Elements of Friction Theory and Nanotribology

Combining the classical theories of contact mechanics and lubrication with the study of friction on the nanometer range, this multi-scale book for researchers and students alike guides the reader deftly through the mechanisms governing friction processes, based on state-of-the-art models and experimental results.

The first book in the field to incorporate recent research on nanotribology with classical theories of contact mechanics, this unique text explores atomic scale scratches, non-contact friction and fishing of molecular nanowires as observed in the laboratory. Beginning with simple key concepts, the reader is guided through progressively more complex topics, such as contact of self-affine surfaces and nanomanipulation, in a consistent style, encompassing both macroscopic and atomistic descriptions of friction, and using unified notations to enable use by physicists and engineers across the scientific community.

ENRICO GNECCO is a Senior Scientist at IMDEA Nanoscience, Madrid, where his research focuses on friction phenomena on the atomic scale, controlled manipulation of nanoparticles and theoretical bases of nanotribology.

ERNST MEYER is Professor of Experimental Physics at the University of Basel and module coordinator in the NCCR for Nanoscale Science. He is a former Chairman of the European Science Foundation collaborative network 'Nanotribo'.

Elements of Friction Theory and Nanotribology

Enrico Gnecco and Ernst Meyer

CAMBRIDGE
UNIVERSITY PRESS

CAMBRIDGE
UNIVERSITY PRESS

University Printing House, Cambridge CB2 8BS, United Kingdom

Cambridge University Press is part of the University of Cambridge.

It furthers the University's mission by disseminating knowledge in the pursuit of education, learning and research at the highest international levels of excellence.

www.cambridge.org
Information on this title: www.cambridge.org/9781107006232

First published 2015

A catalogue record for this publication is available from the British Library

Library of Congress Cataloguing in Publication data

Gnecco, Enrico.
Elements of friction theory and nanotribology / Enrico Gnecco and Ernst Meyer.
pages cm
Includes bibliographical references and index.
ISBN 978-1-107-00623-2
1. Tribology. 2. Friction. 3. Elasticity. I. Meyer, E. (Ernst), 1962– II. Title.
TJ1075.G58 2015
621.8′9–dc23
2014043415

ISBN 978-1-107-00623-2 Hardback

Contents

Preface

Friction permeates every aspect of our life. It accompanies us when we walk and our fingers when they slide on the display of a tablet. Friction produces very annoying results when a chalk is rubbed against a blackboard and may cause tremendous damage when it fails to hold two tectonic plates together and a powerful earthquake is suddenly generated. Friction can also be very useful, when a cat suddenly jumps in front of our car and the brake pedal avoids serious consequences; and even pleasant, when a talented violinist takes up a bow and starts playing his Stradivarius. In any case, friction is certainly not a boring subject, and writing a book about friction is definitely not an easy task.

In spite of an immense amount of experimental data, a general theory of sliding friction between two solid surfaces is still missing. The simple Amontons' law, stating that the friction is proportional to the normal force, has been found to work exceptionally well in a variety of situations. Based on this law, theoretical models with different degrees of complexity have been derived and successfully applied to reproduce real situations. Even if Amontons' law is universally accepted as empirical evidence rather than as a consequence of first principles, the attitude is rapidly changing and it is now possible to prove by analytical means that the friction between two rough elastic surfaces has to be almost proportional to the loading force. A different situation is encountered when studying the drag force accompanying the motion of a solid object in a viscous liquid. Here, the Navier–Stokes law works usually quite well, which made hydrodynamic lubrication an established subject a long time ago. Still, problems arise when the lubricants are confined and the friction can only be investigated, theoretically, using atomic-scale models.

In the past 25 years, significant progress has been achieved in the understanding of the basic principles of sliding friction. This progress was essentially caused by the invention of the atomic force microscope (AFM) and the tremendous growth of computational power. The AFM has allowed us to investigate the motion of nano-asperities driven on solid surfaces with unprecedented space and force resolution. The atomic-scale friction features so measured are found to be in good agreement with a model developed by Ludwig Prandtl sixty years before the AFM was developed. On the other hand, molecular dynamics simulations involving a few hundred thousand atoms can be run nowadays in a reasonable time scale, although the duration of the processes reproduced by these virtual experiments is too short compared to the real measurements. Much more difficult is to explain the different wear processes which usually accompany the sliding. A detailed atomistic description of these phenomena is not feasible even with the fastest supercomputers. At the same time, it is not possible to visualize the structure of a wear scar on the atomic scale, although good progress is being made using transmission electron microscopy and, again, AFM.

Having this in mind, we believe that a 'modern' approach needs to be adopted to explain the fundamental friction theories, as we understand them nowadays, to undergraduate and graduate students in physics or engineering, and to anyone interested in this multidisciplinary and fascinating subject. In this book we have made a rather simple choice, and limited the discussion to theoretical results based on well-posed analytical derivations and numerical calculations, and to experiments aimed to shed light on nanoscale friction and performed in well-defined environmental conditions such as ultra-high vacuum. It was in no way our intention to present long tables of friction coefficients or to introduce purely phenomenological models. For this reason, no attempts to discuss abrasive, adhesive and other forms of wear have been made, with the exception of a few focused investigations on the nanoscale. Similarly, we have not included technical details regarding the chemical composition of contacting surfaces or lubricants, which would have led us too far from our goal.

Classifying and ordering the material is also not easy. A problem that we had to face was unifying the notation, since the same physical quantities are often addressed in different ways by physicists and engineers. Having in mind the various backgrounds of our readers, we have divided the book into four parts. In the first part, the basic theory of elastic contacts is discussed. The influence of friction on normal contacts, partial slips, sliding and rolling of elastic objects with simple geometric shapes is introduced with the minimal assumption that Amontons' law is applicable. The second part of the book focuses on more advanced and not always independent topics such as rough, viscoelastic, adhesive and plastic contacts, thermal and electric effects at the interface between two surfaces, fracture and macroscopic stick–slip. In all these frames, the connection to friction is rather obvious. A particular emphasis is given to the theory recently developed by Bo Persson, which, in our opinion, can explain several phenomena more elegantly than any alternative finite element model. In the third part theoretical models and representative experiments at the basis of modern nanotribology are presented in more detail. Besides atomic-scale sliding friction, we will also discuss manipulation, wear and non-contact friction experiments and the Prandtl–Tomlinson model for atomic-scale stick–slip. The last part of the book is dedicated to the dynamics of viscous fluids and its application to lubrication. This part ends with an overview of important phenomena observed in tiny 'spots' such as capillary condensation, fluid flow between rough surfaces and spreading of liquid droplets on a solid surface. Friction force microscopy, gas viscosity and slip boundary conditions in the Navier–Stokes equation are briefly discussed in separated appendices. In this way, we hope that the main message conveyed by our book is that investigating friction is not a messy task but a rather elegant exercise.

Before starting, we would like to thank all the people who accompanied us in the study of friction and related phenomena. Even if it is not possible to cite all of them, special acknowledgment goes to Hans-Joachim Güntherodt, Alexis Baratoff, Roland Bennewitz, Shigeki Kawai, Marcin Kisiel, Anisoara Socoliuc, Sabine Maier, Karine Mougin, Raphael Roth, Pascal Steiner, Thilo Glatzel, Tibor Gyalog, Martin Bammerlin, Rodolfo Miranda, Carlos Pina, Johannes Gierschner, Reinhold Wannemacher, Pawel Nita, Santiago Casado, Patricia Pedraz, Carlos Pimentel, Robert Szoszkiewicz, Pasqualantonio Pingue, Ruben Perez, Juanjo Mazo, Renato Buzio, Ugo Vibusa and Stefano Brizzolari. We also thank Karyn Bailey, Emily Trebilcock, Roisin Munnelly, Bronte Rawlings and Simon Capelin from Cambridge University Press for assisting us in the publishing process, and Frances Lex for critical comments and improvements to the manuscript. Last but not least, E.G. is immensely grateful to his wife Tatiana and his son Valerio. Without their infinite patience in the uncountable hours spent in front of the screen, this book would have never reached its conclusion.

1
Introduction

The study of friction, wear and lubrication between two surfaces in relative motion is called *tribology*. This term is derived from the Greek verb 'tribos', which means 'to rub'. On one hand tribology aims at a scientific foundation of these phenomena. On the other hand it aims at a better design, manufacture and maintenance of devices which are affected by these 'annoyances'. Tribology has a very important economical outcome. According to one of the first reports on this issue, tribological problems accounted for 6% of the Gross Domestic Product in industrialized countries in the 1960s [160]. This percentage may have increased by now. Tribological problems are found in pinions, pulleys, rollers and continuous tracks, in pin joints and electric connectors, and may cause more failure than fracture, fatigue and plastic deformation. On the other hand, friction is highly desirable, or even essential, in power transmission systems like belt drives, automobile brakes and clutches. Friction can also reduce road slipperiness and increase rail adhesion. Before starting our rather theoretical description of tribology, it is important to recall the milestones that have marked the progress in this subject from the dawn of civilization.

1.1 Historical notes

More than 40 000 years ago a complex process such as the generation of frictional heat from the lighting of fire was already well known. Nowadays the same process is studied by a branch of tribology, which is known as 'tribochemistry' and is focusing, more generally, on friction-induced chemical reactions. The early use of surface lubricants to reduce friction is unambiguously proven by a famous painting from ancient Egypt, in which a 'prototribologist' supports the work of a few dozen slaves by pouring oil in front of the heavy sled that they are pulling (Fig. 1.1). More than four thousand years later Leonardo da Vinci (1452–1519) started a systematic investigation of tribology, as documented by his drawings (Fig. 1.2). Leonardo's

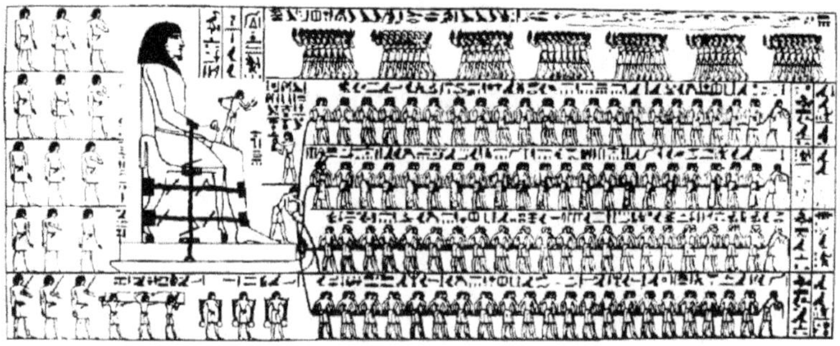

Figure 1.1 Transportation of an Egyptian colossus from the gravestone of Tehuti-Hetep (ca. 1880 B.C.). Note the officer at the feet of the statue lubricating the ground in front of the sled.

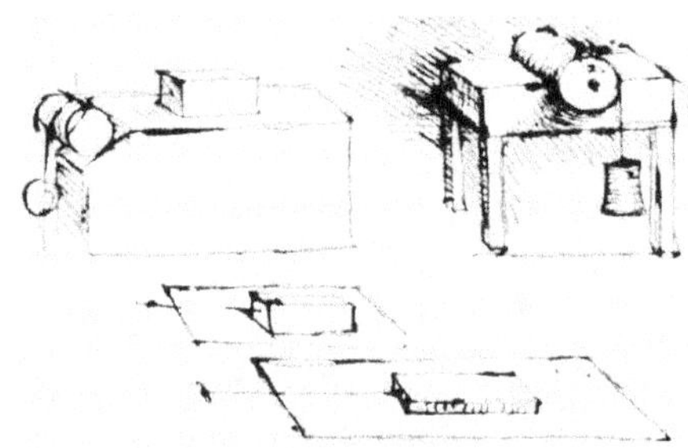

Figure 1.2 Original sketches of the friction experiments performed by Leonardo.

intuition and perseverance resulted in the formulation of the first friction law, which states the proportionality between friction and normal force. Nevertheless, this key observation quickly sank into oblivion until the French physicist Guillaume Amontons rediscovered it in 1699. The Swiss Leonhard Euler, possibly the greatest mathematician of the eighteenth century, was the first person who clearly distinguished between static and kinetic friction. Euler also made an attempt to relate friction to microscopic processes by speculating that friction is ultimately caused by the interlocking of rigid irregularities. A few years later, Charles de Coulomb, best known for his work on electricity and magnetism, observed that the kinetic friction is almost independent of the sliding velocity, whereas the static friction may vary depending on the time of stationary contact of the surfaces.

A turning point in the history of tribology was the theory of frictionless contact of non-conformal elastic solids. This theory was developed by the German physicist Heinrich Hertz in 1882 when he was only 23 years old, and forms the basis of modern contact mechanics. The Hertz theory was extended to include the contribution of adhesive forces by Kenneth Johnson and coworkers almost 90 years later. The difference between apparent and real contact areas was pointed out only in 1942 by Frank Bowden and David Tabor. They also proposed that the friction between two clean metal surfaces originates from the formation and rupture of cold weld junctions and concluded that, if the deformation of the junctions were entirely plastic, the coefficient of friction should be around 0.33, as indeed is measured in many metal pairings. The relation between friction and roughness was further investigated, among others, by John Archard (1957), Greenwood and Williamson (1966) and Bo Persson (2002), who proved, under more and more realistic conditions, that the real contact area is approximately proportional to the normal force.

A key process, when two surfaces slide past each other, is the so-called stick–slip. Stick–slip is caused by elastic instabilities and, in the context of atomic-size contacts, it was first modeled by Ludwig Prandtl in 1928, and experimentally substantiated sixty years later by atomic force microscopy (AFM). The slip can be thermally activated, which leads to characteristic variations of friction with temperature and driving velocity. Thermal activation is also important in capillary condensation and plastic flow, and may lead to various 'ageing' effects which are the focus of numerous theoretical and experimental investigations nowadays. On geological scales, stick–slip is also a key mechanism in earthquakes, as first recognized by Brace and Byerlee in 1966.

The advent of experimental techniques allowing one to measure friction down to the nanoscale and of fast computers allowing one to simulate the atomic interactions between two sliding surfaces resulted in the rise of the so-called 'nanotribology'. While the stick–slip motion of nanometer sized asperities can be readily investigated by AFM, other techniques such as the quartz crystal microbalance and the surface force apparatus have allowed researchers to measure the friction between adsorbate films and substrates or, respectively, between two atomically flat surfaces with intercalated lubricant films. On the other hand, molecular dynamics simulations are throwing light, more and more accurately, on the atomic origins of friction.

Without lubricants, almost no machine made of metal would work, and the Industrial Revolution would not have occurred. The theory of hydrodynamic lubrication was pioneered by Euler, Bernoulli, Poiseuille, Navier and Stokes between 1730 and 1845. It was the last mentioned who discovered that the frictional drag on a spherical particle slowly moving in a fluid is proportional to the velocity of the sphere. A series of key experiments was conducted by Gustave Hirn, who observed that the friction in a bearing is proportional to the sliding velocity and to the viscosity of the lubricant oil. An interpretation of his results, based on hydrodynamic lubrication and not on the more established concept of interlocking asperities, was first given by Nikolai Petrov in 1883, whereas the theory of fluid mechanics was fully established by Osborne Reynolds. Even if the Reynolds theory is still widely used in the design of modern lubricated machinery, this theory breaks down when the separation between the two sliding surfaces becomes comparable to their roughness. Systematic investigations of this problem were performed by Richard Stribeck, who introduced a curve which still holds his name (1902). The concept of boundary lubrication was introduced in 1922 by the biologist William Hardy while studying friction on solid surfaces covered by fatty molecules with hydrocarbon chains of different lengths.

As mentioned above, plastic flow plays an important role in contact ageing. The theory of plasticity, from the Greek verb 'plassein', meaning 'to shape', is also important in determining the stability of soils. Criteria for the yielding of these materials were proposed in early works by Coulomb and the Scottish engineer William Rankine, whereas the first scientific studies of plasticity in metals started only in 1864, when the French engineer Henri Tresca (whose name is also associated with the construction of the Eiffel Tower) published his famous criterion for yielding. This criterion was improved by Richard von Mises in 1913 and fundamental investigations of plasticity flourished in Germany in the early twentieth century under the leadership of Prandtl, who introduced the concept of plastic flow. The theory of plasticity is supported nowadays by powerful computer simulations, which are essential to control technological processes such as the rolling of strips or the extrusion of rods and tubes.

Our overview would not be complete without mentioning wear processes. Wear was well known to our ancestors, who exploited it to create artistic sculptures and useful tools by rubbing dense stones against softer ones in different ways. In spite of its importance, the variety and complexity of wear phenomena make the development of general physics laws interpreting wear processes quite challenging. Related to wear (and to friction) is the study of fracture mechanics, which was initiated by the British engineer Alan Griffith during World War I. The Griffith's criterion, which is based on simple energetic considerations, can elegantly explain the failure of brittle materials. Fracture dynamics is not fully understood and is nowadays a subject of beautiful theoretical and experimental investigations.

2

Dry friction and damped oscillators

In this chapter we introduce the two categories of friction forces experienced by a rigid object sliding on a solid surface or moving in a viscous fluid. These forces have a different nature. Sliding friction increases with the normal force and is usually independent of the velocity. Viscous friction depends on the shape of the object and is proportional to the velocity, provided that this is low enough. Furthermore, while an object in a fluid can be set into motion by an arbitrarily low force, this is not the case if the same object lies on a solid surface, since a static friction force needs to be overcome in this case. Static friction allows us to join objects together using screws. It also has a key role in the propulsion and braking of vehicles and in transmission belts. Sliding (or kinetic) friction is important in pivots and collar bearings, not to mention uncountable situations in everyday life. Viscous friction can be exploited in mechanical dampers to mitigate the effects of forced oscillations. Since the theory of these oscillations is of pivotal importance in physics and engineering, it will be recalled in this chapter, whereas a detailed description of various situation involving viscous drag is provided in the last part of the book.

2.1 Amontons' law

In order to start and to keep moving a solid block on a solid surface, different *friction forces* F_{fric} have to be overcome and opposed. The *static friction* F_{s} corresponds to the minimum tangential force required to initiate sliding. The *kinetic friction* F_{k} perfectly balances the tangential force needed to maintain the sliding at a given (average) speed. These forces are intrinsically different. The static friction does not do any work, while the kinetic friction equals the dissipative work done at the interface divided by the distance covered by the block.

According to *Amontons' law* [5], the friction force is proportional to the normal force F_{N} acting on the block:

$$F_{\mathrm{fric}} = \mu F_{\mathrm{N}}. \tag{2.1}$$

Table 2.1 *Typical coefficients of static and kinetic friction.*

Physical situation	μ_s	μ_k
Rubber on concrete	1.0	0.8
Steel on steel	0.74	0.57
Aluminum on steel	0.61	0.47
Glass on glass	0.94	0.4
Copper on steel	0.53	0.36
Wood on wood	0.25–0.5	0.2
Wood on wet snow	0.14	0.1
Metal on metal (lubricated)	0.15	0.06
Wood on dry snow	–	0.04
Teflon on teflon	0.04	0.04
Ice on ice	0.1	0.03
Synovial joints in humans	0.01	0.003

Furthermore, it is independent of the nominal area of contact. The ratio between F_{fric} and F_N is the *coefficient of friction* μ, and it is usually different for static and kinetic friction. The static friction coefficient depends on the time of stationary contact (so-called *contact history*), on the elastic and geometric properties of the contacting surfaces and on the way in which the driving forces are applied. On the other hand, the kinetic friction coefficient is much better defined once the temperature, humidity, velocity and surface properties are reproducible. The values of the friction coefficient are usually lower than one, and the static coefficient μ_s is always equal to or larger than the kinetic coefficient μ_k.

As far as we are concerned with macroscopic contacts, we will also accept the validity of *Coulomb's law* and assume that, under dry conditions, F_k is independent of the sliding velocity. This is not the case at very low or very high velocities, where thermal effects or, respectively, inertial effects become important.

A representative list of friction coefficients is given in Table 2.1. For lubricated metal surfaces typical values of μ_s are in the range of 0.1–0.3. Higher values are observed after prolonged sliding if the lubricant film is worn off. For common engineering surfaces the friction coefficient does not depend significantly on the surface roughness, unless the surfaces are extremely smooth or rough. Amontons' law is also modified in the presence of strong adhesive forces.

Suppose now that a block rests on a plane inclined by an angle α, as in Fig. 2.1. If α is slowly increased, the block will start moving when

$$\tan\alpha = \mu_s. \tag{2.2}$$

This value defines the *angle of friction* (or *angle of repose*) α_c. If $\mu_s = 0.1$ the angle of friction is about 6°. Thus, the coefficient of static friction can be simply estimated by measuring α_c.

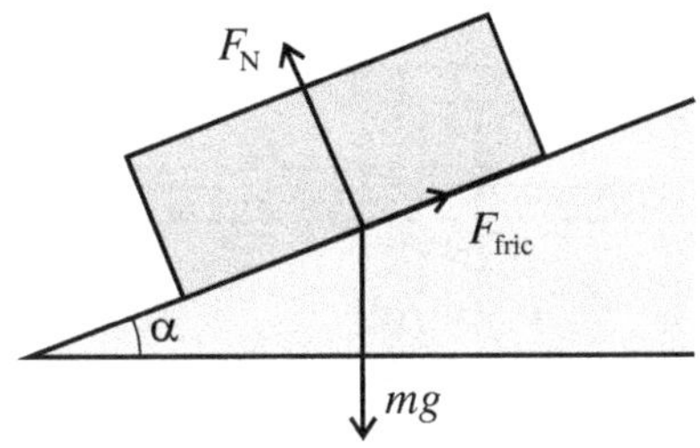

Figure 2.1 Forces on a solid block of mass m resting on an inclined plane ($g = 9.8$ m/s^2 is the acceleration due to gravity).

2.2 Applications to representative mechanical systems

Amontons' law is essential to understand the operation of machines and common mechanical parts. Here, we will discuss a few examples to illustrate its usefulness. We will not distinguish between static and kinetic friction coefficients since the values to be used are clear from the context.

Screws

A *screw* is able to convert a torque into a linear force. A square-thread screw with a mean radius R and pitch[1] b can be seen as a plane inclined by an angle α such that $\tan\alpha = (b/2\pi R)$, and wrapped around a cylinder. Thus, the screw will be self-locking if $\alpha < \alpha_c$, where α_c is defined by Eq. (2.2). Due to this possibility, the applications of screws for holding objects together are uncountable.

Screws can in principle also be used in power transmission, although they are not very efficient in this case. Exploiting the analogy with the inclined plane, it can be indeed demonstrated that the *efficiency* of a screw, i.e. the ratio between the useful work done and the energy transferred to a mechanism, is

$$\eta = \frac{\tan\alpha}{\tan(\alpha + \alpha_c)}.$$

The maximum efficiency η_{max} is achieved when $\alpha = 45° - \alpha_c/2$ and is equal to

$$\eta_{max} = \frac{1 - \sin\alpha_c}{1 + \sin\alpha_c}.$$

If $\mu = 0.1$, a value of $\eta_{max} \approx 0.82$ is reached when $\alpha \approx 42°$. If $\alpha = \alpha_c(\approx 6°)$, the efficiency drops to 0.49. These low values, compared to other transmission mechanisms such as belt drives (see below), explain why 'lead screws' are rarely used for transferring large amounts of power.

[1] The pitch of a screw is the rise corresponding to a rotation of 360°.

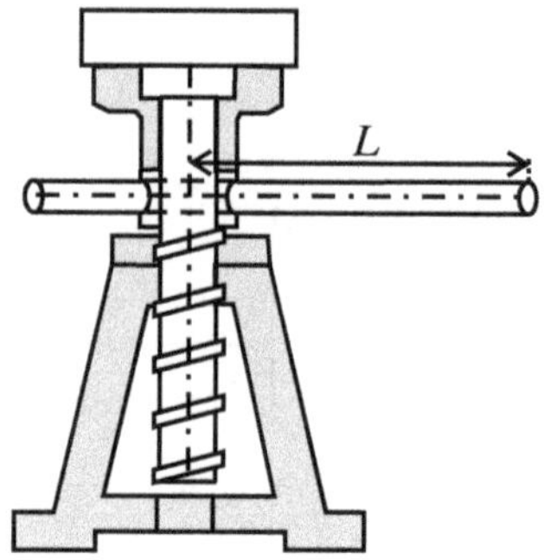

Figure 2.2 A jackscrew.

Jackscrews

If a screw is used as a lifting machine, we can define its *mechanical advantage* MA as the ratio between the load which can be lifted by the screw and the horizontal force F_{ext} applied on it. It is not difficult to see that

$$\text{MA} = \frac{1}{\tan(\alpha + \alpha_{\text{c}})}.$$

The mechanical advantage can be increased by one or two orders of magnitude if the force is applied to the end of a long horizontal bar connected to the screw, as in Fig. 2.2. In this case

$$\text{MA} = \frac{L/R}{\tan(\alpha + \alpha_{\text{c}})}, \tag{2.3}$$

where L is the length of the bar. If $L = 20R$, $\alpha = \alpha_{\text{c}}$ and $\mu = 0.1$, the 'jackscrew' will be able to lift a weight almost 100 times larger than the applied force.

If a V-thread with angle of inclination β is used, the previous formulas are still valid, provided that the coefficient of friction μ in the definition (2.2) is replaced by $\mu/\cos\beta$. In this case the value of α_{c} increases, and the mechanical advantage is reduced. Nevertheless, V-threads are easier to manufacture and for this reason they are much more common than square-threads.

Pivots, collars and clutches

Pivots and *collar bearings* are commonly used to support an axial load acting on a rotating shaft. Pivots are placed at the end of the shaft, whereas collars can be located at any position (Fig. 2.3). If the pressure is uniform it is easy to see that the frictional torque acting on a flat pivot of radius R (Fig. 2.3(a)) is

$$M_{\text{fric}} = \frac{2}{3}\mu F_{\text{N}} R. \tag{2.4}$$

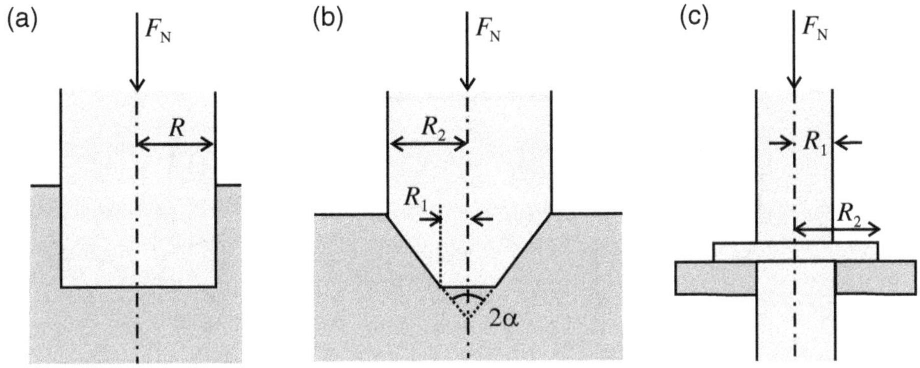

Figure 2.3 (a) A flat pivot, (b) a conical pivot and (c) a collar bearing.

For a truncated conical pivot:

$$M_{\text{fric}} = \frac{2\mu F_{\text{N}}}{3 \sin \alpha} \frac{(R_2^3 - R_1^3)}{(R_2^2 - R_1^2)}, \tag{2.5}$$

where R_1 and R_2 are, respectively, the inner and outer radii of the pivot, and α is the half-angle of the cone (Fig. 2.3(b)). Equation (2.5), with $\alpha = 90°$, can be also applied to collar bearings (Fig. 2.3(c)).

However, since the wear rate at a given pressure is proportional to the sliding velocity and hence to the distance r from the axis of the shaft, a bearing will be more and more damaged at increasing values of r and the pressure distribution will consequently change with time. This means that the formulas (2.4) and (2.5) are strictly valid only for brand new elements. A good agreement with observations is found assuming that the wear rate becomes uniform. Since the wear rate is proportional to pr, where p is the pressure, it can be proven that, in this case, the frictional torque is

$$M_{\text{fric}} = \frac{1}{2} \mu F_{\text{N}} R$$

for a flat pivot, and

$$M_{\text{fric}} = \frac{\mu F_{\text{N}}}{2 \sin \alpha} (R_1 + R_2) \tag{2.6}$$

for a truncated conical pivot. Equations (2.5) and (2.6) can be also applied to plate clutches (with $\alpha = 0$) and to conical clutches connecting two shafts rotating at different speed.

Belt drives

Consider two pulleys with radius R connected by a flexible elastic belt (Fig. 2.4). The initial tension in the belt is T_0. If a torque M_{ext} is applied to one of the pulleys, it

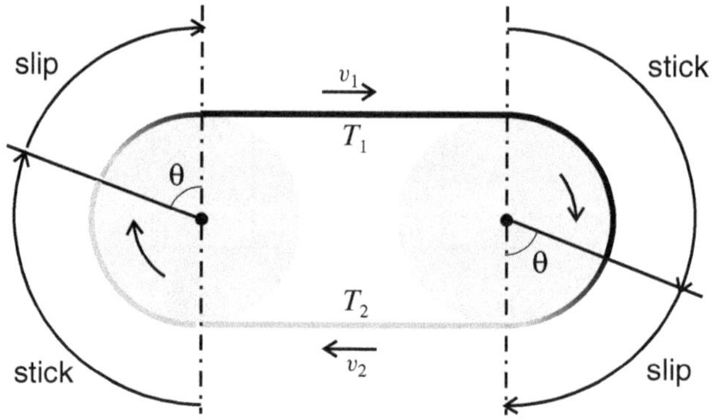

Figure 2.4 Belt transmission.

will be transmitted to the second one. This causes a difference between the tensions T_1 and T_2 on the tight side and slack side of the belt, which is given by

$$T_1 - T_2 = M_{\text{ext}}/R.$$

The tension increases over an arc $R\theta$ around the driving pulley, where the belt slips. The length of this arc is determined by the *capstan equation*

$$\mathrm{e}^{\mu\theta} = T_1/T_2,$$

which is obtained by integrating the tension variation over an infinitesimal arc $\mathrm{d}\theta$:

$$\mathrm{d}T/T = \mu\,\mathrm{d}\theta$$

(μ is the coefficient of friction between belt and pulley). The slip arc is located next to the point where the belt runs out of the driving pulley. A corresponding arc is located around the driven pulley, and this pulley runs slower than the driving pulley in proportion to the transmitted torque. If v is the mean speed acquired by the belt, the transmitted power is

$$P = (T_1 - T_2)v = 2T_0 v \frac{\mathrm{e}^{\mu\theta} - 1}{\mathrm{e}^{\mu\theta} + 1}.$$

Note that the efficiency of a belt is typically very high (~0.9). Since transmission belts do not require lubrication and are relatively cheap, they have found numerous applications ranging from automotive engines to transportation of heavy materials.

The previous formulas are also valid for a V-grooved belt, provided that μ is replaced by $\mu/\sin\beta$, where β is the half-angle of the groove profile. In this case the length of the slip arc can be reduced significantly. For this reason V-grooved belts are the most common choice for applications to power transmission.

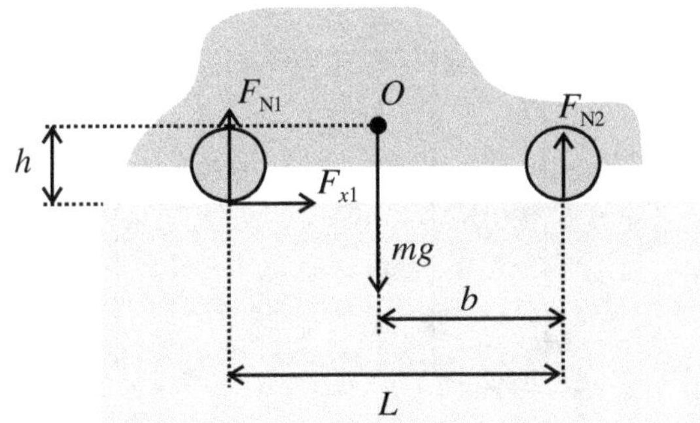

Figure 2.5 Accelerating car on a level road.

Propulsion and braking of vehicles

The processes of accelerating and braking a car of mass m are schematically represented in Fig. 2.5. Suppose that the car is accelerated by a driving couple applied to the rear axle. In this case, the forward acceleration a is determined by the equations

$$F_{N1} + F_{N2} = mg, \qquad F_{x1} = ma,$$
$$F_{x1}h = F_{N1}(L - b) - F_{N2}b,$$

where b and h are the horizontal and vertical distances of the center of mass of the car from the point of contact between the front wheels and the road, L is the distance between the front axle and the rear axle (so-called 'wheelbase') and $g = 9.8$ m/s^2 is the acceleration due to gravity. Slip ('wheelspin') will be avoided if the static friction on the rear wheels $F_{x1} < \mu F_{N1}$. The previous equations imply that the maximum forward acceleration is [325, section 4.12]

$$a_{\max} = \frac{\mu b g}{L - \mu h}. \tag{2.7}$$

For instance, if $L = 2.5$ m, $b = 1.1$ m, $h = 0.5$ m and $\mu = 1$, the acceleration $a_{\max} \approx 0.55g$, meaning that the car will reach a speed of 100 km/h in about five seconds. Note that the acceleration increases the load on the rear wheels and decreases the load on the front wheels according to the relations

$$F_{N1} = \frac{m}{L}(gb + ah), \qquad F_{N2} = \frac{m}{L}\left[g(L - b) - ah\right].$$

If the car is front-wheel driven, the maximum acceleration is lower.

On the other hand, if the car is braked, it can be proven in a similar way that the maximum retardation

$$|a_{\max}| = \frac{\mu(L - b)g}{L - \mu h}$$

is reached if the braking couple is applied to the front axle.

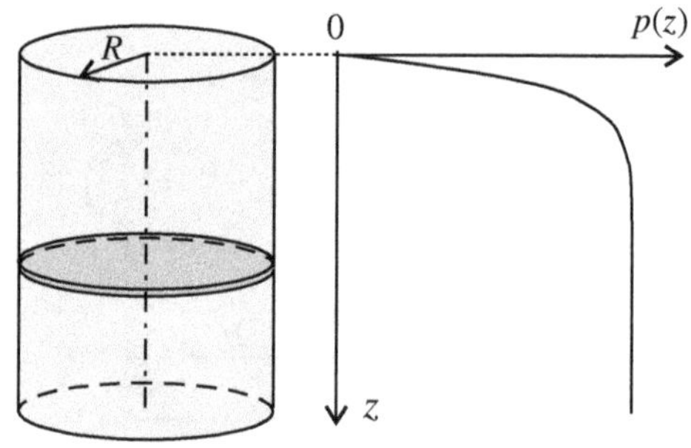

Figure 2.6 Pressure distribution in a container filled with sand.

The hourglass

Friction also explains why the pressure p at the bottom of a cylindrical container of radius R filled with sand is almost independent of the filling height (so that the flow velocity from an hourglass is constant). If the sand has a constant density ρ, the total force F acting on a slice of thickness Δz at a depth z (Fig. 2.6) is equal to the difference between the weight of the slice, $\rho g \pi R^2 \Delta z$, and the friction force at the wall. The friction force is given by

$$F_{\text{fric}} = \mu \pi R^2 \Delta z \cdot p(z).$$

Since $F = \pi R^2 \Delta p$, a simple integration leads to the conclusion that the pressure increases with z as

$$p(z) = \frac{\rho g R}{2\mu}\left(1 - e^{-2\mu z/R}\right).$$

Thus, the pressure approaches a constant value at a depth $z \sim R$.

2.3 Viscous friction

The friction experienced by a solid object sliding on a rigid surface is usually independent of its velocity v. This is not the case if the same object moves in a fluid. In this case a resistive force (*drag*) proportional to v appears, provided that v is not too high. Drag forces can be exploited in mechanical dampers ('dashpots') to slow down other moving components. If a block of mass m is connected to a dashpot, the friction force acting on the block can be written as

$$F_{\text{fric}} = -m\gamma v, \tag{2.8}$$

where γ is the *damping coefficient*. On much smaller scales Eq. (2.8) also applies to the motion of molecules diffusing on a solid surface. In this case γ can be related to electronic and phononic dissipation mechanisms, as discussed in Section 15.8.

Dissipative processes in viscoelastic materials are also governed by Eq. (2.8), as seen in Chapter 9. In several occasions in this book we will be interested in the influence of viscous forces on small oscillations. For this reason it is important to review the fundamental formulas describing the damping of harmonic oscillators.[2] Since we are not concerned with internal degrees of freedom, we can represent the oscillators as point masses.

Damped harmonic oscillators

In a first approximation the motion of a point mass m connected by a spring to a pinning center is described by the equation[3]

$$m\ddot{x} + kx = 0,$$

where k is the spring constant. The particle executes harmonic oscillations with a characteristic *resonance frequency*

$$\omega_0 = \sqrt{k/m},$$

a certain amplitude A and an arbitrary phase α:

$$x(t) = A\cos(\omega_0 t + \alpha).$$

The total energy of the oscillator is simply related to the amplitude as $E = kA^2/2$.

If, in addition to the spring force, a periodic external force $F_{\text{exc}}(t) = F_0 \cos\omega t$ is applied to the particle, a second oscillatory motion is superimposed:

$$x(t) = A\cos(\omega_0 t + \alpha) + B(\omega)\cos\omega t,$$

where $B(\omega) = F_0/m(\omega_0^2 - \omega^2)$. If $\omega \to \omega_0$ the amplitude of the driven oscillations becomes very large. If the excitation frequency $\omega = \omega_0$ the amplitude $B(\omega_0)$ will increase linearly with time till the oscillations cease to be small and non-linear effects appear.

Suppose now that, instead of the periodic force F_{exc}, a friction force $F_{\text{fric}} = -m\gamma\dot{x}$ acts on the oscillator. In this case, two different behaviors can be observed. If $\gamma/2 < \omega_0$ the point mass oscillates with a frequency

$$\omega_1 = \sqrt{\omega_0^2 - (\gamma/2)^2}.$$

[2] The derivations of these formulas can be found in any textbook of classical mechanics, e.g. [175] or [114].

[3] A single dot denotes the first derivative of a physical quantity with respect to time, and a double dot the second derivative.

Most importantly, the amplitude of the oscillations is not constant anymore, but decays exponentially with a decay constant $\gamma/2$:

$$x(t) = Ae^{-\gamma t/2}\cos(\omega_1 t + \alpha).$$

The total energy of the system, averaged over the period of the oscillations, also decays exponentially (with a decay constant γ):

$$\overline{E}(t) = E_0 e^{-\gamma t}.$$

If $\gamma/2 > \omega_0$ the point mass asymptotically approaches the equilibrium position $x = 0$ without oscillating and with two different decay constants $\gamma/2 \pm \sqrt{(\gamma/2)^2 - \omega_0^2}$.

Finally, suppose that the periodic force F_{exc} and the friction force F_{fric} act simultaneously on the point mass. If $\omega \approx \omega_0$, after a transient with a decay constant $\gamma/2$, the system oscillates with an amplitude

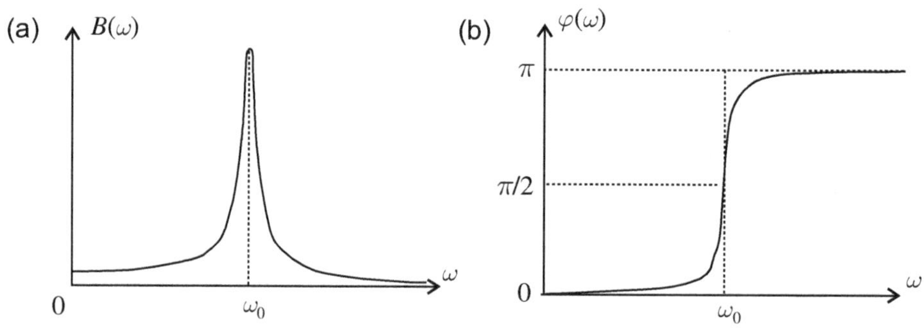

Figure 2.7 Frequency response of a driven damped harmonic oscillator: (a) oscillation amplitude and (b) phase lag.

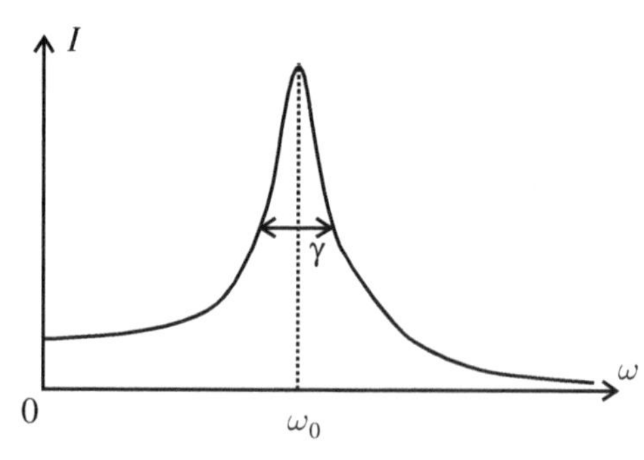

Figure 2.8 Resonance peak of a damped harmonic oscillator.

$$B(\omega) = \frac{F_0}{2m\omega_0\sqrt{(\omega-\omega_0)^2+(\gamma/2)^2}}$$

and a phase lag

$$\varphi(\omega) = \arctan\frac{\gamma}{2(\omega_0-\omega)}.$$

The frequency dependencies of B and φ are shown in Fig. 2.7. If $\omega = \omega_0$ the amplitude has a maximum, but remains finite. The phase lag varies from 0 to π, passing through $\pi/2$ when $\omega = \omega_0$. The intensity of the oscillations, defined as the mean energy adsorbed in the oscillation period, depends on the frequency as

$$I(\omega) \propto \frac{(\gamma/2)^2}{(\omega-\omega_0)^2+(\gamma/2)^2}. \tag{2.9}$$

The curve represented by Eq. (2.9) is a Lorentzian with width γ (Fig. 2.8) and the ratio ω_0/γ is the *quality factor* Q of the oscillator.

Part I

Elastic Contacts

3
Elements of the theory of elasticity

The distribution of the friction forces at the interface between two contacting bodies is significantly influenced by the elastic properties of the two materials. The goal of this chapter is to recall the basic concepts of the theory of elasticity. Here we will first show how the elastic moduli describing the response to simple compressive and shear deformations can be used to relate the stress and strain occurring in an arbitrary deformation of a body. The elastic moduli also define the velocity of the longitudinal and shear sound waves propagated in the bulk and of the Rayleigh waves propagated on the surface of the body. General expressions for the strain energy will be introduced also. If the distribution of the forces on the surface of an elastic body is known, the stress and strain distribution in the bulk are unequivocally determined by the Navier–Cauchy equations. These equations are greatly simplified in plane stress or plane strain problems.

3.1 Strain

When a solid object is deformed, each point in it is subjected to a *displacement*

$$\mathbf{u}(\mathbf{r}) = \mathbf{r}' - \mathbf{r},$$

where $\mathbf{r} \equiv (x, y, z)$ and $\mathbf{r}' \equiv (x', y', z')$ are the vectors defining the position of the point (in a fixed frame of reference) before and after the deformation. The *strain tensor* is defined as

$$\varepsilon_{ij} = \frac{1}{2}\left(\frac{\partial u_i}{\partial x_j} + \frac{\partial u_j}{\partial x_i}\right), \tag{3.1}$$

where x_i and u_i ($i = 1, 2, 3$) are the components of the vectors $\mathbf{r}$ and $\mathbf{u}$, respectively. In engineering, the notation $\gamma_{ij} = 2\varepsilon_{ij}$ is frequently used when $i \neq j$, and the diagonal components of ε_{ij} are simply denoted by ε_i. Note that the definition (3.1) implicitly assumes that the displacement vector $\mathbf{u}$ is small. If this is not the case, additional quadratic terms in the gradients $\partial u_i/\partial x_j$ need to be included. This situation may occur in the deformation of thin rods (Section 7.1).

The sum of the diagonal components of the strain tensor gives the relative variation of the volume V in the deformation:

$$\varepsilon_{kk} = \Delta V/V,$$

where $\varepsilon_{kk} \equiv \varepsilon_{11} + \varepsilon_{22} + \varepsilon_{33}$.

The strain tensor can be separated as

$$\varepsilon_{ij} = \frac{1}{3}\delta_{ij}\varepsilon_{kk} + \left(\varepsilon_{ij} - \frac{1}{3}\delta_{ij}\varepsilon_{kk}\right), \tag{3.2}$$

where $\delta_{ij} \equiv 1$ if $i = j$ and $\delta_{ij} \equiv 0$ if $i \neq j$. The first term on the right hand side of Eq. (3.2) corresponds to a *hydrostatic compression* which does not change the shape of the body. The second term corresponds to a *pure shear* modifying the shape, but not the volume of the body. This term is called *deviatoric strain*.

3.2 Stress

Consider an arbitrary surface element with unit normal **n** inside a deformed body (Fig. 3.1). The components of the surface force **T** acting on this element (the so-called *stress vector* or *traction vector*) can be written as

$$T_i = \sigma_{ij}\, n_j, \tag{3.3}$$

where σ_{ij} is the *stress tensor*. The projections of **T** perpendicular and parallel to the surface element define the *normal stress* $\sigma_{\mathrm{n}} = \mathbf{T} \cdot \mathbf{n}$ and the *shear stress* $\tau = \sqrt{T^2 - \sigma_{\mathrm{n}}^2}$, respectively. In this way the component σ_{xy} of the stress tensor represents the shear in the x direction caused by a traction in the y direction (and similarly for the other combinations of Cartesian coordinates).

If **n** is oriented along one of the three mutually perpendicular *principal axes* of stress, the shear stress along the corresponding *principal plane* must be zero.

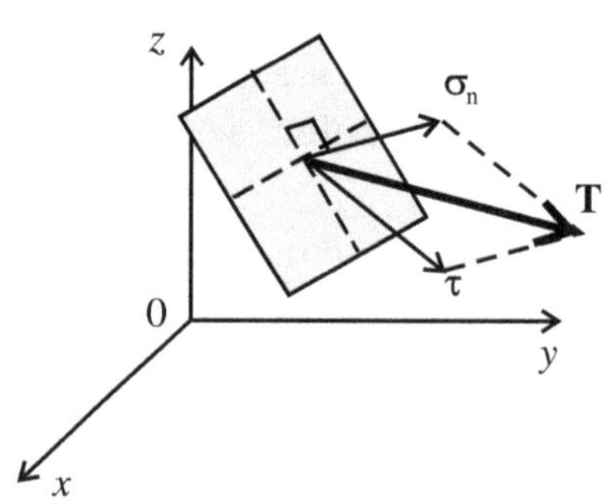

Figure 3.1 Traction vector on an arbitrarily oriented surface.

The *principal normal stresses* are obtained as the solutions σ_i $(i = 1, 2, 3)$ of the characteristic equation

$$\det(\sigma_{ij} - \sigma\delta_{ij}) = 0. \tag{3.4}$$

They are usually ordered so that $\sigma_1 \geq \sigma_2 \geq \sigma_3$. The average value of the normal stresses

$$p = \frac{1}{3}(\sigma_1 + \sigma_2 + \sigma_3)$$

is the *hydrostatic pressure* at the given point. Furthermore, it can be proven that the shear stress reaches its maximum value

$$\tau_{\max} = \frac{1}{2}(\sigma_1 - \sigma_3)$$

along the planes oriented at $\pm 45°$ with respect to the principal axes 1 and 3. If $\tau_{\max}$ exceeds a certain threshold value at any location in the deformed solid the theory of elasticity breaks down and, as will be discussed in Section 12.1, the response of the material becomes plastic.

Strain energy

The definition (3.3) implies that the components of the force **f** acting on the unit volume of an elastically deformed body are [177, section 2]

$$f_i = \frac{\partial \sigma_{ij}}{\partial x_j}.$$

From the work done by **f** during mechanical loading it is easy to see that the density of the strain (or elastic) energy in the body is [177, section 3]

$$U_{\mathrm{el}} = \frac{1}{2}\sigma_{ij}\varepsilon_{ij}. \tag{3.5}$$

Thus, the components of the strain tensor ε_{ij} and the components of the stress tensor σ_{ij} are the first derivatives of the energy density (3.5) with respect to σ_{ij} or ε_{ij}:

$$\varepsilon_{ij} = \frac{\partial U_{\mathrm{el}}}{\partial \sigma_{ij}}, \qquad \sigma_{ij} = \frac{\partial U_{\mathrm{el}}}{\partial \varepsilon_{ij}}.$$

3.3 Isotropic elastic materials

If a solid is *elastic* the relation between the stress tensor σ_{ij} and the strain tensor ε_{ij} is linear:

$$\sigma_{ij} = C_{ijkl}\varepsilon_{kl}.$$

The *stiffness tensor* C_{ijkl} has in general 21 independent components, which become only two if the body is isotropic. Anisotropic materials include single crystals, which are commonly used in microelectromechanical systems (MEMS), but also wood, fiber reinforced composite materials and even polycrystalline metals, if specific textures are formed during manufacture.

Bulk and shear moduli

Taking into account the separability of the strain tensor expressed by Eq. (3.2), the stress–strain relation in an isotropic elastic material can be written as

$$\sigma_{ij} = K\delta_{ij}\varepsilon_{kk} + 2G\left(\varepsilon_{ij} - \frac{1}{3}\delta_{ij}\varepsilon_{kk}\right), \tag{3.6}$$

where K and G are the *bulk modulus* and the *shear modulus* of the material, respectively. Vice versa:

$$\varepsilon_{ij} = \frac{1}{9K}\delta_{ij}\sigma_{kk} + \frac{1}{2G}\left(\sigma_{ij} - \frac{1}{3}\delta_{ij}\sigma_{kk}\right). \tag{3.7}$$

The meaning of K and G becomes clear if the body undergoes a hydrostatic compression or a pure shear. In the first case $\sigma_{ij} = -p\delta_{ij}$ and the relative variation in volume is

$$\Delta V/V = -p/K.$$

In the second case $\varepsilon_{kk} = 0$ so that

$$\tau_{ij} = G\gamma_{ij}. \tag{3.8}$$

Substituting the stress–strain relation (3.6) into the expression (3.5) for the density of strain energy, the latter can be written as

$$U_{\mathrm{el}} = G\left(\varepsilon_{ij} - \frac{1}{3}\delta_{ij}\varepsilon_{kk}\right)^2 + \frac{1}{2}K\varepsilon_{kk}^2. \tag{3.9}$$

The requirement that the elastic energy U_{el} has a minimum in the undeformed equilibrium state of the body implies that the moduli K and G must be positive.

Young's modulus and Poisson's ratio

Consider now an elastic beam undergoing a *simple compression* along its longitudinal (z) direction. In this case the relations between the applied pressure p, the transverse elongation ε_x and the longitudinal compression ε_z of the beam can be written introducing the *Young's modulus* E and the *Poisson's ratio* ν of the material:

$$p = E\varepsilon_z, \qquad \varepsilon_x = -\nu\varepsilon_z. \tag{3.10}$$

The first of the relations (3.10) corresponds to the famous *Hooke's law*, and dates back to 1678. Equation (3.8) can be considered as the equivalent of Hooke's law for shear deformations. For most metals Young's modulus is of the order of 100 GPa. For instance, $E = 210$ GPa for steel. However, E can be as low as 2 MPa in the case of rubber. Poisson's ratio is dimensionless, and takes values around 0.3 for most metals.

Note that the shear modulus can be expressed as a function of E and ν as

$$G = \frac{E}{2(1+\nu)}. \tag{3.11}$$

Other relations among the elastic moduli are

$$K = \frac{E}{3(1-2\nu)}, \qquad E = \frac{9KG}{3K+G}, \qquad \nu = \frac{3K-2G}{2(3K+G)}. \tag{3.12}$$

Using the first of the equations (3.12) the stress–strain relations (3.6) and (3.7) can be written as

$$\sigma_{ij} = \frac{E}{1+\nu}\left(\varepsilon_{ij} + \frac{\nu}{1-2\nu}\delta_{ij}\varepsilon_{kk}\right) \tag{3.13}$$

and

$$\varepsilon_{ij} = \frac{1}{E}\left((1+\nu)\sigma_{ij} - \nu\delta_{ij}\sigma_{kk}\right). \tag{3.14}$$

Since $K > 0$ and $G > 0$ the second of the equations (3.12) implies that E is also positive. From the third equation it follows that ν can only vary between -1 and $1/2$. The limit value $\nu = 1/2$ is approached by *incompressible* materials like rubber. In these materials $G \approx E/3$ and $K \gg G$, meaning that a rubber block is easy to bend but difficult to compress. Negative values of ν are not found in nature, although they have been reported in artificial polymer foams (so-called *auxetic* materials [172]). In this case, the material expands transversally when stretched longitudinally.

Introducing Young's modulus and Poisson's ratio, the strain energy density (3.9) can also be written as

$$U_{\text{el}} = \frac{E}{2(1+\nu)}\left(\varepsilon_{ij}^2 + \frac{\nu}{1-2\nu}\varepsilon_{kk}^2\right). \tag{3.15}$$

Constrained modulus

If the sides of an elastic beam under compression are constrained, only the component ε_z of the strain tensor ε_{ij} is different from zero. In this case the relation between the applied pressure p and the longitudinal compression becomes

$$p = M\varepsilon_z,$$

where

$$M = \frac{(1-\nu)E}{(1+\nu)(1-2\nu)} \tag{3.16}$$

is the *constrained modulus* of the material. For incompressible materials $M \approx K \gg E$. This explains why a piece of rubber which cannot expand freely is much stiffer than if it were not constrained.

3.4 Equilibrium of elastic bodies

If bulk forces are negligible, as we assumed so far, the equilibrium of an elastically deformed body can be described by substituting the relation (3.13) into the general equation $f_i \equiv \partial\sigma_{ij}/\partial x_j = 0$. As a result the displacement vector $\mathbf{u}$ is found to satisfy the *Navier–Cauchy equations*:

$$(1-2\nu)\nabla^2\mathbf{u} + \nabla(\nabla\cdot\mathbf{u}) = 0, \tag{3.17}$$

where ∇ is the gradient and ∇^2 is the Laplacian operator. In the two common situations presented below, Equations (3.17) can be reduced to two dimensions (2D) and greatly simplified.

Plane strain

If a body is deformed perpendicularly to the z direction, the components τ_{xz} and τ_{yz} of the stress tensor and the components ε_z, ε_{xz} and ε_{yz} of the strain tensor are zero. In this case, we can introduce a *stress function* $\varphi(x, y)$ such that [2]

$$\sigma_x = \frac{\partial^2\varphi}{\partial y^2}, \qquad \tau_{xy} = -\frac{\partial^2\varphi}{\partial x\partial y}, \qquad \sigma_y = \frac{\partial^2\varphi}{\partial x^2}. \tag{3.18}$$

The stress function satisfies the biharmonic equation

$$\nabla^2(\nabla^2\varphi) = 0. \tag{3.19}$$

Once the components σ_x and σ_y of the stress tensor are determined from Equations (3.18) and (3.19) with appropriate boundary conditions, the third component σ_z can be calculated as $\sigma_z = \nu(\sigma_x + \sigma_y)$.

Plane stress

If an elastic plate is stretched along its plane perpendicular to the z axis, the stress components σ_z, τ_{xz}, τ_{yz} and the strain components ε_{xz}, ε_{yz} are zero. Again, we

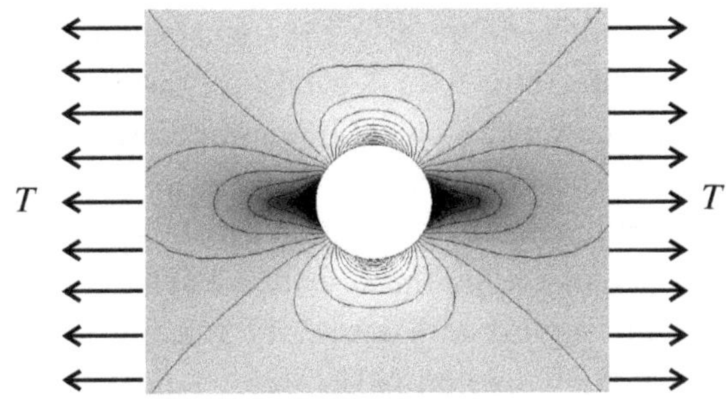

Figure 3.2 Maximum shear stress distribution around a circular hole in an infinite elastic plate under tension.

can introduce a stress function $\varphi(x, y)$ defined by Eq. (3.18) and satisfying the biharmonic equation. Note that in this case the stress distribution is independent of the material constants of the plate.

As an example, consider an infinite plate with a circular hole of radius R, which is subjected to a uniform tension T along the x direction (Fig. 3.2). The resulting stress is distributed at the edge of the hole as $\sigma_\theta = T(1 - 2\cos 2\theta)$, where θ is referred to the direction of the applied force [177, section 13]. The maximum stress $\sigma_{\max}$ corresponds to $\theta = \pm\pi/2$ and it is three times the stress at infinity. The ratio $\sigma_{\max}/T$ defines the so-called *stress concentration factor* S. More generally, if the hole has an elliptical shape with semi-axes a and b, the ratio is [145]

$$S_{\text{ellipse}} = 1 + 2a/b,$$

and tends to infinity if $b/a \to 0$. To reduce this value it is common practice to drill a hole at the tip of a linear slit. For a spherical cavity, the stress concentration factor depends on Poisson's ratio as [177, section 7]

$$S_{\text{sphere}} = \frac{27 - 15\nu}{2(7 - 5\nu)}.$$

We will come back to these problems in the context of fracture mechanics (Section 13).

3.5 Elastic waves

Elastic waves can propagate in an isotropic solid in the form of longitudinal waves or shear waves. The velocity of propagation is given, in the two cases, by the formulas

$$c_{\text{l}} = \sqrt{M/\rho}, \qquad c_{\text{s}} = \sqrt{G/\rho},$$

where M is the constrained modulus, G is the shear modulus and ρ is the density of the material. Note that the ratio c_l/c_s depends only on Poisson's ratio and varies between 0 and $1/\sqrt{2}$. In the seismological literature, longitudinal and shear waves are called *primary* (P) and *secondary* (S) *waves* respectively. Typical values for P-wave velocities in earthquakes are in the range of 5–8 km/s, whereas S-waves are slower. On the surface of a solid the so-called *Rayleigh waves* can also propagate. Their velocity, c_R, is slightly less than c_s by a factor depending on the elastic constants of the material [177, section 24].

4

Normal contacts

The formulas introduced in Chapter 3 can be applied to study the deformation and the stress distribution in two elastic bodies in contact. In this chapter we will first assume that a normal force is concentrated in a single point on the free surface of an elastic half-space, or uniformly distributed on a straight line, an infinitely long strip and also on circular or rectangular areas. The half-space will be also indented by a rigid punch or a rigid cylinder. The normal contact between two elastic spheres or cylinders is discussed within the general theory developed by Hertz. Even if this theory is not applicable if geometric singularities appear, as for a wedge or a cone penetrating into a half-space or in the non-conforming contact formed by a pin in a hole, analytical solutions can be derived also in these cases. Finally, we will show how the pressure distribution in a normal contact is modified by the interfacial friction, which results in the appearance of slip areas. To keep the discussion as short as possible, the derivations of the analytical expressions are not carried out. The interested reader will find them in Johnson's book [156] and references therein.

4.1 Pressure on an elastic half-space

Point loading

The problem of a concentrated force F_N acting normally to the free surface of an elastic half-space (Fig. 4.1) was first solved by Boussinesq [30]. The normal and radial deformations at the surface are inversely proportional to the distance ρ from the point of application of the force:

$$u_z = -\frac{(1-\nu)F_N}{2\pi G\rho}, \qquad u_r = -\frac{(1-2\nu)F_N}{4\pi G\rho}, \tag{4.1}$$

where G and ν are the shear modulus and Poisson's ratio of the material. The normal stress and the component of the shear stress parallel to the free surface (inside the solid) do not depend on the elastic properties of the material:

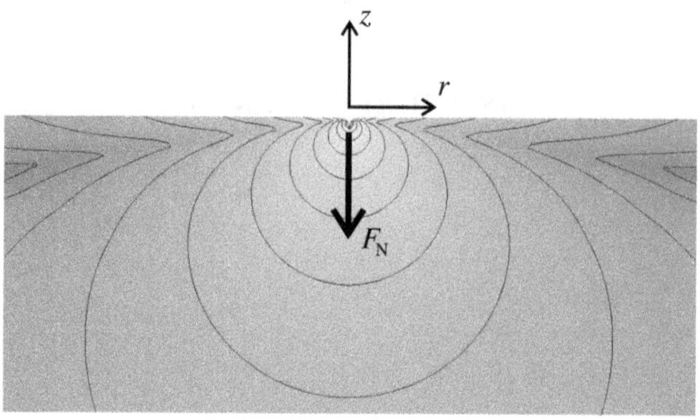

Figure 4.1 Maximum shear stress distribution (logarithmic scale) in an elastic half-space deformed by a concentrated normal force.

$$\sigma_z = -\frac{3F_N}{2\pi}\frac{z^3}{\rho^5}, \qquad \tau_{rz} = -\frac{3F_N}{\pi}\frac{rz^2}{\rho^5}, \tag{4.2}$$

where r is the projection of the position vector onto the free surface. However, this is not the case for the other components of the stress tensor:

$$\sigma_r = \frac{F_N}{2\pi}\left[(1-2\nu)\left(\frac{1}{r^2} - \frac{z}{\rho r^2}\right) - \frac{3zr^2}{\rho^5}\right],$$

$$\sigma_\varphi = -\frac{F_N}{2\pi}(1-2\nu)\left(\frac{1}{r^2} - \frac{z}{\rho r^2} - \frac{z}{\rho^3}\right),$$

where φ is the azimuthal angle around the z axis. In Fig. 4.1 the contours of constant maximum shear stress $\tau_{\max}$ (Section 3.2) are also plotted, assuming that $\nu = 0.33$.

Line loading

The stress field and the deformation resulting from a given force distribution can be obtained by superposition from the results for a concentrated force. If a line load (force per unit length) f_N is uniformly applied along an infinite straight line on the free surface of an infinite half-space, as in Fig. 4.2, the stress in the solid is radially distributed and decreases in intensity as $1/\rho$:

$$\sigma_\rho = -\frac{2f_N}{\pi}\frac{\cos\theta}{\rho}.$$

Both σ_ρ and the maximum shear stress $\tau_{\max} = \sigma_\rho/2$ are constant on a family of circles passing through the point of application of the force. The normal displacement varies logarithmically, and is determined up to an additive constant [93]:

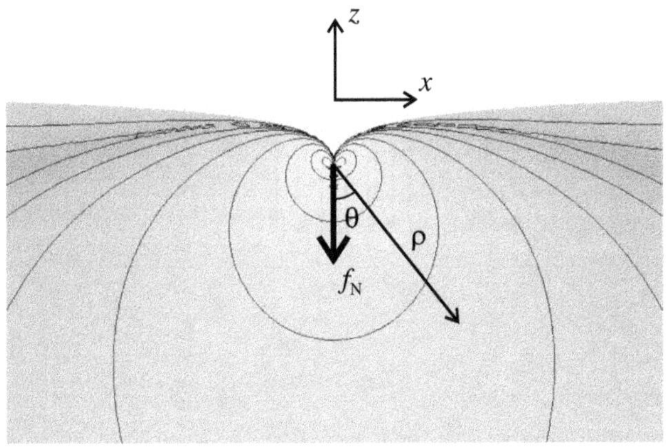

Figure 4.2 Maximum shear stress distribution (logarithmic scale) in an elastic half-space deformed by a normal force distributed along a line (perpendicular to the plane of the figure).

$$u_z = \frac{(1-\nu) f_N}{\pi G} \ln|x| + \text{const.} \tag{4.3}$$

Note that the logarithmic divergence is typical of infinite problems in 2D and is removed if the sample has a finite size. The lateral displacement is constant and given by

$$u_x = \mp \frac{(1-2\nu) f_N}{4G}. \tag{4.4}$$

Strip loading

If a uniform pressure p acts on a strip of half-width a the maximum shear stress τ_{max} is constant on a family of circles passing through the edges of the loaded area [156, section 2.5] (Fig. 4.3). The maximum shear stress τ_{max} peaks on the semi-circle with diameter $2a$, where it reaches the value p/π. The normal displacement u_z can be evaluated from the equation of equilibrium for the gradient

$$\frac{\partial u_z}{\partial x} = -\frac{(1-\nu)}{\pi G} \int_{-a}^{a} \frac{p(s)\, \mathrm{d}s}{x-s}, \tag{4.5}$$

which is obtained by integrating the first of the equations (4.1) on the loaded area and differentiating with respect to z. As a result:

$$u_z = \frac{(1-\nu)}{2\pi G} p \left[(a+x) \ln\left(1+\frac{x}{a}\right)^2 + (a-x)\ln\left(1-\frac{x}{a}\right)^2 \right] + \text{const.} \tag{4.6}$$

Inside the strip the tangential displacement u_x is proportional to the distance x from the axis. Outside the strip u_x is constant and equal to

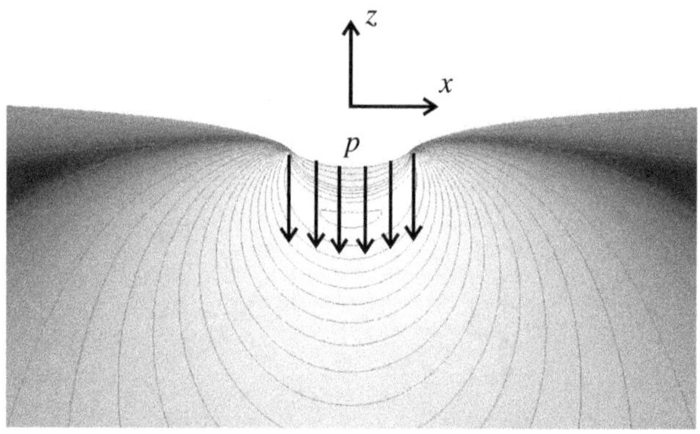

Figure 4.3 Maximum shear stress distribution in an elastic half-space deformed by a normal force uniformly distributed on a strip.

$$u_x = \mp \frac{(1-2\nu)}{2G} pa. \tag{4.7}$$

Again, these conclusions change if the 2D system is finite.

Circular loading

If a uniform pressure p is applied to a circular region of radius a on an elastic half-space, the normal displacement is distributed as in Fig. 4.4. Inside the loaded circle u_z can be expressed in terms of the complete elliptic integrals of the first and second kind,[1] K and E, as

$$u_z = -\frac{2(1-\nu)pa}{\pi G} E\left(\frac{r}{a}\right) \qquad (r < a)$$

[156, section 3.4]. Outside the circle:

$$u_z = -\frac{2(1-\nu)pr}{\pi G} \left[E\left(\frac{a}{r}\right) - \left(1 - \frac{a^2}{r^2}\right) K\left(\frac{a}{r}\right) \right] \qquad (r > a).$$

The radial displacement inside the circle is proportional to r:

$$u_r = -\frac{(1-2\nu)pr}{4G} \qquad (r < a).$$

[1] The functions K and E are defined as

$$K(k) = \int_0^{\pi/2} \frac{d\theta}{\sqrt{1 - k^2 \sin^2\theta}}, \qquad E(k) = \int_0^{\pi/2} \sqrt{1 - k^2 \sin^2\theta}\, d\theta.$$

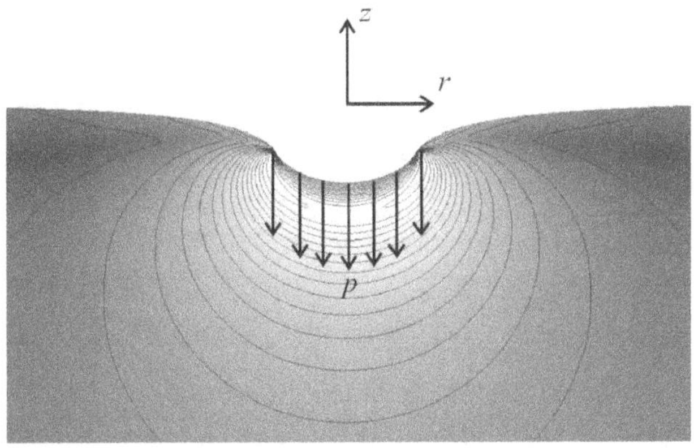

Figure 4.4 Maximum shear stress distribution in an elastic half-space deformed by a normal force uniformly distributed on a circular region.

Outside this circle u_r decreases as if the whole load were concentrated in the center:

$$u_r = -\frac{(1-2\nu)pa^2}{4Gr} \qquad (r > a).$$

Analytical expressions for the stress components in the solid were derived by Love [194]. Along the z axis the normal stress does not depend on the elastic properties and decreases with z as

$$\sigma_z = p\left(1 - \frac{z^3}{(a^2+z^2)^{3/2}}\right).$$

Rectangular loading

The stress distribution resulting from a uniform pressure applied on a rectangular region $a \times b$ is of utmost importance in soil mechanics and foundation engineering. The normal stress below a corner of the rectangle can be found by superposition from Eq. (4.2) [62, section 5.6]:

$$\sigma_z = \frac{p}{4\pi}\left[\frac{2abz\sqrt{f(z)}}{z^2 f(z)+a^2b^2}\frac{f(z)+z^2}{f(z)} + \arctan\left(\frac{2abz\sqrt{f(z)}}{z^2 f(z)-a^2b^2}\right)\right], \qquad (4.8)$$

where $f(z) = a^2 + b^2 + z^2$ and the arctangent takes values between 0 and π. Equation (4.8) can be also used to determine the normal stress below any point P in the rectangle, observing that P can be seen as the common corner point of four rectangular subregions.

4.2 Indentation of an elastic half-space

Instead of considering a localized force or a uniform pressure distribution, we will now assume that an elastic half-space is indented by a rigid object pushed by a

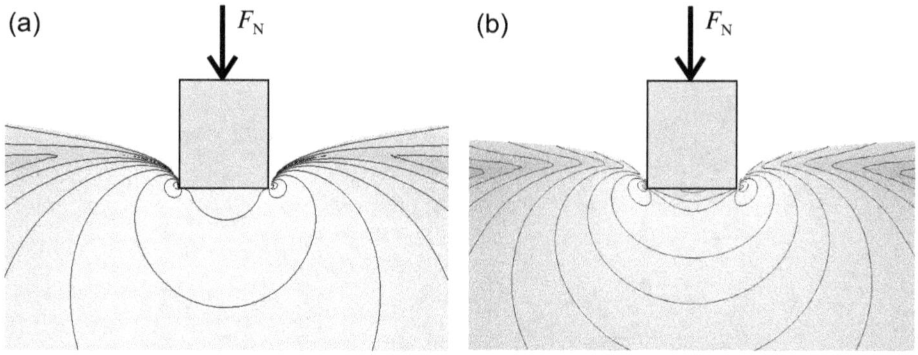

Figure 4.5 Maximum shear stress distribution (logarithmic scale) and deformation in an elastic half-space indented (a) by a rigid flat punch and (b) by a rigid cylinder.

normal force F_N over a finite distance δ. This problem can be solved in a closed form if the indenter is shaped as a punch or a cylinder.

If the indentation is made with a rigid punch of half-width a (Fig. 4.5a) the pressure distribution at the interface is

$$p(x) = \frac{p_0}{\sqrt{1 - x^2/a^2}}, \tag{4.9}$$

where $p_0 = F_N/\pi a$. The normal displacement outside the contact strip is given by the expression

$$u_z(x) = \delta - \frac{(1-\nu)F_N}{\pi G} \ln\left(\frac{x}{a} + \sqrt{\frac{x^2}{a^2} - 1}\right) \qquad (|x| > a)$$

which, as usual, is not applicable to a finite sample. Note that the pressure p becomes infinite at the edges of the strip, where the deformation has an infinite gradient. The maximum shear stress varies as

$$\tau_{max} = -\frac{F_N}{2\pi\sqrt{2a\rho}} \sin\theta,$$

where ρ and θ are polar coordinates centered around the lines $x = \pm a$. If the material is compressible ($\nu < 0.5$) the surface tends also to move towards the center of the strip according to the expression

$$u_x(x) = -\frac{(1-2\nu)(1+\nu)F_N}{\pi E} \arcsin\frac{x}{a}.$$

If the half-space is indented by a rigid cylinder with its axis perpendicular to the surface, as in Fig. 4.5b, the pressure distribution in the contact circle is

$$p(r) = \frac{p_0}{\sqrt{1 - r^2/a^2}}, \tag{4.10}$$

where $p_0 = F_N/2\pi a^2$. In this case the penetration depth is related to F_N as

$$\delta = \frac{(1 - \nu)F_N}{4aG}. \tag{4.11}$$

Outside the circle the normal displacement is

$$u_z(r) = \frac{(1 - \nu)F_N}{2\pi a G} \arcsin \frac{a}{r} \qquad (r > a).$$

Again, the pressure (4.10) becomes infinite at the edge of the contact area, where the deformation has an infinite gradient. The stress distribution in the solid was calculated by Sneddon [309].

4.3 The Hertz theory

In the problems discussed so far the area of contact does not change with the normal force F_N. This is not the case if an elastic sphere is pressed against a half-space. Here, it can be seen by interference techniques that the contact area is a circle, the radius of which increases as $F_N^{1/3}$ [156, section 4.1]. A general solution to these problems was found by Hertz [136].

Contact between elastic bodies

Consider two solid objects whose surfaces touch at a point (Fig. 4.6). If this point is not a singularity for the surface geometries, we can define the radii of curvature R_1', R_2' and R_1'', R_2'' of the two surfaces, and a common tangent plane passing through the point of contact. With a proper choice of the x and y axes along this plane, it is always possible to write the gap between the two surfaces (outside the contact area) as

$$h(x, y) = Ax^2 + By^2.$$

The quantities A and B are related to the radii of curvature by the expressions

$$A + B = \frac{1}{2}\left(\frac{1}{R_1'} + \frac{1}{R_2'} + \frac{1}{R_1''} + \frac{1}{R_2''}\right)$$

and

$$B - A = \frac{1}{2}\left[\left(\frac{1}{R_1} - \frac{1}{R_2}\right)^2 + \left(\frac{1}{R_1'} - \frac{1}{R_2'}\right)^2 + 2\left(\frac{1}{R_1} - \frac{1}{R_2}\right)\left(\frac{1}{R_1'} - \frac{1}{R_2'}\right)\cos 2\alpha\right]^{1/2},$$

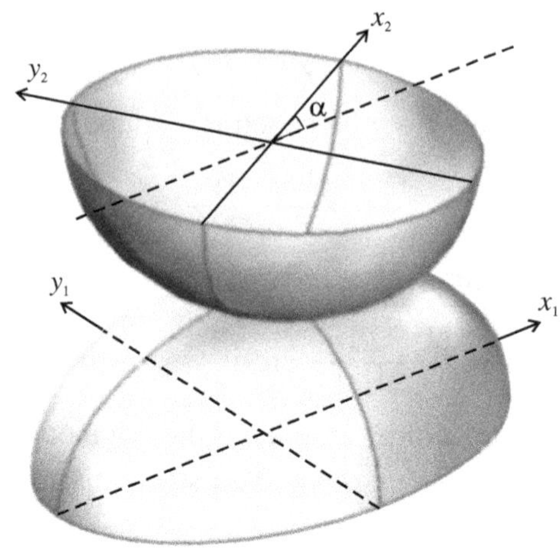

Figure 4.6 Geometry of two elastic bodies with convex surfaces in contact at a point.

where α is the angle formed by the normal sections with radii of curvature R_1 and R_1'. Note that we have implicitly assumed that the two surfaces are not conforming, i.e. they are not likely to form a contact of a size comparable to their dimensions.

If the two bodies are pressed against each other by a force F_N perpendicular to the tangent plane, they will deform and touch over a finite *area of contact* which, in a first approximation, is an ellipse with semi-axes a and b. Furthermore, the bodies will approach each other over a distance δ. The expressions relating these quantities are not trivial. The elastic properties of the two materials appear only in the combination

$$\frac{1}{E^*} = \frac{1-\nu_1^2}{E_1} + \frac{1-\nu_2^2}{E_2}, \tag{4.12}$$

where E_i and ν_i $(i = 1, 2)$ are the Young's modulus and Poisson's ratio of the two materials. The semi-axes a and b are implicitly determined by the relations

$$A = \frac{p_0}{E^*}\frac{b}{\varepsilon^2 a^2}\left(K(\varepsilon) - E(\varepsilon)\right)$$

and

$$B = \frac{p_0}{E^*}\frac{b}{\varepsilon^2 a^2}\left(\frac{a^2}{b^2}E(\varepsilon) - K(\varepsilon)\right),$$

where E and K are the complete elliptic integrals introduced in Section 4.1, $\varepsilon = \sqrt{1 - b^2/a^2}$ is the eccentricity of the ellipse and $p_0 = 3F_\mathrm{N}/2\pi ab$ is the maximum pressure in the contact. Once a and b are known, the penetration depth can be calculated as

$$\delta = \frac{p_0}{E^*} b K(\varepsilon).$$

In spite of the complexity of the problem, two simple relations are valid in general. First, δ increases with F_N as

$$\delta \propto F_N^{2/3}. \tag{4.13}$$

Second, the pressure distribution in the contact ellipse is related to a and b by

$$p(x, y) = p_0 \sqrt{1 - \frac{x^2}{a^2} - \frac{y^2}{b^2}}. \tag{4.14}$$

Note that the pressure drops continuously to zero towards the edge of the contact area and its maximum value, which is reached at the center of the contact area, is 3/2 the average pressure. Useful approximations for a, b and δ have been derived by Hamrock and Dowson using the least squares method [131].

Contact between two spheres

In the case of two elastic spheres with radii R_1 and R_2, the previous expressions are considerably simplified. The quantities A and B are both equal to $1/2R$, where the equivalent radius R of the spheres is defined as

$$\frac{1}{R} = \frac{1}{R_1} + \frac{1}{R_2}. \tag{4.15}$$

In this way the problem can be seen to be completely equivalent to that of a rigid sphere of radius R pressed against an elastic half-space with Young's modulus E^* (Fig. 4.7). The contact area is a circle with radius

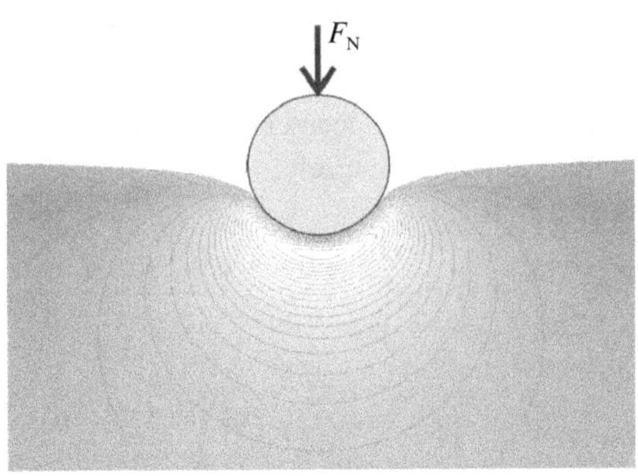

Figure 4.7 Maximum shear stress distribution in an elastic half-space indented by a rigid sphere.

$$a = \left(\frac{3F_N R}{4E^*}\right)^{1/3} \tag{4.16}$$

and the pressure distribution is

$$p(r) = p_0\sqrt{1 - \frac{r^2}{a^2}}, \tag{4.17}$$

where $p_0 = 3F_N/2\pi a^2$. The penetration depth is

$$\delta = \left(\frac{3}{4E^*\sqrt{R}}\right)^{2/3} F_N^{2/3}, \tag{4.18}$$

and is related to the contact radius a by the simple expression

$$\delta = a^2/R. \tag{4.19}$$

According to the previous formulas, if two steel balls with radius of 10 mm are pressed against each other by a force of 1 N, they will approach up to a distance of about 0.2 μm.

The pressure (4.17) causes a vertical displacement

$$u_{z,i}(r) = \frac{(1-\nu_i^2)}{E_i}\frac{\pi p_0}{4a}(2a^2 - r^2) \tag{4.20}$$

of the surface of each sphere inside the loaded circle, and

$$u_{z,i}(r) = -\frac{(1-\nu_i^2)}{E_i}\frac{p_0}{2a}\left((2a^2 - r^2)\arcsin\frac{a}{r} + ar\sqrt{1 - \frac{a^2}{r^2}}\right)$$

out of it. Apart from the normal compression, the mutual contact pressure produces a tangential displacement which, inside the contact circle, is given by

$$u_{r,i}(r) = -\frac{(1-2\nu_i)(1+\nu_i)}{3E_i}\frac{a^2}{r}p_0\left[1 - \left(1 - \frac{r^2}{a^2}\right)^{3/2}\right].$$

Outside this circle the tangential displacement takes the same values as if the load were concentrated in the center of the circle.

Finally, the stress distribution is

$$\sigma_r = -\frac{E}{1+\nu}\frac{u_r}{r} - p(r), \qquad \sigma_\theta = -\frac{E}{1+\nu}\frac{u_r}{r} - 2\nu p(r) \tag{4.21}$$

inside the loaded circle and

$$\sigma_r = \sigma_\theta = -p_0(1+\nu)\left(1 - \frac{z}{a}\arctan\frac{a}{z}\right) + \frac{p_0}{2(1+z^2/a^2)}, \tag{4.22}$$

$$\sigma_z = -\frac{p_0}{\sqrt{1+z^2/a^2}} \tag{4.23}$$

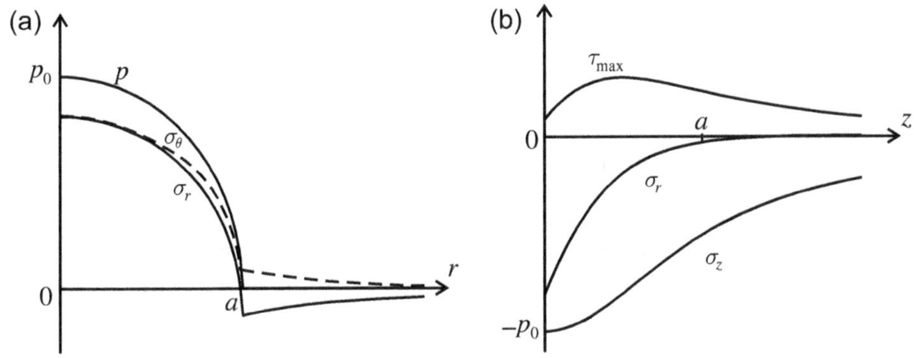

Figure 4.8 Stress distribution accompanying the deformation of two elastic spheres pressed against each other: (a) in the contact area and (b) along the axis of symmetry of the problem (for one sphere only).

along the z axis. The relations (4.21) and (4.22) are plotted in Fig. 4.8 together with the variation of the maximum shear stress τ_{max} along the z axis. Note that τ_{max} peaks beneath the contact area. If $\nu = 0.3$ the peak value is $0.31 p_0$. In the contact circle the stress components are compressive, except for the radial component σ_r at the edge, which tends to $(1 - 2\nu)p_0/3$. This tensile stress may cause cracks when a sphere is pressed into contact.

Collision between elastic spheres

The elastic potential energy U_{el} of two spheres in contact can be determined by observing that the normal force F_N is the first derivative of U_{el} with respect to the penetration depth δ. Using Eq. (4.18) we easily get

$$U_{el}(\delta) = \frac{8}{15} E^* \sqrt{R} \delta^{5/2}. \tag{4.24}$$

The expression (4.24) allows us to solve an interesting problem: how long is the time of contact t_c when the spheres are launched against each other with a relative speed v_0? In the frame of reference of the center of mass, the spheres have a kinetic energy $E_{kin} = mv_0^2/2$, where $m = m_1 m_2/(m_1 + m_2)$ is the reduced mass of the system. If v_0 is small compared to the velocity of sound, the time t_c can be estimated by the law of conservation of energy. As a result [70]:

$$t_c = 2.87 \left(\frac{m^2}{RE^{*2} v_0} \right)^{1/5}.$$

The maximum penetration depth can be also estimated as

$$\delta_{max} = \left(\frac{15 m v_0^2}{16 \sqrt{R} E^*} \right)^{2/5}.$$

If $v_0 = 1$ m/s, the two steel balls considered before would remain in contact for a time $t_c \approx 70$ μs with penetration depth $\delta_{max} \approx 23$ μm.

Contact between two cylinders

If two cylinders are pressed against each other along a generator by a line load F_N, the contact area is a strip.[2] The half-width a of this strip and the pressure distribution in it can be determined by taking the limit $b \to \infty$ in the general equations of the Hertz theory. As a result:

$$a = \sqrt{\frac{4RF_N}{\pi E^*}}, \tag{4.25}$$

where the equivalent R is defined by Eq. (4.15), and

$$p(x) = p_0\sqrt{1 - \frac{x^2}{a^2}}, \tag{4.26}$$

with $p_0 = 2F_N/\pi a$. In this case the penetration depth δ does not follow the general relation (4.13) and cannot be derived from the Hertz theory. The maximum shear stress τ_{max} is distributed as shown in Fig. 4.9. Along the z axis it varies as

$$\tau_{max} = p_0 a\left(z - \frac{z^2}{\sqrt{a^2 - z^2}}\right),$$

independently of Poisson's ratio, and peaks at $0.30p_0$ at $z = 0.78a$.

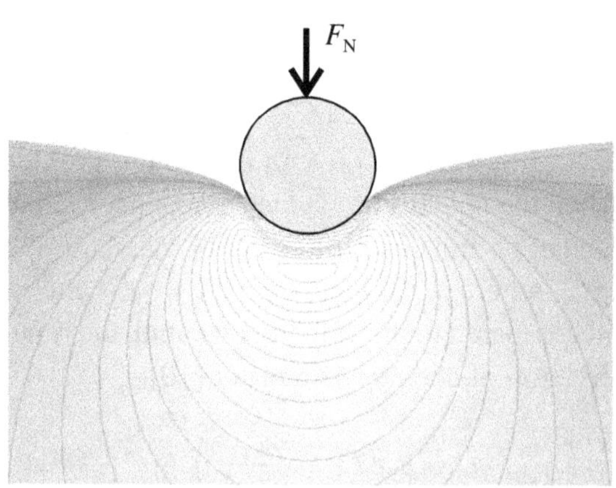

Figure 4.9 Maximum shear stress distribution in an elastic half-space indented by a rigid cylinder.

[2] If two cylinders cross each other perpendicularly, the problem is equivalent to that of a sphere in contact with a plane.

However, real cylinders have a finite size, and the contact stress concentrates at their ends. This problem can be reduced if the cylinders are reshaped in the form of barrels.

Axisymmetric indenters

Suppose now that a rigid indenter with an axisymmetric profile $z(r) = Ar^\beta$ is pressed against an elastic half-space. The resulting deformation was calculated by Sneddon, who derived the load – displacement relation [310]

$$F_\mathrm{N} = \frac{2E^*}{(\sqrt{\pi}A)^{1/\beta}} \frac{\beta}{\beta+1} \left(\frac{\Gamma(\beta/2+1/2)}{\Gamma(\beta/2+1)} \right)^{1/\beta} \delta^{1+1/\beta}. \tag{4.27}$$

In Eq. (4.27) Γ is the gamma function defined as

$$\Gamma(x) = \int_0^\infty t^{x-1} \mathrm{e}^{-x}\, \mathrm{d}x.$$

The Sneddon theory is applied in the Oliver–Pharr model for nanoindentation, discussed in Section 12.8.

4.4 Beyond the Hertz theory

Contact between a wedge or a cone and a half-space

If a wedge or a cone is pressed against a half-space, the contact has a singularity and the Hertz theory cannot be applied. However, the surface deformation can still be determined using the relation (4.5) for the displacement gradients (in the case of a wedge, and a similar relation for a cone). Note that, in order for the deformation to be small, the half-angle α of the indenter has to be sufficiently large. In other words, the formulas introduced below are only valid if the wedge or the cone is blunt. In this case the contact area between a wedge and a half-space turns out to be a strip with half-width [156, section 5.2]

$$a = \frac{F_\mathrm{N}}{E^*} \tan\alpha,$$

where F_N is the normal force per unit length. The pressure distribution in the contact area is

$$p(x) = p_0 \cosh^{-1}(a/x), \tag{4.28}$$

where

$$p_0 = \frac{E^*}{\pi \tan\alpha}.$$

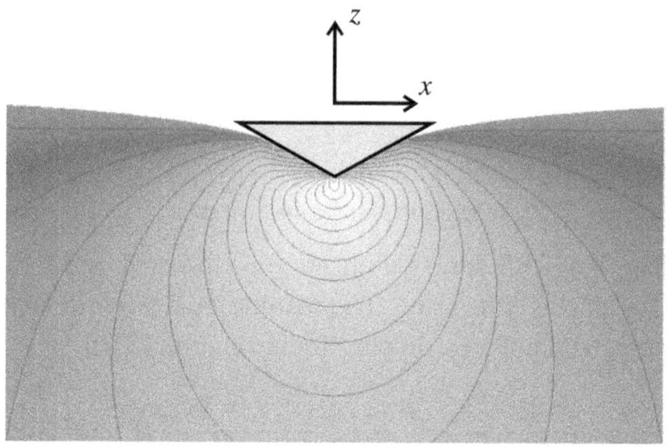

Figure 4.10 Maximum shear stress distribution in an elastic half-space indented by a rigid blunt wedge.

The distribution (4.28) has a logarithmic singularity when $x \to 0$. In spite of that, the maximum shear stress τ_{max} remains finite and peaks at p_0, independently of the applied load, at the apex of the wedge (Fig. 4.10).

Similarly, the contact area between a blunt cone and a half-space [195] is a circle with radius

$$a = \sqrt{\frac{2F_N}{\pi E^*}\tan\alpha},$$

where F_N is the normal force. The pressure distribution $p(r)$ is given again by the relation (4.28), with

$$p_0 = \frac{E^*}{2\tan\alpha}.$$

As for a rigid wedge, p_0 coincides with the peak value of the maximum shear stress τ_{max}, which is reached at the apex of the cone.

Conforming surfaces

Conforming surfaces behave quite differently from the predictions of the Hertz theory. This is due to the fact that the application of a light load can dramatically vary the size of the contact area.

An important example is the problem of a pin in a hole (Fig. 4.11a). This problem was first solved for elastically similar materials by A. Persson [241] and generalized to dissimilar materials by Ciavarella and Decuzzi [57]. For similar materials the contact arc length, 2α, depends on the ratio between the difference ΔR between the radii of the pin and the hole and the normal force F_N according to the relation

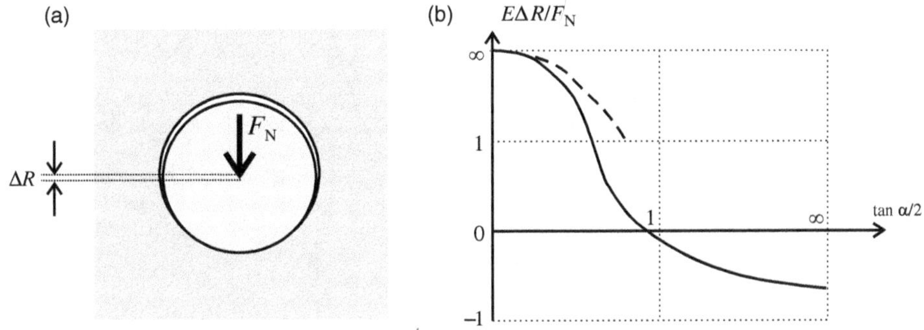

Figure 4.11 (a) Contact of a pin with a conforming hole. (b) The solution for the contact arc according to Eq. (4.29) (continuous curve) and the Hertzian approximation at low load (dashed curve).

$$\frac{E^*\Delta R}{F_N} = \frac{2}{\pi}\frac{1-x^2}{x^2} - \frac{1}{\pi^2 x^2(1+x^2)}\int_{-x}^{x} \log\frac{\sqrt{x^2+1}+\sqrt{x^2-s^2}}{\sqrt{x^2+1}-\sqrt{x^2-s^2}}\frac{ds}{1+s^2} \tag{4.29}$$

with $x = \tan(\alpha/2)$. Equation (4.29) is plotted in Fig. 4.11b together with the Hertzian expression, which is recovered at low load or at large clearance. Note that the contact is stiffer than expected from the Hertz theory. This is also the case for an elastic sphere in a conforming cavity [116].

A pin in a hole is an example of *receding contact*. If the unloaded pin perfectly fits the hole, a gap suddenly appears when the pin is loaded, and the contact area shrinks down.

4.5 Influence of friction on normal contact

We discuss now the influence of the interfacial friction in the contact of two objects pressed *normally* against each other.

Indentation of an elastic body

Consider a rigid punch (with base width $2a$) indenting an elastic half-space. If the static friction were capable of preventing slip completely (see also section 5.2), the pressure distribution would not differ significantly from (4.9) and a tangential traction $\tau = \mu p$ would appear in the contact area, where μ is the coefficient of friction between the punch and the half-space. However, this traction would become infinite at the edges of the contact, a situation which cannot be sustained in practice. As a result, slip must occur. The slip considerably changes the distribution of the traction τ, but not that of the pressure p, in the contact area. The ratio τ/p is

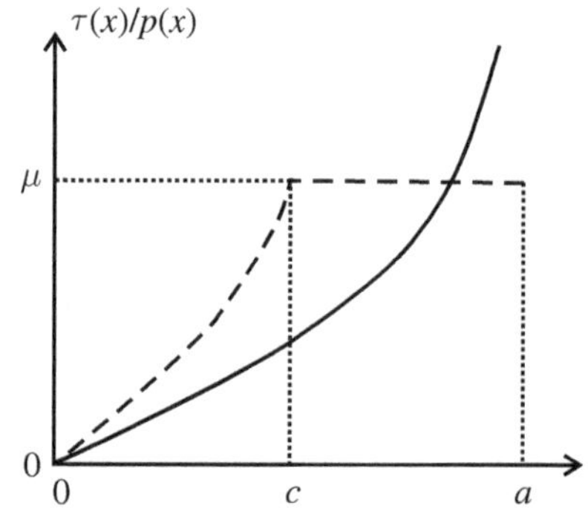

Figure 4.12 Ratio between tangential traction and normal pressure for a rigid flat punch indenting an elastic half-space with no slip (continuous curve) and with slip (dashed curve). Parameter values: $\nu = 0.3$ and $\mu = 0.237$. Adapted from [315] with permission from Cambridge University Press.

plotted in Fig. 4.12 as a function of the distance x from the axis of the strip. Slip occurs out of a thinner strip, the half-width c of which can be implicitly determined using a relation derived by Spence [315]:

$$\frac{E\left(\sqrt{1 - c^2/a^2}\right)}{E(c/a)} = \frac{1 - 2\nu}{2(1 - \nu)\mu} \tag{4.30}$$

(E is the complete elliptic integral of the second kind introduced in Section 4.1).

Similarly, if a rigid cylinder pressed against an elastic half-space completely adhered to it, the pressure distribution would not differ significantly from (4.10), but the friction would become infinite at the edge, meaning that slip must occur out of a circular area.

Hertzian contacts

Interfacial friction plays a role in the contact of two non-conforming surfaces only if the elastic constants of the two materials are different. In this case the influence of the tangential traction on the normal pressure is usually not negligible. If two cylinders are pressed against each other along a generator, a relation similar to (4.30) holds, with the factor multiplying $(1/\mu)$ replaced by the *Dundurs parameter* [156, section 5.4]

$$\beta = \frac{(1 - 2\nu_1)/G_1 - (1 - 2\nu_2)/G_2}{2(1 - \nu_1)/G_1 + 2(1 - \nu_2)/G_2}. \tag{4.31}$$

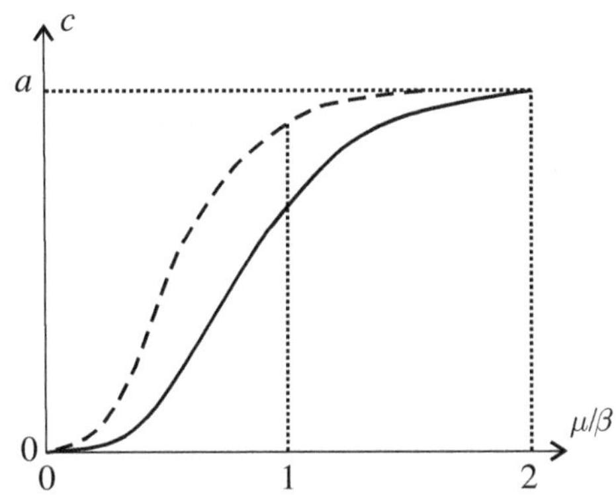

Figure 4.13 Extension of the no-slip area in the contact between two dissimilar cylinders (continuous curve) and two dissimilar spheres (dashed curve) as a function of the ratio μ/β defined in the text.

The half-width c of the contact area is plotted in Fig. 4.13 as a function of μ/β. If two dissimilar spheres are pressed against each other, the radius c of the no-slip circle is given by the relation [115]

$$\frac{a}{2c}\ln\left(\frac{a+c}{a-c}\right) = \frac{\beta}{\mu}E\left(\sqrt{1-c^2/a^2}\right),$$

which is also plotted in Fig. 4.13.

5

Tangential contacts

In this chapter we consider the deformation of elastic bodies under the influence of tangential forces. Again we will first assume that the force is concentrated on a single point on the free surface of an elastic half-space, or uniformly distributed on a straight line or a long strip. If the force is applied to a cylinder or a sphere pushed against the half-space, an infinite traction appears at the edge of the contact area, resulting in the appearance of slip. The slip areas becomes larger at increasing values of the force, till the body is set into motion when the static friction is overcome. The deformation resulting from an object pressed against an elastic half-space and sliding over it will be discussed in the case of rigid spheres or cylinders. Finally, we will briefly mention how oscillating tangential forces and oscillating torques may affect the contact between two elastic materials.

5.1 Traction on an elastic half-space

Point traction

The problem of a concentrated tangential force F_x acting on the free surface of an elastic half-space was solved by Cerruti [50]. The three components of the resulting surface displacement are [177, par. 8]

$$u_x = \frac{F_x}{4\pi G r}\left(2(1-\nu) + \frac{2\nu x^2}{r^2}\right),$$

$$u_y = \frac{\nu}{2\pi G}\frac{xy}{r^3}F_x, \qquad u_z = \frac{1-2\nu}{4\pi G}\frac{x}{r^2}F_x,$$

where G and ν are the shear modulus and Poisson's ratio of the material, and r is the distance from the point of application of the force. The non-zero components of the stress distribution on the free surface are:

$$\sigma_x = -\frac{3x}{r^3}\left[1 + 2\nu\left(\frac{x^2}{r^2} - 1\right)\right]\frac{F_x}{2\pi}, \qquad \sigma_y = -\frac{6\nu x y^2}{r^5}\frac{F_x}{2\pi},$$

$$\tau_{xy} = \frac{y}{r^3}\left[-1 + 2\nu\left(\frac{3x^2}{r^2} - 1\right)\right]\frac{F_x}{2\pi}.$$

As for the normal forces, the deformation resulting from a distribution of tangential forces can be found by superposition.

Line traction

If a tangential line force F_x is uniformly distributed along the y axis, as in Fig. 5.1, the tangential displacement varies logarithmically with the distance x from this axis:

$$u_x(x) = -\frac{(1-\nu)F_x}{2\pi G}\ln|x| + \text{const.} \tag{5.1}$$

The surface ahead of the force is depressed by an amount proportional to F_x whereas the surface behind it rises by the same amount:

$$u_z(x) = \pm\frac{(1-2\nu)F_x}{4G}.$$

The stress in the solid is radially distributed and its contour lines are semicircles passing through the traction line:

$$\sigma_r = -\frac{2F_x}{\pi}\frac{\cos\theta}{\rho}. \tag{5.2}$$

Note that in Eq. (5.2) the angle θ is measured clockwise from the line of action of the force, so that the stress is compressive if $x > 0$ and tractive if $x < 0$.

If the tangential force is oriented parallel to the y axis, the x dependence of the displacement in the force direction does not differ qualitatively from the previous case:

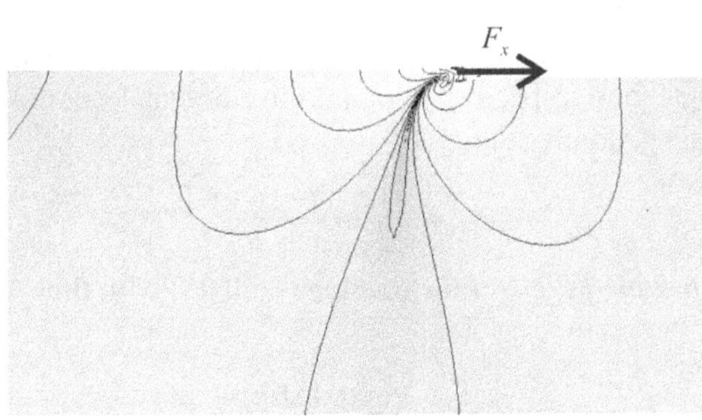

Figure 5.1 Maximum shear stress distribution (logarithmic scale) in an elastic half-space deformed by a tangential force uniformly distributed along a line.

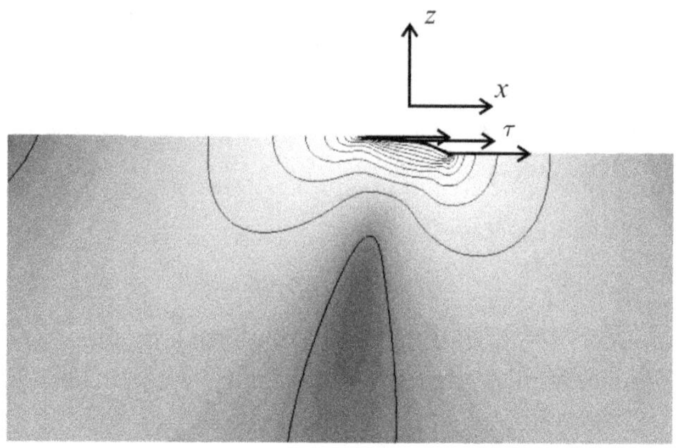

Figure 5.2 Maximum shear stress distribution in an elastic half-space deformed by a tangential force uniformly distributed on a strip.

$$u_y(x) = -\frac{F_y}{\pi G} \ln |x| + \text{const.}$$

However, there is no displacement in the normal direction.

Strip traction

If a tangential stress $\tau(x)$ is applied to an infinite strip of half-width a, as in Fig. 5.2, the displacement u_x is obtained from a relation which is completely analogous to Eq. (4.5):

$$\frac{\partial u_x}{\partial x} = -\frac{1-\nu}{\pi G} \int_{-a}^{a} \frac{\tau(s)\,\mathrm{d}s}{x-s}. \tag{5.3}$$

If τ is uniform, the formulas for the resulting surface displacement are similar to those obtained when a uniform pressure p is applied to the strip, with u_z replaced by u_x, and u_x replaced by $-u_z$: see Eq. (4.6) and (4.7). However, the stress always remains finite when a uniform pressure is applied, whereas the stress component σ_x becomes infinite, and respectively compressive or tensile, at the edges of a strip which is loaded tangentially (Fig. 5.3):

$$\sigma_x = \frac{2\tau}{\pi} \ln \frac{x-a}{x+a}.$$

This may cause *fretting fatigue* if the traction oscillates with time.

5.2 Partial slip

When two elastic objects are pressed against each other and, at the same time, they are moved laterally, the large traction at the edges of the contact area causes

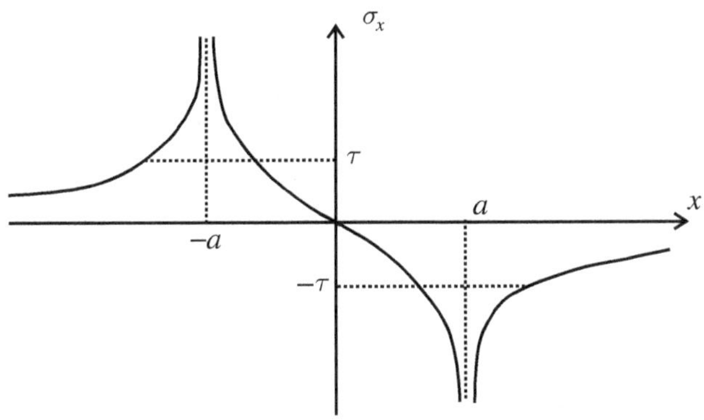

Figure 5.3 Tangential shear stress distribution on the free surface of the half-space.

a *partial slip* which reduces the size of the contact. A similar phenomenon is observed if the objects are twisted about an axis passing through the area of contact.

Cylinder

Suppose that a tangential force (per unit length) F_x is applied to an elastic cylinder, which is pressed along a generator against an elastic half-space by a normal force F_N. The ratio F_x/F_N is supposed to be lower than the coefficient of friction μ, so that the cylinder cannot start sliding. If the cylinder adhered completely to the substrate, the tangential displacement in the contact area would be constant, corresponding to a tangential stress

$$\tau(x) = \frac{\tau_0}{\sqrt{1 - x^2/a^2}},$$

where a is given by (4.25) and $\tau_0 = F_x/\pi a$. However, an infinite value of τ at the edges of the contact cannot be sustained and, in practice, slip must occur out of a strip of half-width c.

In the slip region the tangential stress is

$$\tau_1(x) = \mu p_0 \sqrt{1 - x^2/a^2}, \tag{5.4}$$

where $p_0 = 2F_N/\pi a$. In the stick region, a contribution

$$\tau_2(x) = -\frac{c}{a}\mu p_0 \sqrt{1 - \frac{x^2}{c^2}} \tag{5.5}$$

must be added to (5.4) in order to have equal tangential displacements in the two surfaces. The resulting stress distribution $\tau(x)$ in the two regions is shown

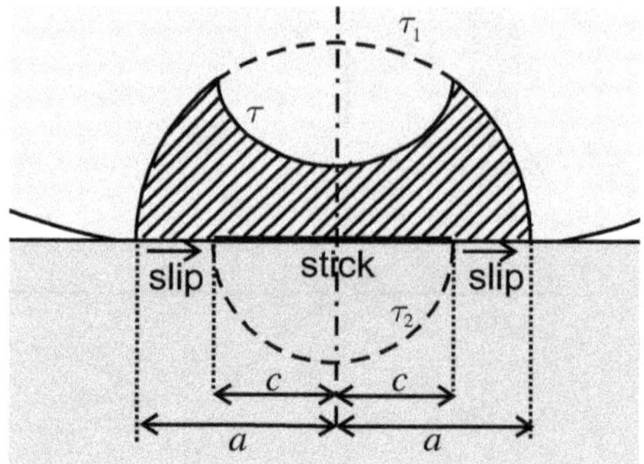

Figure 5.4 Tangential stress distribution in the elastic contact between a cylinder and a half-space (dashed area) in the presence of a normal force F_N and a tangential force $F_x < \mu F_N$.

in Fig. 5.4. Integrating $\tau(x)$ and equating the result to F_x, c can be precisely determined, as first done by Cattaneo [49]:

$$c = a\sqrt{1 - \frac{F_x}{\mu F_N}}. \tag{5.6}$$

Sphere

If a tangential force F_x is applied to an elastic sphere pressed against an elastic half-space the tangential stress in the contact area is

$$\tau(x) = \frac{\tau_0}{\sqrt{a^2 - r^2}},$$

where a is given by (4.16) and $\tau_0 = F_x/2\pi a^2$. The corresponding displacement is proportional to F_x:

$$\delta_x = \frac{\pi(2-\nu)}{4G^*}\tau_0 a, \tag{5.7}$$

where

$$\frac{1}{G^*} = \frac{2-\nu_1}{G_1} + \frac{2-\nu_1}{G_2} \tag{5.8}$$

is the effective shear modulus of the materials in contact. As for a cylindrical contact, slip is unavoidable. The radius of the circular stick region is

$$c = a\left(1 - \frac{F_x}{\mu F_N}\right)^{1/3}, \tag{5.9}$$

and the expression for the lateral displacement becomes

$$\delta_x = \delta_0\left(1 - c^2/a^2\right), \tag{5.10}$$

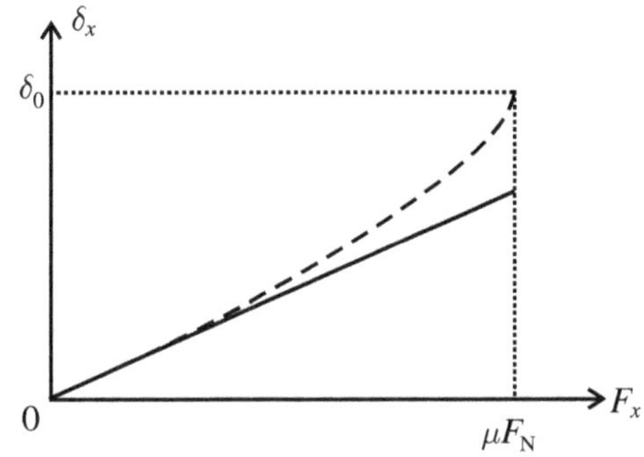

Figure 5.5 Tangential displacement of the contact between a sphere and a half-space in the presence of a normal force F_N and a tangential force $F_x < \mu F_N$: without slip (continuous curve) and with slip (dashed curve).

where $\delta_0 = 3\mu F_N/16aG^*$. The quantity δ_x is plotted in Fig. 5.5 as a function of F_x. Note that at low values of F_x Eq. (5.10) is well approximated by the linear relation (5.7) for no slip.

Torsional traction

Suppose now that a rigid cylinder with radius a adheres to the surface of an elastic half-space and is twisted about its axis by a torque M_z. In this case the contact area turns by an angle [156, section 3.9]

$$\beta = \frac{3M_z}{16Ga^3}. \tag{5.11}$$

The azimuthal displacement in this area is $u_\varphi = \beta r$ and the circumferential stress is

$$\tau_\varphi(r) = \frac{3M_z r}{4\pi a^4\sqrt{1 - r^2/a^2}}. \tag{5.12}$$

Since the normal displacement is zero, the pressure distribution is not modified by the twist.

If an elastic sphere is pressed against an elastic half-space and twisted about the normal axis through the center of the contact circle, the twist angle is given, in principle, by Eq. (5.11). However, the tangential stress would become infinite at the edge of the contact circle so that, once again, slip must occur. The radius c of the stick region depends in a non-trivial way on the parameter β [197]:

$$\beta = \frac{3\mu F_N}{4\pi a^2}\left(1 - \frac{c^2}{a^2}\right)\left(\frac{1}{G_1} + \frac{1}{G_2}\right)\left(K(\sqrt{1 - c^2/a^2}) - E(\sqrt{1 - c^2/a^2})\right),$$

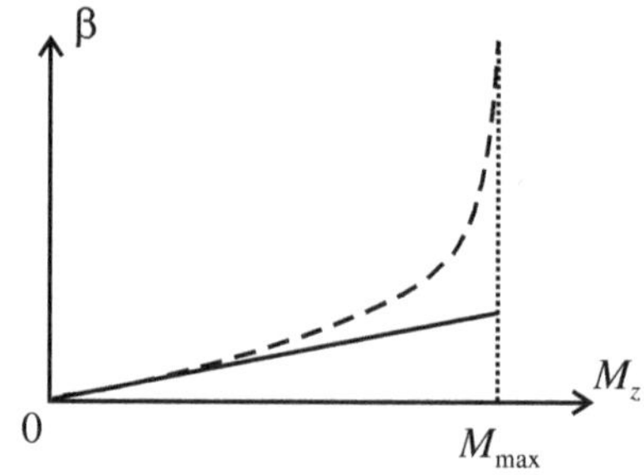

Figure 5.6 Twist angle of a circular region subjected to a torque: without slip (continuous curve) and with slip (dashed curve).

where K and E are the complete elliptic integrals of the first and second kind (see Section 4.1). The corresponding relation between β and M_z is plotted in Fig. 5.6. The maximum value that M_z can reach, before the stick region shrinks to only one point and complete slip occurs, is

$$M_{\text{max}} = (3\pi/16)\mu F_{\text{N}} a.$$

5.3 Sliding of elastic objects

Sliding cylinder

If a cylinder slides on an elastic half-space, with its axis parallel to the free surface, the tangential stress caused by the kinetic friction is

$$\tau(x) = \tau_0 \sqrt{1 - x^2/a^2}. \tag{5.13}$$

In Eq. (5.13) a is the half-width of the contact strip and $\tau_0 = 2\mu F_{\text{N}}/\pi a$, where μ is the friction coefficient (see Section 4.3). The maximum shear stress distribution produced by the combined effect of the normal pressure and the tangential traction is shown in Fig. 5.7 for the case $\mu = 0.2$ and $\nu = 0.33$. Compared to the results without friction (Fig. 4.9), τ_{max} peaks at a point which is closer to the contact area. If μ exceeds a critical value, the peak is located on this area. Note that we have implicitly assumed that the influence of friction on the shape and size of the contact area and on the pressure distribution can be neglected. This hypothesis can be rigorously verified [37].

Sliding sphere

Similarly to a cylinder, if a rigid sphere slides on an elastic half-space a tangential stress

$$\tau(x) = \tau_0 \sqrt{1 - r^2/a^2}$$

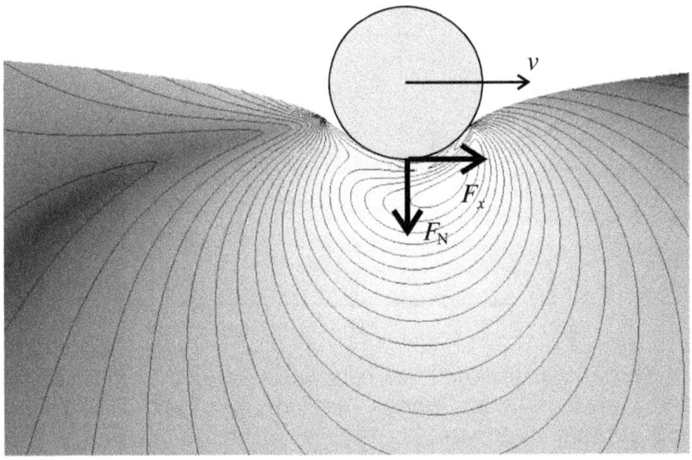

Figure 5.7 Maximum shear stress distribution in an elastic half-space under the combined pressure and tangential traction of a rigid cylinder sliding on it (friction coefficient: $\mu = 0.2$).

appears, where a is the radius of the contact circle and $\tau_0 = 3\mu F_N/2\pi a^3$. This traction causes a lateral displacement

$$u_x = \frac{\pi\tau_0}{32Ga}\left[4(2-\nu)a^2 - (4-3\nu)x^2 - (4-\nu)y^2\right], \qquad u_y = \frac{\pi\tau_0}{32Ga}2\nu xy$$

inside the circle. Expressions for the surface displacement and the stress distribution have been derived [153, 130]. Again, the maximum shear stress τ_{max} peaks at a point on the contact area if μ exceeds a critical value depending on Poisson's ratio ν.

Sliding punch

If a rigid punch of half-width a slides on an elastic half-space, the influence of the interfacial friction on the pressure distribution is quantified by a relatively simple formula [156, section 2.8]:

$$p(x) = \frac{p_0\cos(\pi\gamma)}{\sqrt{1-x^2/a^2}}\left(\frac{a+x}{a-x}\right)^{\gamma},$$

where $p_0 = F_N/\pi a$ and the parameter γ is related to the friction coefficient by

$$\tan(\pi\gamma) = -\frac{\mu(1-2\nu)}{2(1-\nu)}. \tag{5.14}$$

Comparing with the formula (4.9) obtained in the static case, the pressure is reduced on the front half of the punch and increased on the rear, although the variation is very small (Fig. 5.8).

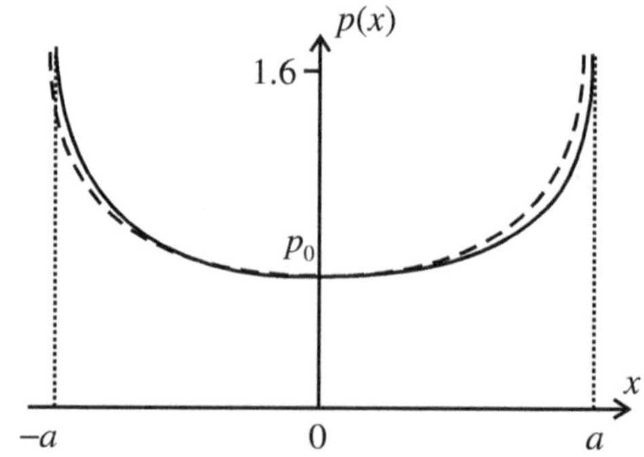

Figure 5.8 Pressure at the interface between an elastic half-space and a rigid punch sliding on it: without friction (dashed curve) and with friction ($\mu = 0.5$, $\nu = 0.3$, continuous curve).

5.4 Influence of oscillating forces

Suppose that two elastic spheres in contact are subjected to a tangential force F_x oscillating with an amplitude F_0, in addition to a steady normal force F_N. In this case, the lateral displacement of the spheres depends on the contact history, as shown in Fig. 5.9. The first application of F_x causes a micro-slip out of a circle with a radius c defined by Eq. (5.9) with $F_x = F_0$, whereas in the subsequent unloading phase a reversed slip penetrates more and more into the contact. When $F_x = 0$ the reversed slip extends on a circle of radius c' ($> c$) given by [156, section 7.4]

$$c' = a\left[\frac{1}{2}\left(1 + \frac{c^3}{a^3}\right)\right]^{1/3}.$$

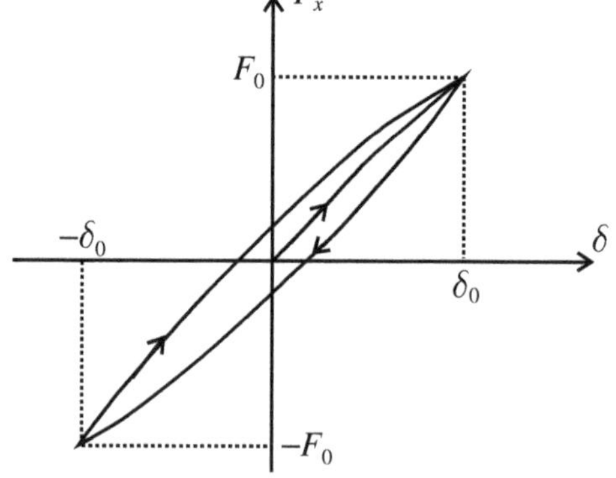

Figure 5.9 Force–displacement cycle in a circular contact subjected to a steady normal force F_N and an oscillating tangential force of amplitude F_0.

When the tangential motion is completely reversed ($F_x = -F_0$) the radius of the no-slip area is again equal to c, and so on. The energy dissipated in a full cycle corresponds to the area of the loop in Fig. 5.9 and can be precisely calculated [221]. If the oscillations are small, it is proportional to the cube of the oscillation amplitude:

$$E_{\rm diss} \approx \frac{1}{36a\mu F_{\rm N} G^*} F_0^3,$$

where G^* is defined by Eq. (5.8). However, a quadratic dependence is usually observed in the experiments. This result may be attributed to internal damping, variations of the friction coefficient and surface roughness [154]. Mindlin and Deresiewicz have also estimated the energy dissipated when the oscillating force forms an angle α with the normal direction [220]. As a result, $E_{\rm diss} \neq 0$ only if $\tan\alpha > \mu$. If this is not the case, no slip or energy loss occur, as substantiated by experiments on steel spheres [154].

Hysteresis and energy dissipation are also observed if an oscillating torque of amplitude M_0 is applied to a circular contact [69]. In this case, for small values of M_0 ($\ll \mu F_{\rm N} a$):

$$E_{\rm diss} \approx \frac{3M_0^3}{16Ga^4 \mu F_{\rm N}}.$$

6
Elastic rolling

This chapter is dedicated to the deformation of two elastic bodies rolling over each other, and the influence of friction on it. We will start by discussing the contact between two cylinders and distinguish between elastically similar and dissimilar materials. The complex problem of rolling contact between three-dimensional objects can be addressed, in a first approximation, with the linear theory developed by Kalker. Specific topics such as the rolling of a sphere in a groove and the deformation of aircraft and automotive tires will conclude the chapter.

6.1 Steady elastic rolling

Creep ratio

Consider the steady rolling of two elastic cylinders over each other. In the absence of deformation and reciprocal sliding the contact points move with a common speed v_0 directed along the x axis. However, if a normal force is applied, a friction force appears, causing a tangential strain ε_{xi} (Section 5.1) and partial slip (Section 5.2) in each cylinder. As a result, the relation between the velocities v_1 and v_2 of the contact points can be written as [156, section 8.1]

$$\frac{v_1 - v_2}{v_0} = \xi_x + (\varepsilon_{x1} - \varepsilon_{x2}), \tag{6.1}$$

where ξ_x is the so-called called *creep ratio.*

Elastically similar cylinders

The influence of the interfacial friction on the rolling of two cylinders with the same elastic properties was first investigated by Carter [48]. As an example, one may think to the wheel of a vehicle which is braked. If the tangential force is less than the static friction, a slip area is detached from the contact. The semi-width c of

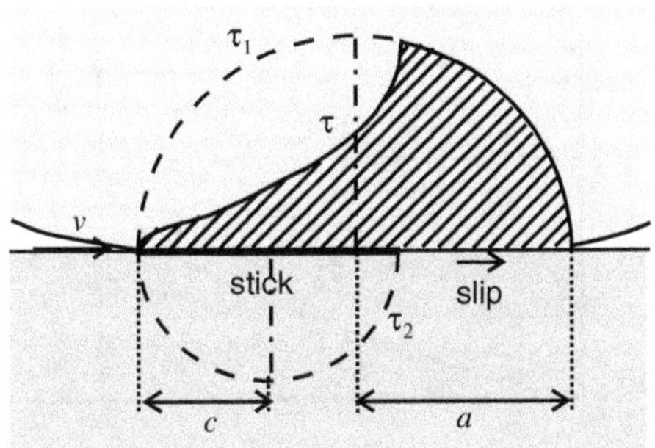

Figure 6.1 Tangential stress distribution accompanying the rolling of elastically similar cylinders in the presence of friction.

the remaining stick area is given by Eq. (5.6). However, the stick area is now shifted next to the leading edge, as shown in Fig. 6.1. This is to satisfy the condition that the direction of slip must oppose the traction in the slip area [156, section 8.3]. The tangential stress $\tau(x)$ is obtained as the sum of the contributions (5.4) and (5.5), with the distribution $\tau_2(x)$ centered in $(c - a)$.

In the stick area the tangential strains in each surface have opposite signs and their moduli are equal to

$$\varepsilon_x = \frac{2(1-\nu^2)}{aE}\mu p_0(a-c),$$

where $p_0 = 2F_N/\pi a$. Since in this area $v_1 = v_2$, the creep ratio, as estimated by Eq. (6.1), is

$$\xi_x = \frac{\mu a}{R}\left(1 - \frac{c}{a}\right),$$

where R is the equivalent radius of the two cylinders and we have used the expression (4.25) for the contact half-width a. The relation between ξ_x and the tangential force F_x is plotted in Fig. 6.2. If $F_x \to 0$ the creep ratio is independent of μ and varies linearly with F_x as

$$\xi_x \approx aF_x/2RF_N.$$

Dissimilar materials

Suppose now that the elastic properties of the two cylinders are different. If the friction were capable of preventing slip entirely, the distribution of tangential stress $\tau(x)$ in the contact could be estimated from the equilibrium equation of the displacement gradient (5.3). Substituting in Eq. (6.1), with the left hand side equal to zero, we would get:

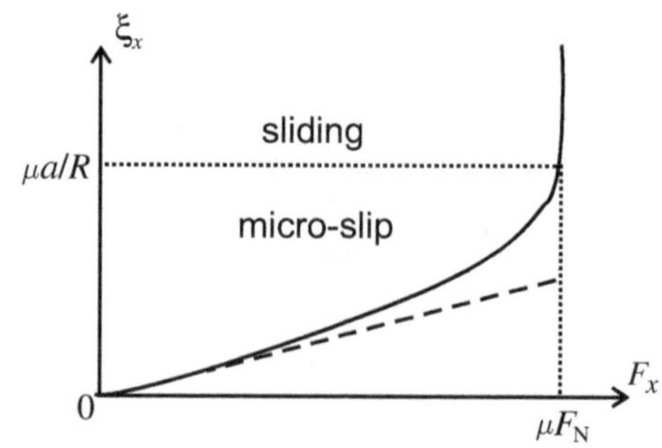

Figure 6.2 Creep ratio in the contact between two elastically similar cylinders rolling over each other as a function of the tangential force.

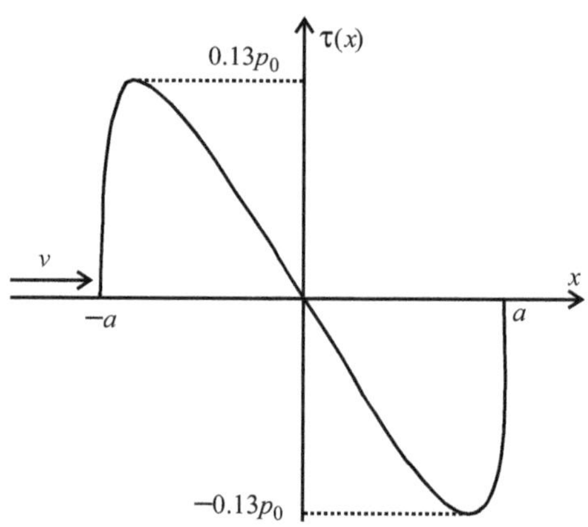

Figure 6.3 Tangential stress distribution in the rolling contact of dissimilar cylinders without slip when $\beta = 0.3$ and $\mu = 0.1$.

$$\pi\beta p(x) + \int_{-a}^{a} \frac{\tau(s)\,\mathrm{d}s}{x - s} = \frac{1}{2}\pi E^* \xi_x,$$

where β is the Dundurs parameter (4.31). Neglecting the effect of $\tau(x)$ on the Hertzian normal pressure (4.26), the creep ratio is given by

$$\xi_x = 2\beta a/\pi R.$$

The corresponding traction distribution [156, section 8.2],

$$\tau(x) = \frac{\beta}{\pi} p_0 \sqrt{1 - \frac{x^2}{a^2}} \ln\left(\frac{a + x}{a - x}\right), \tag{6.2}$$

is plotted in Fig. 6.3.

However, since the ratio $\tau/p \to \infty$ when $x \to \pm a$, slip is once again unavoidable. From numerical analysis [19] two stick areas are expected, separating three areas where slip occurs in alternate directions. The amplitude of the stick areas

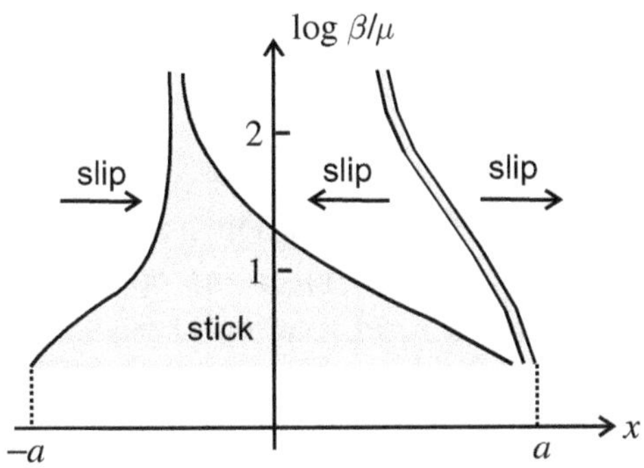

Figure 6.4 Stick and slip areas in the rolling contact of dissimilar cylinders for different values of the ratio β/μ. Adapted from [19] with permission from Elsevier.

depends on the ratio β/μ as shown in Fig. 6.4. This means that a rolling resistance, represented by a frictional torque M_{fric}, appears. Note that M_{fric} is low not only when μ is small, but also when μ is large, since in this case slip is prevented. The maximum value of M_{fric} is expected when $\beta \approx 5\mu$. The corresponding *rolling resistance coefficient*

$$\mu_r = M_{\text{fric}}/F_{\text{N}}R \tag{6.3}$$

is very small ($\sim 10^{-4}a/R$).

A similar analysis for two spheres with different elastic properties results in the following expression for the tangential stress in the rolling contact without slip [156, section 8.2]:

$$\tau(r) = \frac{\beta}{\pi} p_0 \left(-\frac{a}{r}\sqrt{a^2 - r^2} + \frac{1}{r}\int_r^a \frac{\rho^2}{\sqrt{\rho^2 - r^2}} \ln\frac{\rho + r}{\rho - r}\, \mathrm{d}\rho \right).$$

The corresponding creep ratio would be $\xi_x = \beta a/\pi R$ but, again, slip must occur at the edge of the contact circle.

6.2 Three-dimensional rolling

The rolling of 3D elastic bodies is complicated by the relative rotation about the normal axis (*spin*), which couples to the tangential forces F_x and F_y and occurs with an angular velocity $\Delta\Omega$. The contact is formed on an elliptical region with semi-axes a and b given by the Hertz theory. If the friction is large enough to prevent slip completely, the tangential forces and the twisting torque M_z are approximately related to the creep ratios ξ_x, ξ_y and the *spin parameter* $\psi = \Delta\Omega\sqrt{ab}/V$ by linear equations:

$$\frac{F_x}{Gab} = C_{11}\xi_x, \qquad \frac{F_y}{Gab} = C_{22}\xi_y + C_{23}\psi, \qquad \frac{M_z}{G(ab)^{3/2}} = C_{32}\xi_y + C_{33}\psi, \tag{6.4}$$

where the non-dimensional coefficients C_{ij} depend on the eccentricity of the contact ellipse as calculated by Kalker [161]. In the opposite case of complete slip, the relative motion consists of a rigid rotation about a so-called *spin pole* with coordinates

$$x_{\mathrm{p}} = -a\xi_y/\psi, \qquad y_{\mathrm{p}} = a\xi_x/\psi,$$

which may lie inside or outside the contact area. The forces F_x and F_y and the torque M_z can be estimated numerically [161].

The problem of 3D rolling in the case of partial slip can be solved by dividing the contact area into independent thin strips parallel to the rolling direction. The 2D theory is then applied to each of the strips. If only a longitudinal force F_x is present, the stick zone is expected to have a lemon shape (see the shaded region in Fig. 6.5) obtained from the reflection of the leading edge in the straight line [162]

$$x = \frac{G}{2(1-\nu)}\frac{a}{\mu p_0}\xi_x,$$

(p_0 is the maximum contact pressure value as given by the Hertz theory). The creep ratio can be estimated from the tangential force using a non-trivial relation, which reduces to the first of the equations (6.4) with

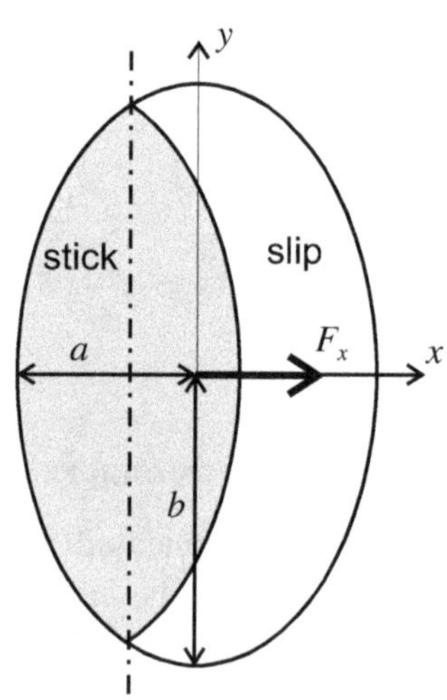

Figure 6.5 Stick and slip areas in an elliptical rolling contact in the presence of a longitudinal force.

$$C_{11} = \frac{\pi^2}{4(1-\nu)}$$

if $\mu \to \infty$. However, the strip model works well only if $b \gg a$. If this is not the case, the interaction between strips cannot be neglected and numerical methods must be used [161].

6.3 Sphere in a groove

The problem of a sphere with radius R rolling in a conforming groove is very important in bearing technology. The contact area is an elongated ellipse, the semi-axes a and b of which are determined by the Hertz theory. The points on the sphere surface have different linear velocities, and the creep ratio depends on the distance y from the axis of the groove as $\xi(y) = \xi_0 - y^2/2R^2$ [156, section 8.5]. The Carter theory discussed in Section 6.1 predicts that slip occurs backwards if $y < R\sqrt{2\xi_0}$ and forwards in the opposite case. The value of ξ_0 is determined by the condition that the total traction force is zero. As a result, the resisting torque M_{fric} can be calculated as a function of the 'conformity parameter' $\Gamma = b^2E^*/4\mu p_0 R$. The result is showed in Fig. 6.6. In the case of close conformity ($\Gamma \gg 1$) a limit value corresponding to a rolling resistance coefficient [134]

$$\mu_r = 0.08(b/R)^2\mu \tag{6.5}$$

is approached. Since the ratio b/R is not negligible, the value of μ_r can be relatively high.

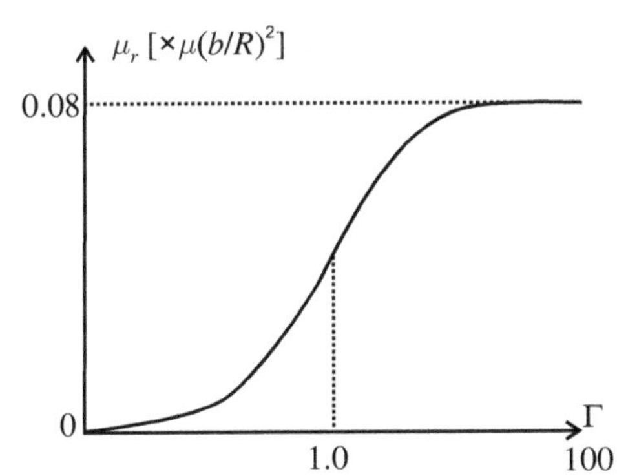

Figure 6.6 Resisting torque on a sphere rolling in a conforming groove as a function of the conformity parameter defined in the text. Adapted from [134] with permission from SAGE Publications.

6.4 Tire mechanics

Aircraft tires

The tire of an airplane is well approximated by a toroidal thin membrane with internal pressure. If the tire is pressed against a rigid plane, the contact area is an ellipse with semi-axes

$$a = \sqrt{(2R - \delta)\delta}, \qquad b = \sqrt{(w - \delta)\delta},$$

where R and w are the radius and the width of the tire and δ is the vertical deflection (Fig. 6.7). The pressure distribution in the contact is uniform and equal to the inflation pressure. The creep ratio can be estimated as the difference between the length of the chord $\overline{AB}$ and the arc $\widehat{AB}$:

$$\xi_x = -\delta/3R. \tag{6.6}$$

Although the strain out of the contact region has been neglected, the relation (6.6) is in good agreement with experimental observations.

Automotive tires

A car tire, due to its stiffer tread and its cross-section shape, forms a roughly rectangular contact area of length $2a$. In this case the pressure is concentrated in the center of the contact.

If the car turns around a corner or its wheels are slightly skewed, a transverse surface traction $\tau(x)$ appears in the contact region, where x is the direction of motion. Considering only the carcass deformation, the lateral displacement $u(x)$ satisfies in both cases the equilibrium equation

$$u(x) - \lambda^2 \frac{\mathrm{d}^2 u}{\mathrm{d}x^2} = \frac{\tau(x)}{k_c},$$

where k_c is the carcass stiffness per unit length [340].

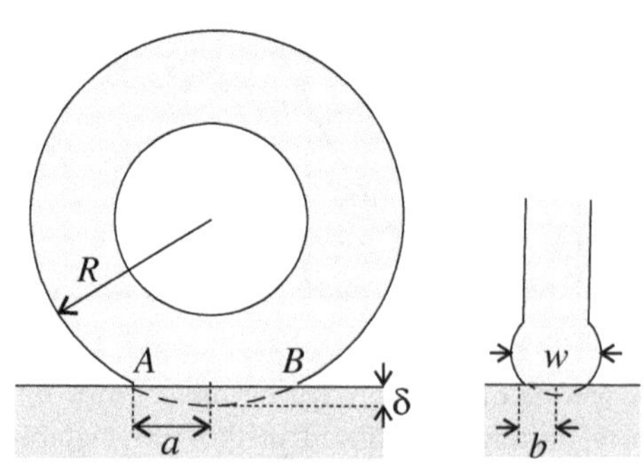

Figure 6.7 Contact between a thin inflated membrane and a rigid plane.

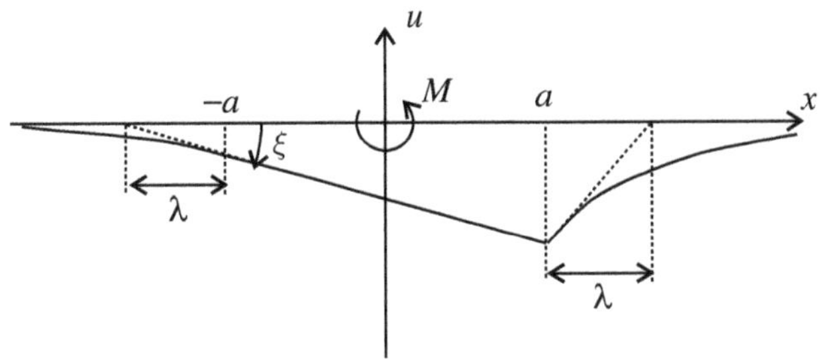

Figure 6.8 Transverse deformation of a car tire slightly skewed with respect to the plane of rolling.

When the wheels are slightly skewed the lateral displacement in the contact region is simply given by

$$u(x) = \text{const.} - \xi x, \tag{6.7}$$

where ξ is the creep ratio. Outside this area there is no traction and $u(x)$ decays exponentially with a characteristic length $\lambda = \sqrt{T/k_c}$, where T is the uniform tension carried by the carcass (Fig. 6.8). From the continuity of the displacement gradient $\mathrm{d}u/\mathrm{d}x$ it follows that the constant in Eq. (6.7) is equal to $-\lambda\xi$. The traction distribution $\tau(x)$ can be integrated over the contact area to give the total *cornering force*

$$F_y = -2k_c\xi a^2\left(\frac{\lambda}{a}+1\right)^2$$

and the self-aligning torque

$$M = -2k_c\xi a^3\left(\frac{1}{3}+\frac{\lambda}{a}+\frac{\lambda^2}{a^2}\right).$$

Note that the discontinuity in $\mathrm{d}u/\mathrm{d}x$ at the trailing edge ($x = a$) corresponds to a concentrated traction, which causes a *sideslip*.

When the car turns around the corner, the deflected shape in the stick region is parabolic:

$$u = \text{const.} + \psi x^2/a$$

and the transverse force (so-called *camber thrust*) is

$$F_y = -2k_c\psi a^2\left(\frac{1}{3}+\frac{\lambda}{a}+\frac{\lambda^2}{a^2}\right).$$

In this case no self-aligning torque is present.

7

Beams, plates and layered materials

Our brief review of the theory of elasticity would not be complete without discussing the deformation of beams, plates and thin overlayers. Cantilever beams are key components in atomic force microscopy, where they are used to detect forces in the nanonewton range. Beams and curved plates present characteristic elastic instabilities when compressed beyond a well-defined threshold. A related phenomenon is also observed in the theory of stick–slip which is described elsewhere. The vibrations of beams and plates will be also discussed briefly.

7.1 Elastic deformation of beams

Bending of beams

Consider a solid beam (oriented along the x direction) which is slightly bent around the y axis by a linear force distribution $p(x)$ as in Fig. 7.1. The deflection $u(x)$ of the beam is described by the *Euler–Bernoulli equation* [177, section 20]:

$$EI\frac{\mathrm{d}^4u}{\mathrm{d}x^4} - p(x) = 0, \tag{7.1}$$

where I is the area moment of the beam's cross-section with respect to the z axis,

$$I = \int y^2\,\mathrm{d}y\,\mathrm{d}z,$$

and E is the Young's modulus of the material. If the beam has a rectangular section with width w and thickness d, $I = wd^3/12$. For a circular cross-section with radius R, $I = \pi R^4/4$.

The bending moment and the shear force in the beam are simply given by the second and third derivatives of $u(x)$:

$$M_y = -EIu''(x), \qquad F_z = -EIu'''(x). \tag{7.2}$$

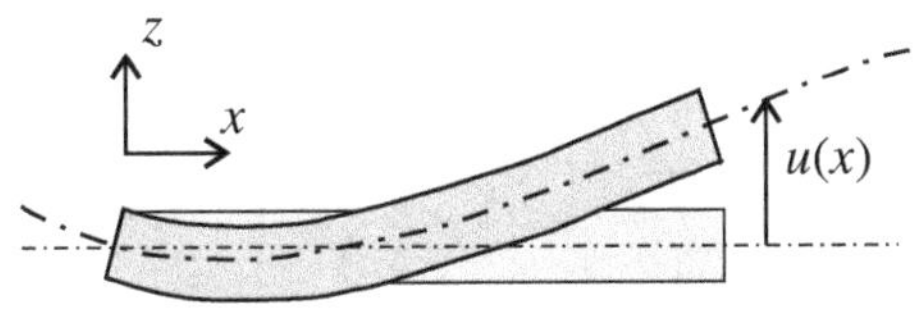

Figure 7.1 Bending of an Euler–Bernoulli beam. The y axis is oriented perpendicularly to the plane in the figure.

The equations (7.2) allow us to write the boundary conditions on the beam in a very simple form. If one end of the beam is *clamped*, both the deflection u and the slope u' are zero. If the end is *supported*, the deflection u and the bending moment are zero, and the last condition implies that $u'' = 0$. Finally, at a free end, both force and bending moment are zero: $u''' = 0$ and $u'' = 0$.

Using Eq. (7.1) the deformation of a cantilever beam of length L, clamped at one end and bent by a force F concentrated at the other end, turns out to be

$$u(x) = \frac{F}{6EI} x^2 (3L - x).$$

If the beam is bent by a uniform force distribution p such as its own weight:

$$u(x) = \frac{p}{24EI} x^2 (x^2 - 4Lx + 6l^2).$$

Note that Eq. (7.1) dates back to the 1750s, but it was not applied on a large scale till the construction of the Tour Eiffel and the first Ferris wheel in Chicago in the late nineteenth century. After that, the Euler–Bernoulli equation has quickly become a cornerstone of structural and mechanical engineering.

Torsion of beams

Suppose now that one end of a straight beam with arbitrary cross-section, oriented along the z axis, is fixed and a torque M is applied to the other end of the beam. In this case the twist angle per unit length is [177, section 16]

$$\theta = M/C, \tag{7.3}$$

where C is the *torsional rigidity* of the beam. The torsional rigidity can be determined using the relation [275]

$$C = 4G \int \psi \, \mathrm{d}x \, \mathrm{d}y, \tag{7.4}$$

where G is the shear modulus of the material and the function $\psi(x, y)$ is the solution of the Poisson equation $\nabla^2 \psi = -1$ with the boundary condition $\psi = 0$. Note that in this way the problem becomes formally identical to the bending of a uniformly loaded membrane, Eq. (7.8), and also to the viscous flow through a pipe,

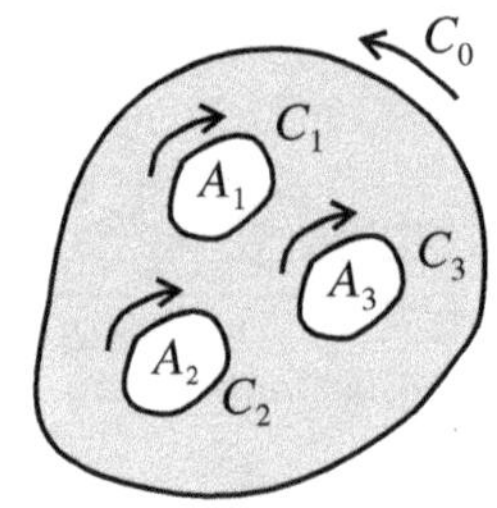

Figure 7.2 A multiply connected cross-section of a beam undergoing torsion.

Eq. (22.6). The function $\psi(x, y)$ corresponds to the vertical deformation of the membrane or, respectively, to the fluid velocity. Using this analogy, it can be easily seen that a cylinder with circular cross-section has a torsional rigidity $C = G\pi R^4/2$, while a long thin plate with width w and thickness $d \ll w$ has a torsional rigidity $C = Gwd^3/3$.

However, Eq. (7.4) is only valid for singly connected cross-sections. If the cross-section of the beam is multiply connected, one has to add the quantity $4G\Sigma_k \psi_k A_k$ to the right hand side of Eq. (7.4), where A_k are the areas enclosed by the curves limiting the cross-section (Fig. 7.2) and ψ_k are the constant values taken by the function ψ on these curves. In this way, it can be proven that a cylindrical pipe has a torsional rigidity

$$C = \frac{\pi G}{2}(R_{\text{ext}}^4 - R_{\text{int}}^4).$$

Dynamic beam equation

The dynamics of a vibrating beam is governed by the time-dependent Euler–Bernoulli equation:

$$EI\frac{\partial^4 u}{\partial x^4} + \rho\frac{\partial^2 u}{\partial t^2} - p(x) = 0, \tag{7.5}$$

where ρ is the mass density per unit length of the beam. In view of applications to AFM (Section 18.2) we present the solution of Eq. (7.5) for the deformed shape of a cantilever beam of length L with a clamped end:

$$u_n(x) \propto \cosh k_n x - \cos k_n x + \frac{\cos k_n L + \cosh k_n L}{\sin k_n L + \sinh k_n L}(\sin k_n x - \sinh k_n x),$$

where the wave numbers k_n ($n = 1, 2, \ldots$) are the solutions of the characteristic equation

$$\cosh k_n L \cos k_n L + 1 = 0.$$

The corresponding resonance frequencies are

$$\omega_n = k_n^2\sqrt{EI/\rho}.$$

The first three modes of vibration of the cantilever beam are shown in Fig. 7.3.

Suppose now that a vertical spring k^* is added to the free end of the cantilever, coupling it to a rigid surface. In this case the boundary conditions at $x = L$ are

$$u''(L) = 0, \qquad u'''(L) = \frac{k^*}{EI}u(L)$$

and the equation for k_n becomes

$$\sinh k_nL \cos k_nL - \sin k_nL \cosh k_nL = \frac{(k_nL)^3 k_N}{3k^*}(1 + \cos k_nL \cosh k_nL),$$

where $k_N = Ewd^3/4L^3$ is the normal spring constant of the beam; k_N is the ratio between a concentrated force applied at $x = L$ and the deflection $u(L)$ of the free beam [279]. The effect of the spring on the shape of the first mode is shown in Fig. 7.4 for increasing values of the stiffness k^*.

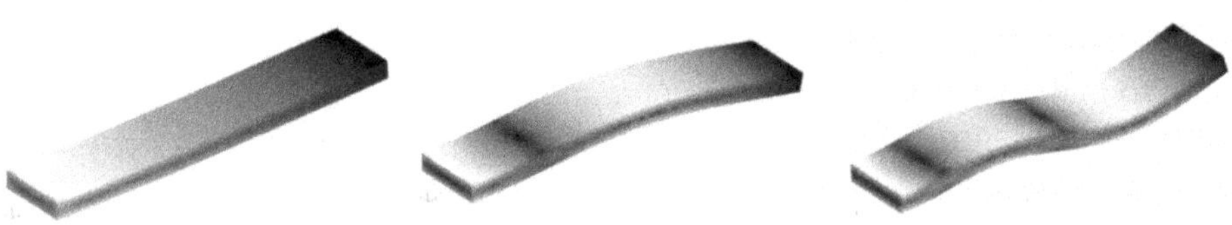

Figure 7.3 Mode shapes for the first three modes of vibration of a cantilever beam (the clamped end is the one on the left).

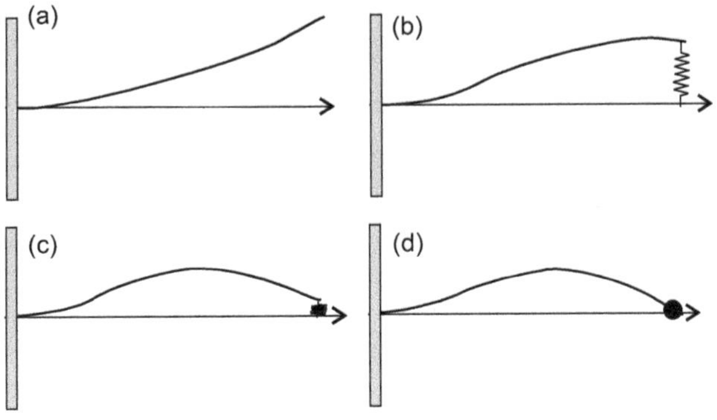

Figure 7.4 Shape of the first vibration mode of a clamped cantilever beam coupled to a rigid surface by a spring with stiffness $k^* = 0, 10, 100, \infty$ (a to d).

7.2 Plate theory

A plate has a thickness d which is much smaller than its planar dimensions. In this case the problem of elastic deformation is reduced to 2D.

Bending of plates

The equilibrium of a plate deformed by a force distribution $p(x, y)$ perpendicular to its surface is described by the equation [193]

$$D\nabla^2(\nabla^2 u) - p(x, y) = 0, \tag{7.6}$$

where $\nabla^2 \equiv \partial^2/\partial x^2 + \partial^2/\partial y^2$ is the 2D Laplacian,

$$D = \frac{Ed^3}{12(1-\nu^2)}$$

is the *bending stiffness* of the plate and ν is the Poisson's ratio of the material. Equation (7.6) can be obtained by minimizing the expression for the elastic energy of the plate, which follows from Eq. (3.15) after equating to zero the components of the displacement vector in the xy plane and all components of the stress tensor except σ_x, τ_{xy} and σ_y [177, sections 11 and 12].

The boundary conditions for Eq. (7.6) are very complicated unless the edges of the plate are clamped or supported (Fig. 7.5(a)). A clamped edge remains horizontal, so that the vertical displacement u and the slope $\partial u/\partial n$ along the direction **n** which is normal to the contour of the plate are both zero (Fig. 7.5(b)). If the edge is supported the deflection and the bending moment are zero. This condition translates into the relations

$$u = 0, \qquad \frac{\partial^2 u}{\partial n^2} + \nu \frac{d\theta}{dl}\frac{du}{dn} = 0,$$

where **l** is the unit vector tangent to the plate edge, and θ is the angle between the x axis and **n**.

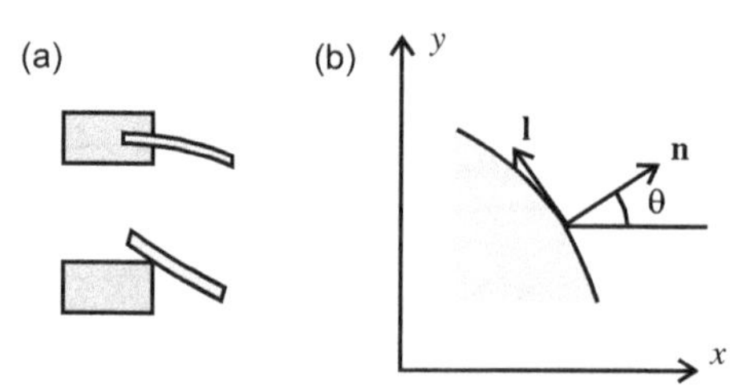

Figure 7.5 (a) Clamped and supported edges. (b) Top view of the edge of a plate.

As an example, if a concentrated force F_N is applied to the center of a circular plate of radius R, the vertical deformation of the plate is

$$u(r) = \frac{F_N}{16\pi D}\left(f(\nu)(R^2 - r^2) - 2r^2 \ln \frac{R}{r}\right), \tag{7.7}$$

where $f(\nu) = 1$ if the edges are clamped, and $f(\nu) = (3+\nu)/(1+\nu)$ if the edges are supported. If the plate is deformed by a uniform pressure p such as its own weight, the solution of Eq. (7.6) is

$$u(r) = -\frac{p}{64D}(R^2 - r^2)r^2$$

if the edges are clampled, and

$$u(r) = -\frac{p}{64D}\left(\frac{(5+\nu)R^2}{1+\nu} - r^2\right)(R^2 - r^2)$$

if the edges are supported.

Membranes

The equilibrium of a *membrane*, i.e. of a thin plate subjected to large stretching forces applied at its edges, is described by a simple equation. If the stretching is isotropic,

$$T\nabla^2 u + p = 0, \tag{7.8}$$

where T is the tension per unit length of the edge. Solving Eq. (7.8), we can easily see that a circular membrane of radius R bent by a uniform pressure p undergoes a parabolic deformation:

$$u(r) = -\frac{p(R^2 - r^2)}{4T}.$$

Vibrations of plates and membranes

The equation for the free oscillations of a plate is

$$\rho\frac{\partial^2 u}{\partial t^2} + \frac{D}{d}\nabla^2(\nabla^2 u) = 0, \tag{7.9}$$

where ρ is the mass density of the material. An exact solution of Eq. (7.9), and an estimation of the resonance frequencies in a closed form is possible only in few cases. For instance, a rectangular plate with supported edges resonates at the (angular) frequencies

$$\omega_{mn} = \sqrt{\frac{D}{\rho d}}\pi^2\left(\frac{m^2}{a^2} + \frac{n^2}{b^2}\right),$$

Figure 7.6 Shapes of the first three modes of vibration of a circular plate.

where m and n are integers. The corresponding mode shapes are:

$$u_{mn}(x, y) \propto \sin \frac{m\pi x}{a} \frac{n\pi y}{b}.$$

For a circular plate of radius R with clamped edges,

$$\omega_n = \sqrt{D/\rho d} k_n^2,$$

where the wave numbers k_n are the solutions of the characteristic equation

$$J_0(kR)I_1(kR) + I_0(kR)J_1(kR) = 0$$

(J_α and I_α are, respectively, the Bessel function of order α of the first kind and the modified Bessel function of order α of the second kind). The corresponding mode shapes,

$$u_n(r) \propto J_0(k_n r) - \frac{J_0(k_n R)}{I_0(k_n R)} I_0(k_n r),$$

are shown in Fig. 7.6 for $n = 1, 2, 3$.

The resonance frequencies of a circular membrane with radius R are the solutions of the equation

$$J_n(\omega_{mn} R \sqrt{\rho d/T}) = 0$$

and the mode shapes are

$$u_{mn}(r, \varphi) \propto J_n(\omega_{mn} r \sqrt{\rho d/T}) \sin(n\varphi + \text{const.}).$$

7.3 Elastic instabilities

The deformation of a beam subject to a longitudinal compression is stable only if the applied force remains below a certain threshold [177, par. 21]. For instance, if both ends of a slender column with length L and area moment of inertia I are clamped, the value of the critical force is

$$F_c = 4\pi^2 EI/L^2. \tag{7.10}$$

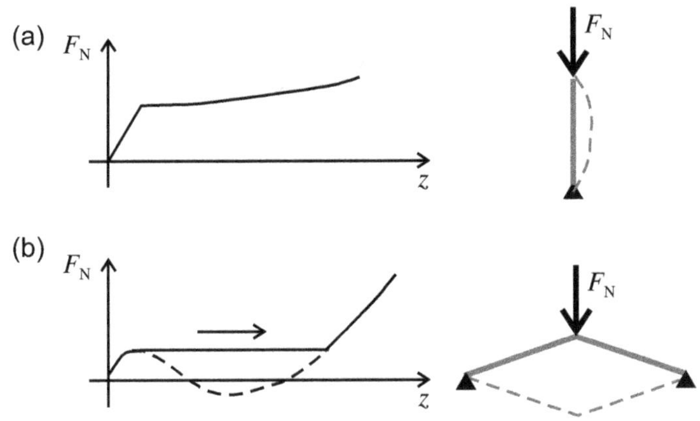

Figure 7.7 Examples of (a) Euler buckling and (b) limit point instability.

This value becomes four times smaller if the ends are hinged, and 16 times smaller if one end of the column is free and the other end is clamped. Once the threshold is exceeded, the column will bend significantly (*buckling*). However, we should note that, if the compression is applied for a very short time $\tau < L/c$, where c is the velocity of sound, the column can sustain much higher loads before buckling (so-called *dynamic buckling*).

Since the values of F_c are proportional to the moment of inertia, a tubular section will be much more efficient than a solid one. Under its own weight, a vertical column will be stable only if $L < L_c = 1.98(EI/\lambda g)^{1/3}$, where λ is the mass density per unit length [163]. Elastic instabilities also occur when a beam is subjected to torsion. If the beam has a circular cross-section, the critical torsion angle per unit length is $\theta_c = 8.98EI/C$, where C is the torsional rigidity of the beam. Buckling can be also observed in bicycle wheels, if the spoke tension is increased beyond a safe level, or in rail tracks excessively heated by the Sun.

The phenomenon of *elastic instability* was discovered by Leonhard Euler in 1757 and is due to a bifurcation appearing in the solution of the equation of static equilibrium. In the cases discussed above the buckled configuration is adjacent to the original one (Fig. 7.7(a)). This is not the case in structures experiencing 'limit point instabilities', and suddenly jumping into very different stable configurations (Fig. 7.7(b)). This second type of instability also occurs in atomic-scale stick–slip, as discussed in Section 15.1.

7.4 Shells

In contrast to thin plates, *shells* have a curved shape in their undeformed state. If a shell has a thickness d and a radius of curvature R, the range of action of a

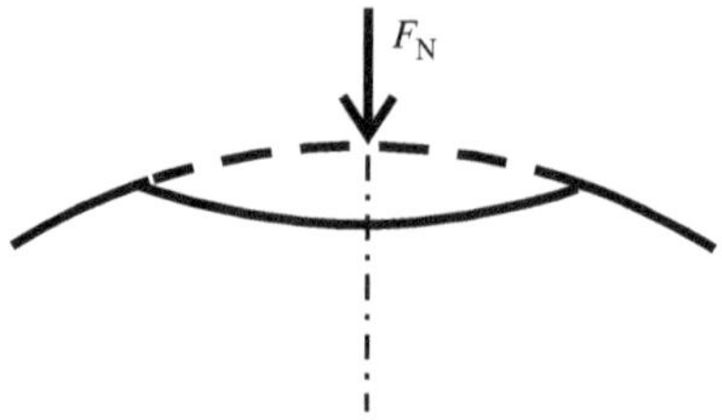

Figure 7.8 Buckling of a thin shell indented by a concentrated force.

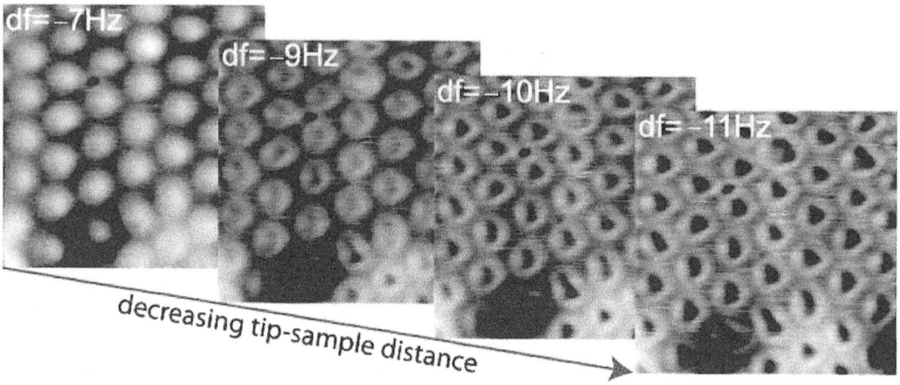

Figure 7.9 Four AFM images (9.6 nm in size) acquired with increasing normal force on the Moiré pattern formed by a graphene monolayer on Ru(0001).

concentrated force F_N is an area $\sim dR$. This result follows from the condition that the sum of the stretching energy and the bending energy of the shell has a minimum [177, section 15]. The corresponding deflection is in the order of $F_N R/Ed^2$. If the normal force is too large, a shell can buckle (Fig. 7.8). The depth of the bulge so formed is $\sim F_N^2 R^2/E^2 d^5$.

An example of shell deformation on the nanoscale is shown in Fig. 7.9 [167]. Here, a regular array of ultrathin 'domes' formed by a graphene monolayer grown on a Ru(0001) surface have been imaged by AFM with increasing normal forces. When the probing tip is retracted, the deformation is fully reversed and no buckling is observed.

7.5 Indentation of elastic plates

The problem of the contact between a rigid object and a thin elastic plate has received only limited attention so far. If a rigid cylinder of radius R is pressed against a plate of length $2L$ with a force F_N per unit length, as in Fig. 7.10, the half-width of the contact strip is [156, section 5.8]

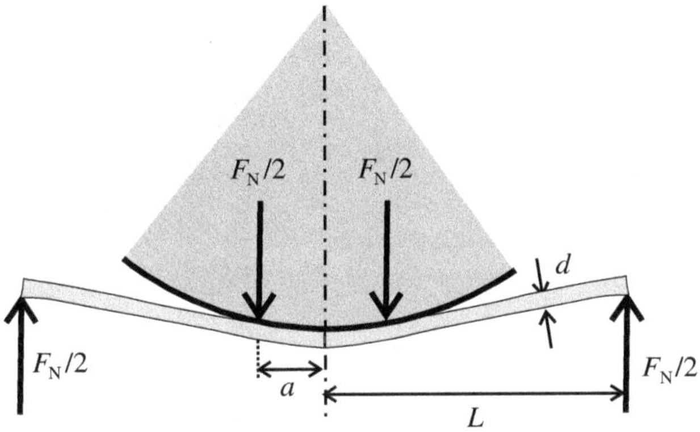

Figure 7.10 Indentation of a thin elastic plate by a rigid cylinder.

$$a = L - \frac{128D}{RF_N}, \tag{7.11}$$

where D is the bending stiffness of the plate. The pressure is concentrated at the edges of the strip and is zero elsewhere. However, these conclusions are only valid if the plate thickness $d \ll a \ll R$.

In a similar way, if a plate is indented by a sphere, the pressure is concentrated on a ring corresponding to the edges of the contact circle. In this case, in order to get realistic distributions, the shear stiffness of the plate cannot be ignored.

7.6 Indentation of thin elastic layers

Suppose that a rigid substrate covered by a thin elastic layer of thickness d is indented by a rigid cylinder of radius R with its axis parallel to the surface of the substrate. For instance we may think of a roller covered by rubber, widely used in processing machinery. The contact strip has a half-width a, which is supposed to be much smaller than d, and we will distinguish between compressible and incompressible layers.

In the first case (Fig. 7.11a) the indentation depth is $\delta = a^2/2R$ and the pressure distribution is [156, par. 5.8]

$$p(x) = p_0(1 - x^2/a^2),$$

with $p_0 = 3F_N/2a$. The length a is obtained from the normal force F_N as

$$a = \left(\frac{\alpha(1 - \beta\nu)RdF_N}{G}\right)^{1/3}.$$

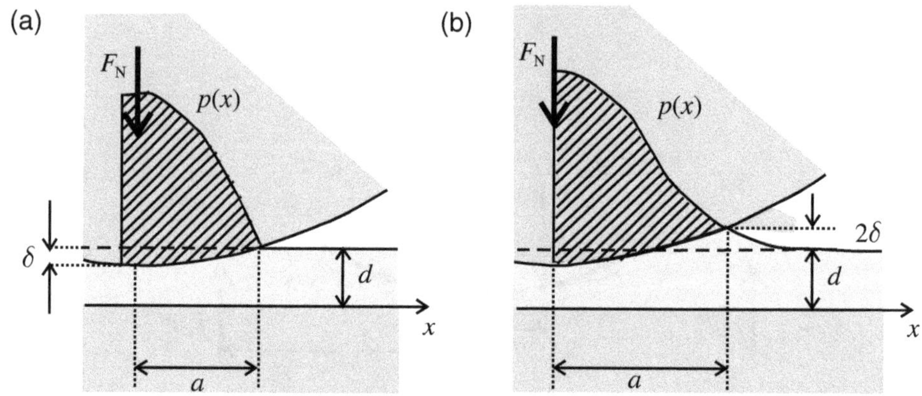

Figure 7.11 Indentation of (a) a compressible and (b) an incompressible elastic layer by a rigid cylinder.

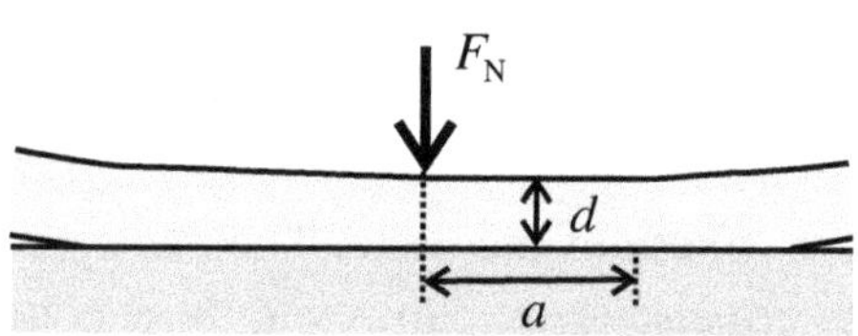

Figure 7.12 Elastic layer lifted off by a concentrated force.

where $\alpha = 3$ and $\beta = 1$ if the layer can slip on the substrate without friction, whereas $\alpha = 6$ and $\beta = 2$ if the layer is bound to the substrate. If the layer is incompressible, part of it is necessarily squeezed out, as shown in Fig. 7.11(b). In this case:

$$a = \left(\frac{45Rd^3F_N}{2E}\right)^{1/5}, \qquad p(x) = \frac{48F_N}{45a}\left(1 - \frac{x^2}{a^2}\right)^2$$

and the indentation depth is $\delta = a^2/6R$.

If the layer is free to lift off the substrate under the action of a concentrated force (Fig. 7.12), the receding contact strip has a half-width [103]

$$a = d\left(1.845\frac{(1-\nu_s^2)E_l}{(1-\nu_l^2)E_s}\right)^{1/3},$$

where the indices s and l refer to the substrate and the layer respectively. Note that, in this case, a is independent of the normal force.

Part II
Advanced Contact Mechanics

8
Rough contacts

In this chapter we discuss the influence of surface roughness in contact problems. After introducing the basic definitions and analyzing early contact models, the theory developed by Bo Persson is presented in more detail. This theory correctly describes the linear dependence of the contact area on the normal force at low loads, and makes important predictions on the load dependence of the average separation between two rough surfaces in contact. Other popular early models are discussed for comparison. The chapter ends with a brief discussion of the contact between wavy surfaces, where simple analytical expressions can be derived.

8.1 Surface roughness

Atomically flat surfaces are very rare. Examples include graphite and mica cleaved along atomic planes. Soft rubber can also be made flat elastically by a contact pressure. Apart from these cases, we can reasonably assume that the contact between two solid surfaces at normal operating loads is discontinuous and the real contact area is only a small fraction of the apparent contact area (Fig. 8.1).

Height distribution

The height profile $h(\mathbf{r})$ of a surface at a point $\mathbf{r} \equiv (x, y)$ above a flat reference plane can be routinely measured by optical and stylus techniques or, on the nanoscale, by AFM. Many surfaces, e.g. those prepared by fracture or by bombardment with small particles, have an approximately Gaussian height distribution

$$P(h) = \frac{1}{\sqrt{2\pi}\, h_{\rm rms}} \exp\left(-\frac{h^2}{2h_{\rm rms}^2}\right). \tag{8.1}$$

In this case the root mean square $h_{\rm rms} = \sqrt{\langle h^2 \rangle}$ coincides with the half-width of the distribution. However, if the surface is polished, the height distribution deviates

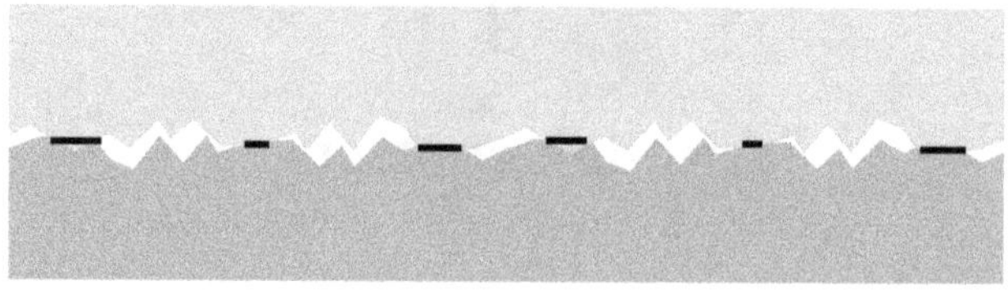

Figure 8.1 Contact between two rough surfaces.

rapidly from (8.1) since the protruding asperities are smoothed out more than the deeper regions of the surface.

Surface roughness power spectrum

The surface roughness can be characterized by the *height–height correlation function*:

$$C(\mathbf{r}) = \langle h(\mathbf{r})h(0)\rangle\,, \tag{8.2}$$

where the angular bracket $\langle\ldots\rangle$ stands for ensemble averaging.[1] Note that in the definition (8.2) it is implicitly assumed that the statistical properties of the surface are translationally invariant. The ratio between the height–height correlation functions in the x and y directions (the so-called *Peklenik number*) is a measure of the anisotropy degree of the surface.

The Fourier transform of the height–height correlation function is the *surface roughness power spectrum* [231]:

$$S(\mathbf{k}) = \frac{1}{(2\pi)^2}\int C(\mathbf{r})e^{-i\mathbf{kr}}\,d^2r.$$

Vice versa:

$$C(\mathbf{r}) = \int S(k)e^{i\mathbf{kr}}\,d^2k,$$

and it is not difficult to see that

$$h_{\text{rms}}^2 = 2\pi\int_0^\infty kS(k)\,dk.$$

As an example, the surfaces prepared by cooling a glassy material from a temperature above T_g (see Section 11.1) have a minimum roughness in the nanometer range, which is caused by thermally excited capillary waves. In this case, it can be proven that [213]

$$S(k) = \frac{1}{4\pi^2}\frac{k_B T_g}{\rho g + \gamma k^2 + Dk^4}, \tag{8.3}$$

[1] The definition (8.2) is not unique and other expressions can be found in the literature.

where ρ is the mass density, g is the acceleration of gravity, γ is the surface tension, D is the bending stiffness of the material and $k_B = 1.38 \times 10^{-23}$ m^2·kg/s^2·K is Boltzmann's constant. Neglecting the gravity term, the rms roughness of the surface on an area of linear size L is given by

$$h_{\text{rms}}^2 \approx \frac{k_B T_g}{2\pi\gamma} \ln \frac{\sqrt{\gamma/D}}{k_L},$$

where $k_L = 2\pi/L$.

It is interesting to observe that a randomly rough surface with a given power spectrum $S(k)$ over an $L \times L$ square area can be generated numerically as

$$h(\mathbf{r}) = \sum_{\mathbf{k}} \frac{2\pi}{L} \sqrt{S(\mathbf{k})} e^{i(\mathbf{kr}+\varphi(\mathbf{k}))}, \tag{8.4}$$

where the components of the vectors $\mathbf{k}$ are integer multiples of k_L and $\varphi(\mathbf{k})$ are random variables uniformly distributed between 0 and 2π [265].

Self-affinity

Many surfaces of interest are *self-affine*. This means that the surface looks the same, independently of the level of magnification ζ which is applied to observe it. However, the magnification is usually not the same along the xy plane and the z direction (Fig. 8.2). Assuming that the two levels of magnification are ζ and ζ^H respectively, the (isotropic) height–height correlation function of a self-affine surface turns out to follow the power law

$$C(r) \propto r^{2H}. \tag{8.5}$$

The *Hurst coefficient* H has typical values between 0.5 and 0.9 and is related to the fractal dimension D_f of the surface, which ranges from 2 to 3, as $H = 3 - D_f$ [10].

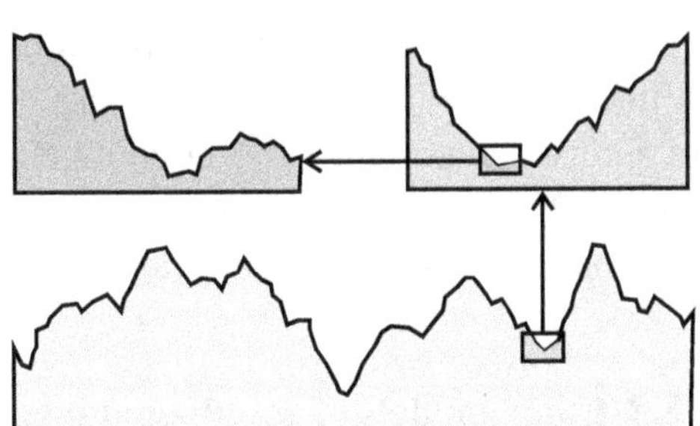

Figure 8.2 A self-affine profile does not change its statistical properties if scaled down by a factor ζ in the lateral direction and ζ^H in the normal direction.

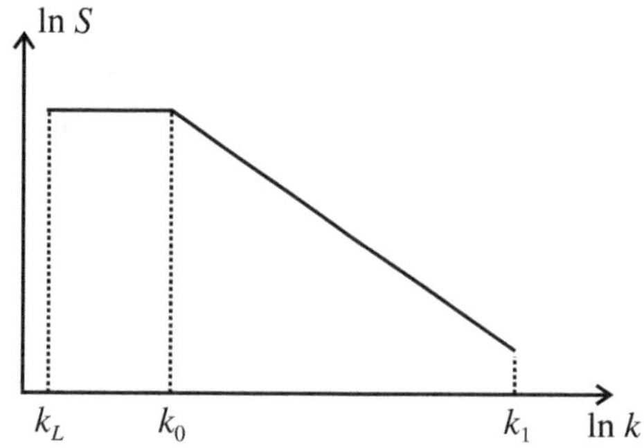

Figure 8.3 Roughness power spectrum of a self-affine surface.

The relation (8.5) breaks down at the linear size L of the surface and, on the other hand, at a length scale a of the order of the smallest components of the surface material at which plastic yield occurs. The ultimate values of the cut-off correspond to interatomic distances, but on a clean rubber surface a is usually of the order of a few μm.

The power spectrum of a perfect self-affine surface follows the power law

$$S(k) = S_0 \left(\frac{k}{k_0}\right)^{-2(H+1)} \tag{8.6}$$

between a 'roll-off' value k_0 and $k_1 = 2\pi/a$ [22]. This means that the slope of the $S(k)$ curve on a logarithmic scale (Fig. 8.3) can be used to determine the fractal dimension of the surface. Below k_0 the spectrum $S(k)$ remains approximately constant (and equal to S_0) down to the value $k_L = 2\pi/L$, which is the smallest possible wave vector. The value of S_0 depends on the rms of the surface height as

$$S_0 = \frac{H}{\pi[1 + H - (k_L/k_0)^2 H]} \left(\frac{h_{\text{rms}}}{k_0}\right)^2.$$

As an example, asphalt pavements are self-affine surfaces with fractal dimension $D_f \approx 2.2$ and $\lambda_0 \equiv 2\pi/k_0 \sim 1$ cm [250]. Surfaces prepared by fracture of brittle materials can be self-affine down to interatomic distances. On the other hand, the surfaces resulting from a glassy material slowly cooled below the glass transition temperature T_g are not self-affine on any length scale, as seen from Eq. (8.3). Still, most surfaces are approximately self-affine in a finite length range.

8.2 Early models of rough contacts

The contact between two rough surfaces with height profiles $h_1(\mathbf{r})$ and $h_2(\mathbf{r})$ is equivalent to the contact of a rigid rough substrate with surface profile $h = h_1 + h_2$ and a flat elastic block with Young's modulus E and Poisson's ratio ν such that

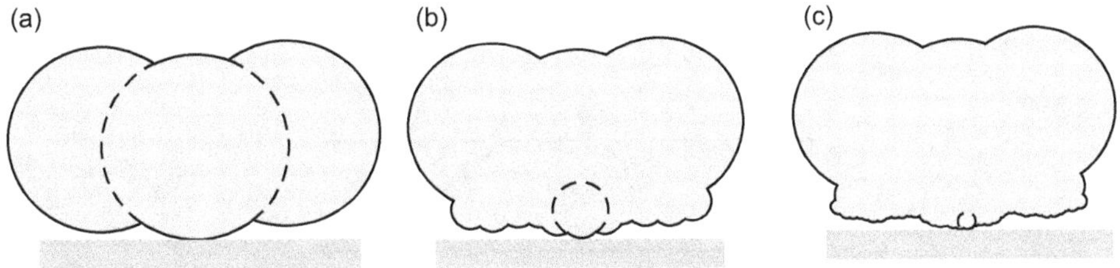

Figure 8.4 Contact between two surfaces at different levels of magnification according to the Archard model.

$(1 - \nu)/E = 1/E^*$, where E^* is the effective elastic modulus defined in Section 4.3 [156]. For this reason, in the rest of this chapter we will only consider contacts of the second kind.

The Archard model

In the Archard model a rough surface is approximated by a series of hierarchically superimposed spheres [6], as shown in Fig. 8.4. Archard proved that the relation between the real contact area A and the normal force F_N for the three geometries with increasing roughness shown in the figure is $A \propto F_N^\alpha$, where $\alpha = 4/5$, 14/15 and 44/45 respectively. This suggests that, in the case of a complex real surface, the exponent $\alpha \approx 1$. In this way, A is expected to be proportional to the load, in line with Amonton's law (2.1).

The Greenwood–Williamson model

A more realistic model, which is also consistent with Amonton's law, has been proposed by Greenwood and Williamson [122]. In the GW model a rough surface is considered as an ensemble of N independent spherical caps with the same radius R (Fig. 8.5). At a given separation d the real contact area is

$$A(d) = N \int_d^\infty A_1(h - d) P(h) \, dh \tag{8.7}$$

and the normal force is

$$F_N(d) = N \int_d^\infty F_1(h - d) P(h) \, dh, \tag{8.8}$$

where A_1 and F_1 are, respectively, the area of the contact formed by each cap and the normal forces required to compress the cap, and $P(h)$ is the height distribution of the caps. Since the caps are supposed to be independent, the theory breaks down when A is no longer smaller than the apparent contact area. If the deformation is elastic, Eqs. (4.19) and (4.18) imply that

Figure 8.5 The Greenwood–Williamson model.

$$A_1(h) = \pi R h, \qquad F_1(h) = \frac{4}{3} E^* \sqrt{R} h^{3/2}.$$

In the case of an exponentially decaying height distribution

$$P(h) \propto \mathrm{e}^{-\lambda h}, \tag{8.9}$$

where λ is a constant, the integrals (8.7) and (8.8) can be easily calculated. As a result, the GW model predicts that the real contact area is proportional to F_N:

$$A(F_\mathrm{N}) = \frac{\sqrt{\pi R \lambda}}{E^*} F_\mathrm{N}.$$

A proportionality relation holds also, in a first approximation, in the more realistic case of a Gaussian height distribution (8.1), assuming that only the rapidly decaying part of the distribution is involved in the contact.

If the contact is fully plastic it is not difficult to see that, for each cap, the penetration depth δ and the radius of the contact area a are related by $\delta \approx a^2/2R$, so that $A_1(h) \approx 2\pi R h$. Since the average pressure $p \approx 3Y$ (see Section 12.8), we get $F_1(h) \approx 6\pi R Y h$ and also, in this case, we can solve the integrals (8.7) and (8.8) and conclude that the area of real contact is proportional to the normal force:

$$A(F_\mathrm{N}) \approx \frac{F_\mathrm{N}}{3Y} \approx \frac{F_\mathrm{N}}{H},$$

where H is the indentation hardness (Section 12.6).

The Bush–Gibson–Thomas theory

In the Bush–Gibson–Thomas (BGT) theory the spheres of the GW model are replaced by paraboloids with randomly distributed curvatures and heights [40]. This means that the surface roughness occurs on different length scales. Also in this case the contact area A turns out to be proportional to the normal force F_N, provided that A remains well below the apparent area of contact A_0 or, equivalently, that F_N is very low. The relation between A and F_N can be expressed via the surface roughness power spectrum $S(k)$ introduced in Section 8.1 as

$$A(F_N) = \sqrt{\frac{\pi}{G_1}\frac{F_N}{2}}, \tag{8.10}$$

where

$$G_1 = \frac{\pi}{4}E^{*2}\int_{k_0}^{k_L} k^3 S(k)\,dk. \tag{8.11}$$

8.3 The Persson theory

In the models presented in Section 8.2 the asperities are treated as independent. In other words these models ignore the fact that, when a given asperity is compressed, the deformation field extends far away and influences the contact of other asperities. Furthermore, the fractal nature of real surfaces is not properly treated. As a result, these theories are only applicable when the contact area is extremely small. This is not the case for the contact theory developed by Bo Persson [250], which takes into account any roughness length scale and is particularly accurate in the case of complete contact.

The starting point of the Persson theory is the apparent contact area on the length scale λ, $A(\lambda)$, which is defined as the projection of the contact area formed when the original surface is smoothed on all length scales below λ. The ratio $\zeta \equiv L/\lambda$ can be considered as the 'magnification' of the surface (Fig. 8.6). For a self-affine surface described by the power spectrum in Fig. 8.3 the magnification ζ can also be defined as the ratio λ_0/λ, where $\lambda_0 = 2\pi/k_0$. For a computer generated surface, the surface profile at the magnification ζ is obtained by restricting the sum in Eq. (8.4) to the values $k < \zeta k_0$.

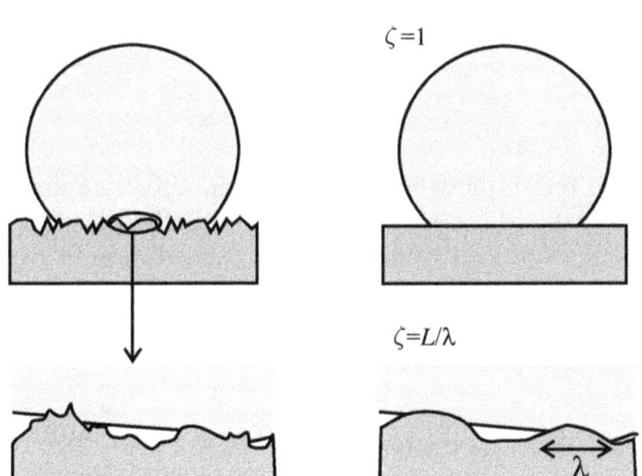

Figure 8.6 Contact between an elastic sphere and a rough hard substrate at the magnification $\zeta = 1$ and $\zeta = L/\lambda$.

Macroasperity contact

An important parameter in the Persson theory is the critical magnification ζ_c at which the contact area breaks up, as shown in Fig. 8.7. This value is approximately given by the condition

$$A(\zeta_c) = A_0(1 - p_c),$$

where the percolation threshold $p_c \approx 0.6$ [262]. The islands formed when the splitting occurs may be called 'macroasperity' contact regions.[2] The concept of macroasperity regions will be applied in Section 11.2 and in Section 24.3 in the context of heat transfer and fluid flow between two rough surfaces in contact.

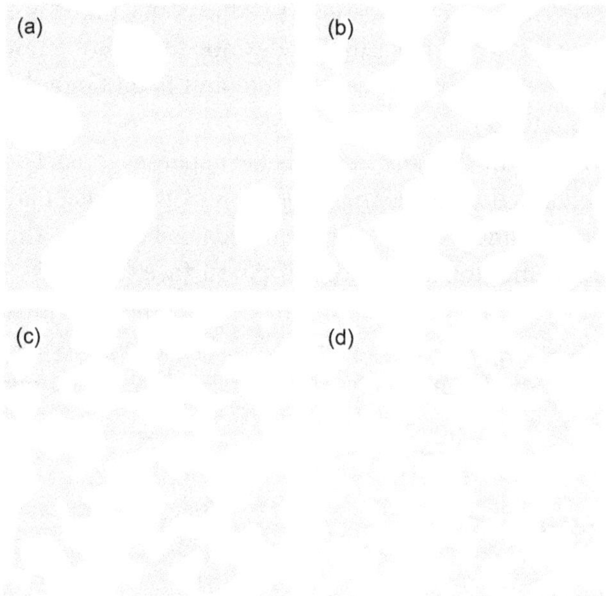

Figure 8.7 Contact region between a rough rigid substrate and an elastic block at different magnifications $\zeta = 3, 9, 12, 648$ (a–d). When $\zeta \approx 12$ the non-contact region percolates. Reproduced from [262] with permission from IOP Publishing.

[2] Assumming a Gaussian height distribution, the average radius of the macroasperities is [251]

$$R = \frac{1}{\zeta_c k_0}\sqrt{0.688\frac{\int x \exp[-(d/h_{\rm rms} + x)^2/2]\,dx}{\int \exp[-(d/h_{\rm rms} + x)^2/2]\,dx}},$$

where d is the separation between the surfaces and $h_{\rm rms}$ is the rms amplitude of the summit height fluctuations.

Diffusion equation for the contact area

Assuming complete contact it can be proven that the stress distribution in the contact area at the magnification ζ, $P(\sigma, \zeta)$, satisfies a diffusion-like equation of the form [246]

$$\frac{\partial P}{\partial \zeta} = G'(\zeta)\frac{\partial^2 P}{\partial \sigma^2}. \tag{8.12}$$

In Eq. (8.12) ζ plays the role of time and the stress σ replaces the spatial coordinate. The 'diffusion coefficient' G' is not constant but depends on $\zeta = k/k_L$, since it is the derivative of

$$G(\zeta) = \frac{\pi}{4}E^{*2}\int_{k_0}^{\zeta k_L} k^3 S(k)\,\mathrm{d}k. \tag{8.13}$$

A key assumption made by Persson is that Eq. (8.12) remains approximately valid also in the case of partial contact.

Equation (8.12) must satisfy the boundary conditions

$$P(0, \zeta) = 0, \tag{8.14}$$

which simply states that, in the absence of adhesion, the surfaces detach when the local stress vanishes, and

$$P(\infty, \zeta) = 0, \tag{8.15}$$

meaning that the stress at the interface cannot become infinitely large. In this way Eq. (8.12) can be solved and, as a result, one finds that the pressure distribution at the interface depends on σ and ζ as

$$P(\sigma, \zeta) = \frac{1}{2\sqrt{\pi G(\zeta)}}\left(\mathrm{e}^{-(\sigma-p)^2/4G} - \mathrm{e}^{-(\sigma+p)^2/4G}\right), \tag{8.16}$$

where p is the squeezing pressure.

The projected contact area at the magnification ζ is simply obtained as

$$A(\zeta) = A_0 P(\zeta), \tag{8.17}$$

where

$$P(\zeta) = \int P(\sigma, \zeta)\,\mathrm{d}\sigma.$$

The area $A(1)$ coincides with the nominal contact area A_0. A pressure p is uniformly distributed on this area, so that[3] $P(\sigma, 1) = \delta(\sigma - p)$, as shown in Fig. 8.8(a).

[3] The Dirac delta $\delta(x)$ is defined by the conditions $\delta(x) = \infty$ if $x = 0$ and $\delta(x) = 0$ otherwise. Furthermore,

$$\int_{-\infty}^{\infty} \delta(x)\,\mathrm{d}x = 1.$$

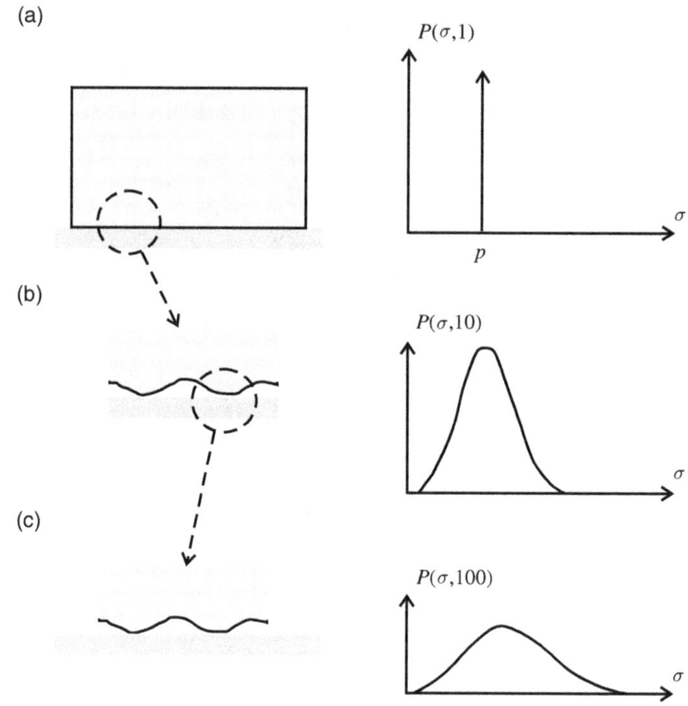

Figure 8.8 Stress distribution in the contact region between a rigid self-affine surface and an elastic flat substrate at increasing magnification ζ.

At higher magnification non-contact regions appear. Since the stress σ goes continuously to zero at the boundaries between contact and non-contact regions, the stress distribution extends down to $\sigma = 0$, as seen in Fig. 8.8(b). Considering that the average stress is equal to p, a corresponding tail must appear in the region $\sigma > p$. If the magnification is further increased, the distribution broadens (Fig. 8.8(c)), similarly to the diffusion of a fluid initially concentrated in a single point.

From Eq. (8.16), after some simplifications we find that the normalized contact area depends on the pressure as follows:

$$P(\zeta) = \frac{1}{\sqrt{\pi G(\zeta)}} \int_0^p e^{-\sigma^2/4G(\zeta)}\, d\sigma \equiv \operatorname{erf}\left(\frac{p}{2\sqrt{G(\zeta)}}\right). \qquad (8.18)$$

The dependence of the projected contact area on ζ is shown in Fig. 8.9(a) on a logarithmic scale. The function $A(\zeta)$ decreases monotonically, introducing shorter and shorter roughness wavelength components and, had the short distance cut-off a not been introduced, it would vanish asymptotically. The dependence of the contact area on the squeezing pressure p (at the highest magnification $\zeta = 1$), as given by Eq. (8.18), is plotted in Fig. 8.9(b).

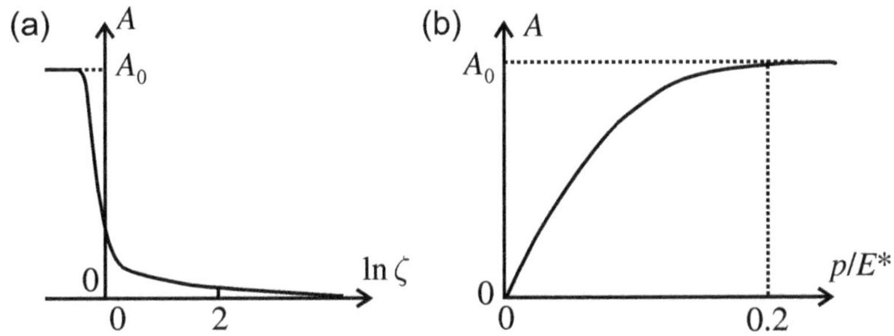

Figure 8.9 Apparent contact area in the Persson theory as a function (a) of the logarithm of the magnification ζ and (b) of the squeezing pressure. Adapted from [250] with permission from Elsevier.

It is interesting to observe that, for high values of p, when the two solids are in complete contact, the second term on the right hand side of Eq. (8.16) is negligible and the stress distribution (at an arbitrary magnification) becomes a Gaussian centered around p with half-width $\sqrt{2G}$:

$$P(\sigma, \zeta) = \frac{1}{2\sqrt{\pi G(\zeta)}} e^{-(\sigma - p)^2/4G(\zeta)}.$$

On the other hand, if $p \ll \sqrt{G}$:

$$P(\zeta) \approx \frac{p}{\sqrt{\pi G(\zeta)}}. \tag{8.19}$$

In this case the contact area is proportional to the normal force, with a coefficient of proportionality $2/\pi$ smaller than in the BGT theory:

$$A(F_N) \approx \frac{F_N}{\sqrt{\pi G_1}},$$

where $G_1 \equiv G(\zeta = 1)$, consistently with the definition (8.11).

8.4 Advanced concepts in the Persson theory

The space separating two solid surfaces has a key importance in many physical processes such as contact resistivity, heat transfer and optical interference. It also determines the frictional response of the surfaces, for instance when the space is filled by a lubricant fluid.

Elastic energy and total contact area

Suppose that a rubber block is pressed against a rough hard substrate. If the contact is complete, it can be proven that the strain energy U_{el} and the total contact area A_{con} depend on the surface roughness power spectrum $S(k)$ as [260]

$$U_{\mathrm{el}} = \frac{\pi}{2} E^* A_0 \int_{k_0}^{k_1} k^2 S(k)\, \mathrm{d}k, \tag{8.20}$$

and

$$A_{\mathrm{con}} \approx A_0 \left(1 + \pi \int_{k_0}^{k_1} k^3 S(k)\, \mathrm{d}k \right), \tag{8.21}$$

where k_0 and k_1 are the smallest and largest values of the surface roughness wave vector.[4]

Partial contact can be approximately accounted for by 'weighting' the integrands in Eqs. (8.20) and (8.21) with the factor $P(k)$ given by (8.18), with $\zeta = k/k_L$. A multiplicative correction factor of the order of one also appears in the expression for the strain energy [248].

Average separation between rough surfaces

If the squeezing pressure is not too large, Persson theory predicts an exponential increase of the normal force with decreasing average separation δ [252]. Indeed, since

$$p(\delta) = -\frac{1}{A_0} \frac{\mathrm{d}U_{\mathrm{el}}}{\mathrm{d}\delta}, \tag{8.22}$$

we can use the expression (8.19) for the dependence of the contact area on the applied load to conclude that, at low loads,

$$p(\delta) \approx \beta E^* \exp\left(-\frac{\delta}{\delta_0} \right). \tag{8.23}$$

In Eq. (8.23) the parameter β depends on the surface roughness but not on the pressure nor on the elastic properties of the solids. The characteristic length δ_0 is of the order of the rms surface roughness h_{rms} and can be approximately estimated as [344]

$$\delta_0 \approx C \int_{k_0}^{k_1} \frac{k^2 S(k)}{\sqrt{I(k)}}\, \mathrm{d}k,$$

where $C \approx 0.4$ and

$$I(k) = \int_{k_0}^{k} k'^3 S(k')\, \mathrm{d}k' \tag{8.24}$$

(note that $I = 4G/\pi E^{*2}$). If the dependence $\delta(p)$ is known from experiments, Eq. (8.22) can be used to estimate the strain energy U_{el} stored in the contact region.

[4] We have assumed that $k_L = k_0$.

For a typical self-affine fractal surface (with $D_f < 2.5$) it can be proven that [252]

$$\beta \approx 0.4 k_0 h_{rms}$$

and the characteristic length depends on the roughness and the Hurst coefficient as

$$\delta_0 \approx \sqrt{\frac{2(1-H)}{\pi H}} h_{rms} \left[r(H) - \left(\frac{k_0}{k_1} \right)^H \right], \tag{8.25}$$

where[5] $r(H) \sim 1$. For asphalt pavements $D_f \approx 2.2$ (see Section 8.1) and, if $k_0 \ll k_1$, the length $\delta_0 \approx 0.4 h_{rms}$. Equation (8.23) is in good agreement with experiments [20], but it is quite different from the predictions of the BGT and GW theories. In those cases, it is expected that $p(\delta) \propto \delta^{-\alpha} \exp(-\beta\delta^2)$, where $\alpha = 1$ or 5/2 respectively [344]. For arbitrary pressure values the average separation can be estimated with the formula [344]

$$\delta(p) = \int_{k_0}^{k_1} \frac{k^2 S(k)}{\sqrt{I(k)}} \int_p^\infty \frac{1}{p'} [C + 3(1-C) P^2(k)] \mathrm{e}^{-p'^2/\pi E^{*2} I(k)} \, \mathrm{d}p' \, \mathrm{d}k. \tag{8.26}$$

Average separation at different magnifications

In Sections 24.2 and 24.3 we will also make use of the average distance $\delta(\zeta)$ between two contacting surfaces at magnification ζ (Fig. 8.10). This quantity is given by Eq. (8.26) with properly modified lower integration limits [344]. We will also need the 'incipient' distance $\delta_{inc}(\zeta)$, which is defined as the average

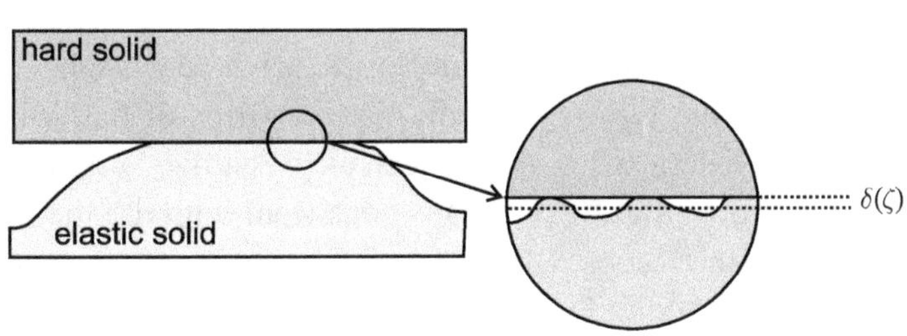

Figure 8.10 Apparent separation between two surfaces at magnification ζ.

[5] The complete definition is

$$r(H) = \frac{H}{2(1-H)} \int_1^\infty \frac{\mathrm{d}x}{x^{1/2(1-H)} \sqrt{x-1}}.$$

separation between the surfaces which appear to come into contact as a result of an infinitesimal decrease in the magnification ζ and is obtained from $A(\zeta)$ and $\delta(\zeta)$ as

$$\delta_{\rm inc}(\zeta) = \delta(\zeta) + \delta'(\zeta)\frac{A(\zeta)}{A'(\zeta)}. \tag{8.27}$$

Both $\delta(\zeta)$ and $\delta_{\rm inc}(\zeta)$ are monotonically decreasing functions of ζ.

Indentation of thin elastic layers

The Persson theory can also be applied to the indentation of a rigid substrate coated by a thin elastic layer. This problem has been already considered in Section 7.6, where the substrate was supposed to be flat. If the substrate is rough it is expected that, for small magnifications, the contact mechanics is independent of the film coating whereas, for very large magnifications, the film can be seen as infinitely thick [254]. The transition occurs at the magnification where the roughness wavelength is of the order of the film thickness.

8.5 Contact of wavy surfaces

Consider now the contact between an elastic half-space and a rigid substrate with a sinusoidal profile of amplitude h_0 and periodicity λ. It can be proven that the contact is complete only if the mean pressure p is larger than $p_{\rm c} = \pi E^* h_0/\lambda$, where $E^* = E/(1-\nu^2)$ [156, section 13.2]. If this is not the case, the contact will occur on parallel strips of half-width a such that the ratio between the real and the apparent contact is [341]

$$\frac{A}{A_0} = \frac{2a}{\lambda} = \frac{2}{\pi}\arcsin\sqrt{\frac{p}{p_{\rm c}}}. \tag{8.28}$$

If $p \ll p_{\rm c}$ the strips are independent, and Eq. (8.28) is consistent with the Hertz theory. In the opposite limit $p \to p_{\rm c}$ only narrow strips of half-width $\lambda/2 - a$ remain out of contact. These strips can be seen as pressurized cracks in an infinite solid, and it can be shown that Eq. (8.28) is consistent with the theory of fracture mechanics introduced in Section 13. The dependence of the contact area on the squeezing pressure is plotted in Fig. 8.11(a).

If the surface is periodic in both the x and y directions, in the case of low pressure the contact is formed by an array of elliptical areas, and is again described by the Hertz theory. At high pressure the small areas of separation will be elliptical and resemble pressurized cracks. If h_0 and λ are the same in both directions, the radius a of the contact circles and the radius b of the penny-shaped cracks in the two limit cases are given by [158]

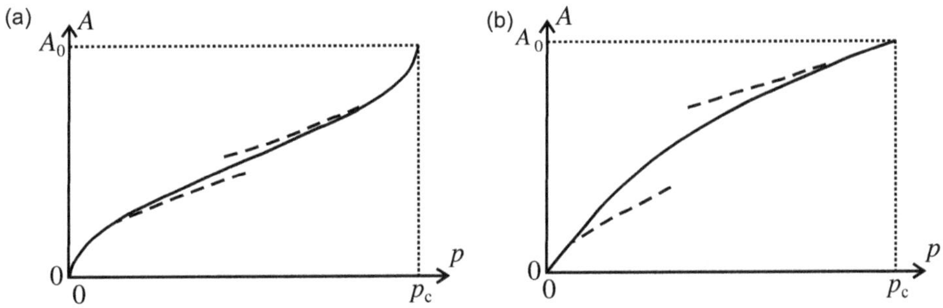

Figure 8.11 (a) Extension of the contact area between a 1D wavy surface and an elastic half-space (continuous curve). The dashed curves represent the dependencies expected from the Hertz theory ($p \to 0$) and from fracture mechanics ($p \to p_c$). (b) The same for a 2D wave surface with equal periodicities. Adapted from [158] with permission from Elsevier.

$$a = \lambda \left(\frac{3}{8\pi} \frac{p}{p_c} \right)^{1/3} \qquad (p \ll p_c), \tag{8.29}$$

$$b = \frac{\lambda}{\pi} \sqrt{\frac{3}{2} \left(1 - \frac{p}{p_c} \right)} \qquad (p \to p_c), \tag{8.30}$$

where $p_c = 2\pi E^* h_0/\lambda$. In Fig. 8.11(b) Eqs. (8.29) and (8.30) are compared with a numeric solution of the problem. The theoretical predictions have been substantiated by experiments on rubber [156, section 13.2].

9

Viscoelastic contacts

The contacts formed by viscoelastic bodies present very peculiar features. The chapter starts examining the stress–strain relations observed in viscoelastic materials and the ways to model them from first principles. Viscoelastic indentation is discussed in the framework of the Radok correspondence principle, which is also extended to the Persson theory. This theory is applied to rubber friction, where an important relation between the friction coefficient and the so-called loss function can be derived. The discussion ends with the rolling resistance of viscoelastic materials and its dependence on a characteristic 'Deborah number'.

9.1 Stress–strain relation

In *viscoelastic* materials like polymers the relation between stress and strain is time-dependent. Since the value of Poisson's ratio ν in these materials is usually larger than 0.4, they can be treated as incompressible. The typical response of a viscoelastic material, when a finite tensile stress σ_0 is applied and suddenly released after a certain time t_1, is shown in Fig. 9.1. The strain $\varepsilon(t)$ responds almost instantaneously (elastically) to the applied stress, and then keeps growing (viscously) at an exponentially decreasing rate. If the material is able to flow it will also acquire a steadily increasing creep strain. When the stress is removed a sudden elastic response is followed by an exponential decay of $\varepsilon(t)$. A non-zero value of strain, caused by creep, is reached asymptotically.

The response $\varepsilon(t)$ to a step function $\sigma(t)$ of height σ_0 defines the *creep compliance* $C(t)$ of the material: $\varepsilon(t) = C(t)\sigma_0$. More generally, if the function $\sigma(t)$ has an arbitrary shape:

$$\varepsilon(t) = \int_{-\infty}^{t} C(t-t')\frac{d\sigma}{dt'}dt'.$$

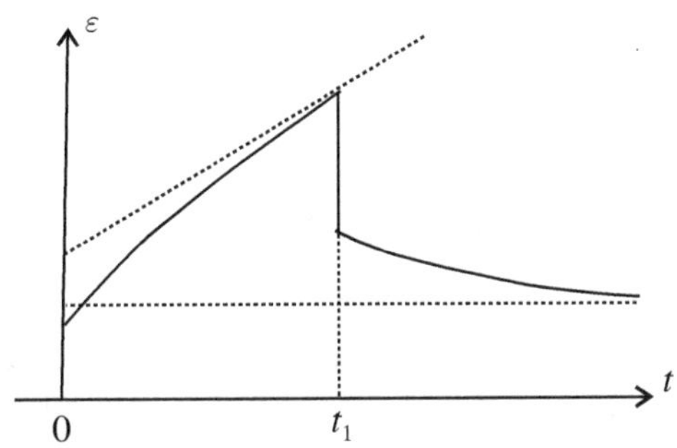

Figure 9.1 Strain variation in a viscoelastic material under the action of a constant stress applied for a period t_1.

Similarly, the *viscoelastic modulus* $E(t)$ is defined by the convolution

$$\sigma(t) = \int_{-\infty}^{t} E(t-t') \frac{\mathrm{d}\varepsilon}{\mathrm{d}t'} \, \mathrm{d}t'. \tag{9.1}$$

If $\varepsilon(t)$ is a step function of height ε_0, the corresponding stress $\sigma(t) = E(t)\varepsilon_0$.

The Fourier transforms of stress and strain,

$$\sigma(\omega) = \frac{1}{2\pi} \int_{-\infty}^{\infty} \sigma(t) \mathrm{e}^{\mathrm{i}\omega t} \, \mathrm{d}t,$$

and similarly for $\varepsilon(\omega)$, are related by the equation

$$\sigma(\omega) = \varepsilon(\omega) E(\omega),$$

where

$$E(\omega) = -\mathrm{i}\omega \int_{0}^{\infty} E(t) \mathrm{e}^{\mathrm{i}\omega t} \, \mathrm{d}t. \tag{9.2}$$

Note that $E(\omega)$ is not the Fourier transform[1] of $E(t)$. The real and imaginary parts of $E(\omega)$ are the *storage modulus* $E'(\omega)$ and the *loss modulus* $E''(\omega)$ respectively. The meaning of the last quantity can be understood if we calculate the energy $\Delta\mathcal{E}$ dissipated in the deformation of a volume V of the material:

$$\Delta\mathcal{E} = \int \sigma_{ij} \dot{\varepsilon}_{ij} \, \mathrm{d}t \, \mathrm{d}V = -\frac{V}{2\pi} \int \omega E''(\omega) \, |\varepsilon(\omega)|^2 \, \mathrm{d}\omega.$$

Alternatively, the energy loss can be written as

$$\Delta\mathcal{E} = \frac{V}{2\pi} \int \omega \mathrm{Im} \left(\frac{1}{E(\omega)} \right) |\sigma(\omega)|^2 \, \mathrm{d}\omega. \tag{9.3}$$

[1] The definition (9.2) is motivated by the fact that this quantity, and not the Fourier transform, is directly accessible in rheological experiments.

The imaginary part of $1/E(\omega)$ is called the *loss function*. Similar definitions apply to the viscoelastic shear modulus $G(t) \approx E(t)/3$.

9.2 Constitutive models

As seen in Section 9.1, viscoelastic behavior has elastic and viscous components. An elastic component can be modeled as a spring following Hooke's law $\sigma = E\varepsilon$. A viscous component can be modeled as a dashpot with a stress–strain rate relation $\sigma = \eta\dot{\varepsilon}$. Based on these assumptions, various models have been proposed.

Maxwell model

In the *Maxwell model* [209] a viscoelastic material is described by a spring and a dashpot connected in series as in Fig. 9.2(a). The variations of stress and strain are related as:

$$\dot{\varepsilon} = \frac{\sigma}{\eta} + \frac{1}{E}\dot{\sigma}. \tag{9.4}$$

If a step strain ε_0 is applied, the stress σ suddenly increases and subsequently decays exponentially with a *relaxation time* $\tau = \eta/E$:

$$\sigma(t) = E\mathrm{e}^{-t/\tau}\varepsilon_0.$$

On the other hand, if a step stress σ_0 is applied the strain ε increases linearly with time (so-called *steady creep*):

$$\varepsilon(t) = \frac{\sigma_0}{E}\left(1 + \frac{t}{\tau}\right). \tag{9.5}$$

The relation (9.5) is valid until ε is no longer small and the model breaks down.

The differential equation (9.4) corresponds to the viscoelastic modulus

$$E(\omega) = \frac{E\mathrm{i}\omega\eta}{E + \mathrm{i}\omega\eta}.$$

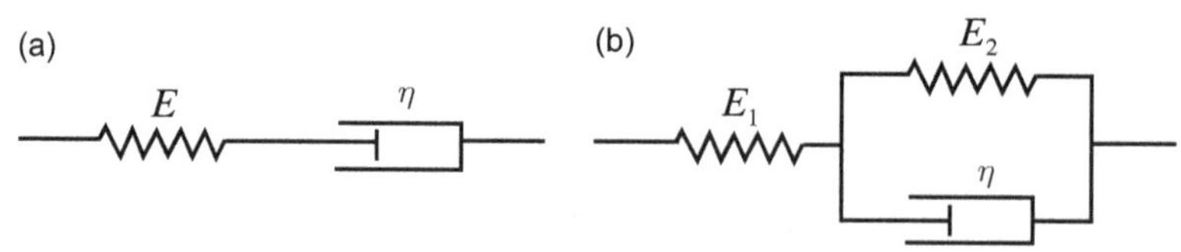

Figure 9.2 (a) Maxwell model and (b) standard solid model for viscoelastic materials.

The storage and loss moduli are, respectively,

$$E'(\omega) = E\frac{(\omega\tau)^2}{1+(\omega\tau)^2}, \qquad E''(\omega) = E\frac{\omega\tau}{1+(\omega\tau)^2}.$$

Standard solid model

The limitations of the Maxwell model can be overcome by adding a second spring in parallel to the dashpot (Fig. 9.2b). In this case

$$\left(1+\frac{E_2}{E_1}\right)\frac{\sigma}{\eta} + \frac{1}{E_1}\dot{\sigma} = \frac{E_2}{\eta}\varepsilon + \dot{\varepsilon}.$$

If a step stress σ_0 is applied, the material will immediately deform to the elastic part of the strain, then gradually approach a final value adding the viscous part (*delayed elasticity*):

$$\varepsilon(t) = \frac{\sigma_0}{E_1} + \frac{\sigma_0}{E_2}\left(1 - \mathrm{e}^{-t/\tau_\varepsilon}\right), \tag{9.6}$$

where $\tau_\varepsilon = \eta/E_2$. If a step strain is applied:

$$\sigma(t) = E_\infty\left(1 + \frac{E_1}{E_2}\mathrm{e}^{-t/\tau_\sigma}\right)\varepsilon_0, \tag{9.7}$$

where $\tau_\sigma = \eta/(E_1+E_2)$ and $1/E_\infty = 1/E_1 + 1/E_2$.

In the frequency domain:

$$E(\omega) = E_\infty\frac{1+\mathrm{i}\omega\tau_\varepsilon}{1+\mathrm{i}\omega\tau_\sigma}. \tag{9.8}$$

The frequency dependence of the real and imaginary parts of $E(\omega)$, as described by Eq. (9.8), is shown in Fig. 9.3. In the 'rubbery' region at low frequencies the solid is soft, whereas it is quite stiff in the 'glassy' region at high frequencies. The storage

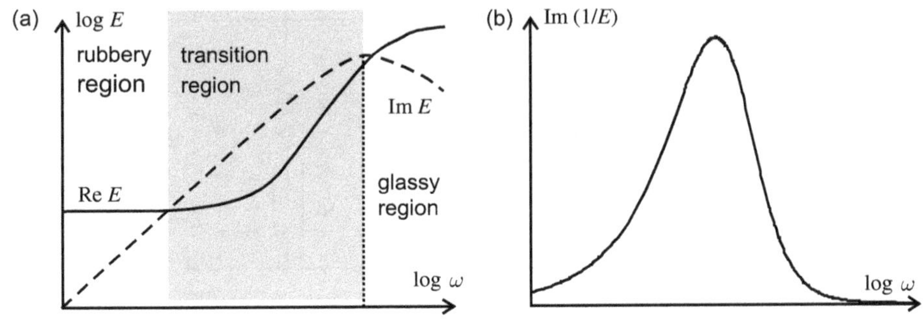

Figure 9.3 (a) Real part (continuous curve) and imaginary part (dashed curve) of the viscoelastic modulus $E(\omega)$ given by (9.8). (b) Loss function $\mathrm{Im}(1/E(\omega))$.

modulus E' tends to the two limit values E_∞ and E_1 respectively at low and high frequencies. The loss function is quite large in the intermediate region, peaking at the frequency $\omega_{max} = 1/\tau_\varepsilon$ (Fig. 9.3(b)).

The scheme presented in Fig. 9.2(b) is know as the Voigt form of the standard solid model. Alternatively, one can add a spring in parallel to the Maxwell element in Fig. 9.2(a). In this case the governing equation is

$$\eta\dot{\sigma} + E_1\sigma = \eta(E_1 + E_2)\dot{\varepsilon} + E_1 E_2 \varepsilon$$

corresponding to the storage and loss moduli

$$E'(\omega) = E_1 + E_2\frac{(\omega\tau)^2}{1+(\omega\tau)^2}, \qquad E''(\omega) = E_2\frac{\omega\tau}{1+(\omega\tau)^2}.$$

Maxwell–Wiechert model

The standard solid model is still too simple to reproduce the response of a polymer over a long time. The creep in Fig. 9.1 can be obtained by adding a second dashpot with viscosity η_2 in series to the spring and the Voigt element in Fig. 9.2(b), which introduces an additional linear term $(\sigma_0/\eta_2)t$ on the right hand side of Eq. (9.6). However, in order to reproduce the general response of a real system, a distribution of stiffness and relaxation times must be introduced. This can be done by connecting a spring and several Maxwell elements in parallel, as shown in Fig. 9.4. In this case the relaxation modulus is expressed by the *Prony series*

$$E(t) = E_\infty + \sum_{i=1}^{N} E_i e^{-t/\tau_i},$$

where E_∞ is the steady-state stiffness corresponding to the spring, and the parameters E_i and τ_i characterize each of the N elements.

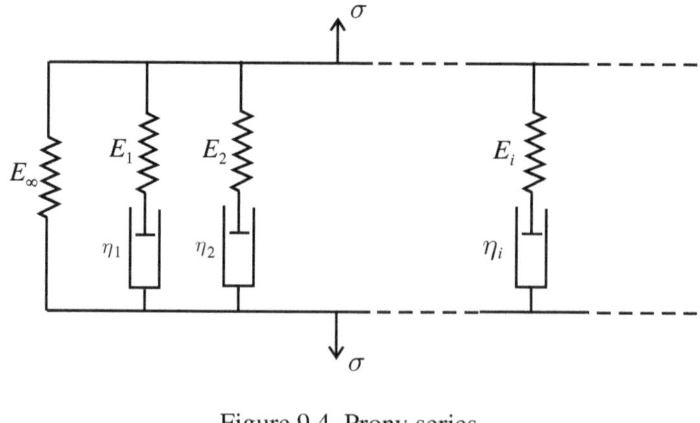

Figure 9.4 Prony series.

9.3 Viscoelastic indentation

Analytical solutions of viscoelastic contact problems can be derived by means of the *correspondence principle* introduced by Radok [280]. According to this principle, if the deformation history of the contacting bodies is known, the effective elastic modulus E^* appearing in the expressions for the pressure distribution in an elastic contact can be replaced by an integral operator:

$$E^* \to \int E^*(t-t')\frac{\mathrm{d}}{\mathrm{d}t'}\cdots \mathrm{d}t',$$

where E^* is now the effective viscoelastic modulus. However, the replacement is only possible if the contact area is growing with time.

In this way it is not difficult to prove that, if a rigid sphere of radius R is pressed into a viscoelastic half-space by a normal force F_N, the contact radius will increase with time as

$$a(t) = \left(\frac{9RF_\mathrm{N}}{16E(t)}\right)^{1/3}. \tag{9.9}$$

If the viscoelastic material is described by the Maxwell model, a will immediately jump to the value $a_0 = (9RF_\mathrm{N}/16E)^{1/3}$ and grow continuously as long as $a \ll R$. At the same time, the pressure will first follow an elastic distribution, with the maximum value in the center of the contact area, and then concentrate more and more towards the edge of the contact, as shown in Fig. 9.5(a) [156, section 6.5]. If the material is described by the standard solid model in the Voigt form, the pressure will keep an approximately elastic distribution all the time, as in Fig. 9.5(b) [346]. Correspondingly, the contact radius will vary from $a_1 = (9RF_\mathrm{N}/16E_1)^{1/3}$ to $a_\infty = (9RF_\mathrm{N}/16E_\infty)^{1/3}$.

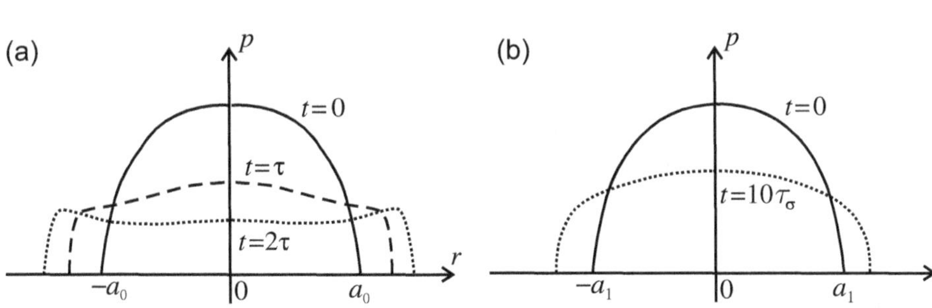

Figure 9.5 Variation of pressure distribution when a rigid sphere is pressed by a step load against a viscoelastic half-space, described by (a) the Maxwell model and (b) the standard solid model. Adapted from [156] with permission from Cambridge University Press.

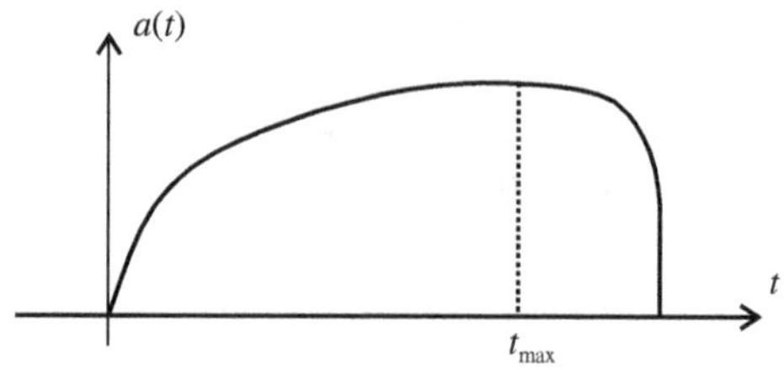

Figure 9.6 Time dependence of the contact radius between a rigid sphere and a viscoelastic half-space under the action of a sinusoidal normal force. Adapted from [156] with permission from Cambridge University Press.

If the contact is first loaded and then unloaded according to a sinusoidal law $F_N = F_0 \sin \omega t$, a will keep growing by creep even when F_N starts to decrease. Only at the time $t_{max} = 3\pi/4\omega$ does the contact area start to decrease (rapidly) to zero. The time dependence of a in this last phase can be estimated as described in [156, section 6.5]. The complete result is shown in Fig. 9.6.

Rough substrates

Suppose now that a viscoelastic block with surface roughness power spectrum $S(k)$ is squeezed against a rough hard substrate. If the block were elastic and the load were very low, we could write the relation (8.19) between the normalized contact area P and the squeezing pressure p at the magnification ζ as

$$E^* P(\zeta) = \frac{2p}{\pi\sqrt{I(\zeta)}}, \tag{9.10}$$

where $I(\zeta)$ is defined by Eq. (8.24). According to the correspondence principle, we can generalize Eq. (9.10) to a viscoelastic solid as

$$\int_{-\infty}^{t} E^*(t-t')\frac{\partial P}{\partial t'}\,dt' = \frac{2p(t)}{\pi\sqrt{I}}.$$

As a result [263]:

$$P(\zeta, t) = \frac{2}{\pi\sqrt{I(\zeta)}}\int_{-\infty}^{\infty} \frac{p(\omega)}{E^*(\omega)} e^{-i\omega t}\,d\omega, \tag{9.11}$$

where $p(\omega)$ is the Fourier transform of $p(t)$. For an arbitrary load, a comparison with Eqs. (8.18) and (8.19) suggests that

$$P(\zeta, t) = \text{erf}\left(\frac{\sqrt{\pi}}{2} P_0(\zeta, t)\right),$$

where P_0 is defined by Eq. (9.11).

If $E^*(t) = E_\infty + (E_1 - E_\infty)e^{-t/\tau}$ and $\sigma(t) = \sigma_0$ for $0 < t < t_1$ and zero otherwise, one gets

$$P_0(\zeta, t) = \frac{2}{\pi\sqrt{I}}\frac{\sigma_0}{E_\infty}\left[1 + \left(\frac{E_\infty}{E_1} - 1\right)e^{-t/\tau}\right]$$

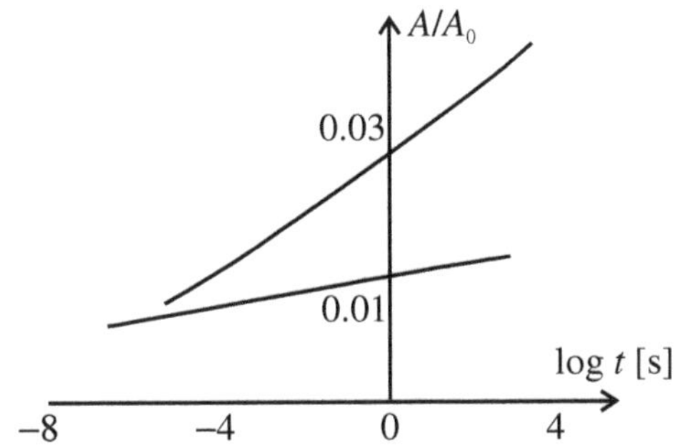

Figure 9.7 Relative contact area of a block of tread rubber (upper curve) and rim rubber (lower curve) squeezed against a steel surface by a pressure $p = 0.1$ MPa. Adapted from [265] with permission from IOP Publishing.

and, for an arbitrarily large load:

$$P(\zeta, t) = \frac{2p}{\pi^2 \sqrt{I(\zeta)}} \mathrm{Re} \int_0^\infty \frac{1 - \mathrm{e}^{-\mathrm{i}\omega t_1}}{-\mathrm{i}\omega} \frac{\mathrm{e}^{-\mathrm{i}\omega t}}{E^*(\omega)} \mathrm{d}\omega. \tag{9.12}$$

The time dependence of the contact area expected from Eq. (9.12) is shown in Fig. 9.7 for two types of rubber. Again, these results are only valid as long as the contact area increases with time.

9.4 Rubber friction

Rubber friction is a complex phenomenon determined by at least four factors: (i) the viscoelastic energy dissipation caused by the surface asperities while sliding, (ii) the opening crack tips, (iii) the shearing of possible contamination films and (iv) wear process. The contribution of adhesion is usually negligible, as shown in Section 10.4. Here, we will focus on the first of these processes, whereas energy dissipation at opening cracks (important on smooth surfaces) will be discussed in Section 13.4.

Suppose first that a rubber block is pressed against a rigid wavy substrate while sliding with a constant velocity v. In this case a qualitative estimation of the kinetic friction coefficient μ_k is possible [259]. The energy dissipation in a contact region with linear size l occurs in a volume $V \sim l^3$ (Fig. 9.8). The stress acting in this volume is $\sigma \sim p \cos \omega_0 t$, where the frequency $\omega_0 \sim v/l$ and p is the average pressure. Substituting into Eq. (9.3) we obtain the energy $\Delta\mathcal{E}$ dissipated in the time $\Delta t = 2\pi/\omega_0$:

$$\Delta\mathcal{E} \sim l^3 p^2 \omega_0 \Delta t \,\mathrm{Im}\left(\frac{1}{E(\omega_0)}\right).$$

Since the product of the friction force $F_\mathrm{k} = \mu_\mathrm{k} F_\mathrm{N}$ and the sliding velocity is equal to the power dissipation $\Delta\mathcal{E}/\Delta t$, and the normal force $F_\mathrm{N} \approx l^2 p$,

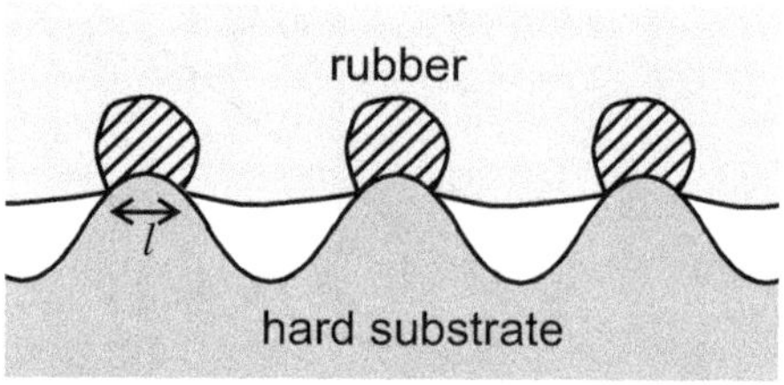

Figure 9.8 The fluctuating stress acting on a sliding rubber block gives rise to energy dissipation via the internal friction of the rubber. Most of the dissipation occurs in the hatched regions.

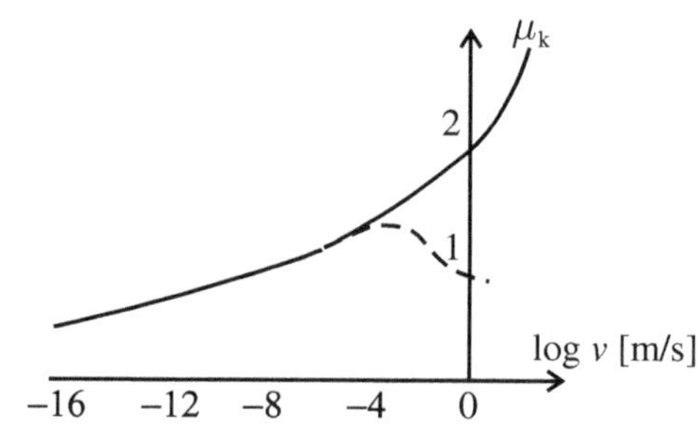

Figure 9.9 Friction coefficient as a function of the sliding velocity for a rubber block sliding on an asphalt road surface (continuous curve). The dashed curve is obtained by introducing the so-called flash temperature (see Section 11.2). Adapted from [251] with permission from IOP Publishing.

$$\mu_k \sim p \, \mathrm{Im}\left(\frac{1}{E(\omega_0)}\right). \tag{9.13}$$

Considering the frequency dependence of the loss function, one can expect a large asperity-induced contribution to the friction if the frequency ω_0 is close to a peak value in Fig. 9.3(b) or in the Prony series (which is not the case in most practical applications).

On an arbitrary rough surface, a precise expression for the friction coefficient can be obtained in the framework of Persson theory [246]:

$$\mu_\mathrm{k} = \frac{1}{2p}\int_{k_0}^{k_1} k^3 S(k) P(k) \int_0^{2\pi} \cos\varphi \mathrm{Im}\, E^*(kv\cos\varphi)\, \mathrm{d}\varphi\, \mathrm{d}k, \tag{9.14}$$

where the normalized contact area $P(k)$ is given by the relation (8.18) with

$$G(k) = \frac{1}{8}\int_{k_0}^{k} k^3 S(k) \int_0^{2\pi} \left|E^*(kv\cos\varphi)\right|^2 \mathrm{d}\varphi\, \mathrm{d}k.$$

The velocity dependence of μ_k, as predicted by Eq. (9.14), is shown in Fig. 9.9. As will be discussed in Section 11.2, this relation is significantly modified at high velocities by taking into account the frictional heating occurring in the contact regions.

9.5 Rolling on viscoelastic bodies

The pressure distribution exerted by a rigid cylinder of radius R rolling on a viscoelastic overlayer of thickness d, as in Fig. 9.10, was calculated by May *et al.* [210]. If the cylinder is pressed against the layer by a normal force F_N and the rolling velocity is v, the contact extends asymmetrically on two strips of width a and b. If the layer is described by the standard solid model in Fig. 9.2(b), with the viscoelastic modulus E replaced by the viscoelastic shear modulus G, the pressure distribution inside the strips is

$$p(x) = \frac{G_{\mathrm{eq}}a^2}{Rd}\left[\frac{1}{2}\left(1-\frac{x^2}{a^2}\right) - \beta \mathrm{D}\left(1+\frac{x}{a}\right) + \beta \mathrm{D}(1+\mathrm{D})\left(1-\mathrm{e}^{(1+x/a)/\mathrm{D}}\right)\right]. \tag{9.15}$$

In Eq. (9.11) $G_{\mathrm{eq}}^{-1} = G_1^{-1} + G_2^{-2}$, $\beta = G_1/G_2$ and the *Deborah number*[2]

$$\mathrm{D} = \frac{\tau_\sigma}{a/v}$$

is the ratio between the relaxation time τ_σ of the material and the time a/v taken to cross the distance a. The lengths a and b can be determined as a function of the parameters β and D from the condition $p(-a) = p(b) = 0$.

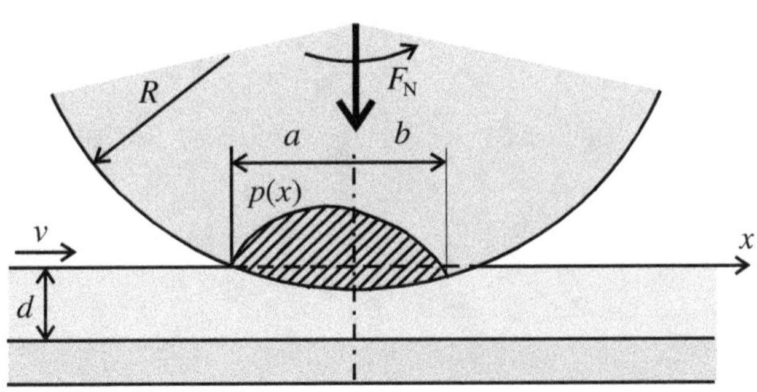

Figure 9.10 Contact formed by a rigid cylinder rolling on a viscoelastic foundation.

[2] This term was introduced by the Israeli engineer Markus Reiner, who was inspired by a biblical song by the prophetess Deborah.

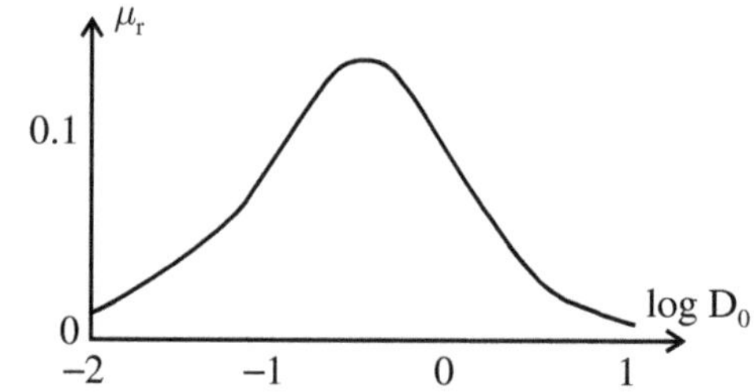

Figure 9.11 Rolling resistance coefficient as a function of the Deborah number for the contact described in Fig. 9.10.

The coefficient of rolling resistance μ_r, defined by Eq. (6.3), can be estimated by observing that the resistive torque is

$$M_{\text{fric}} = -\int_{-a}^{b} xp(x)\,\mathrm{d}x.$$

The torque M_{fric} is plotted in Fig. 9.11 as a function of $\mathrm{D}_0 = \tau_\sigma/(a_0/v)$, where a_0 is the half-width of the contact in the static case (Section 9.3). Note that the maximum value of μ_r is reached when $\mathrm{D}_0 \sim 1$.

10

Adhesive contacts

The adhesion between two elastic surfaces can be described within different frameworks. The JKR model and the DMT models emphasize short-range and long-range forces respectively, and can be considered as extreme cases of the same theory. The intermediate regime has been described by Maugis. The Persson theory makes also significant predictions regarding adhesion. Interestingly, adhesion is expected to play only a minor role in rubber friction. The importance of adhesion in biological systems will also be briefly mentioned.

10.1 The Johnson–Kendall–Roberts model

In the presence of adhesion, the contact between two elastic spheres has a surface energy

$$U_{\text{surf}} = -\Delta\gamma\pi a^2,$$

where $\Delta\gamma = \gamma_1 + \gamma_2 - \gamma_{12}$ is the local change of surface tension upon contact and a is the contact radius (γ_i is the energy of each surface and γ_{12} is the energy of the interface). A displacement with the Hertzian form (4.20) is retained if we add a term

$$p_{\text{adh}}(r) = \frac{p_0'}{\sqrt{1 - r^2/a^2}}$$

to the pressure distribution $p(r) = p_0\sqrt{1 - r^2/a^2}$ given by Eq. (4.17) [157]. Correspondingly, the elastic energy stored in the two bodies becomes

$$U_{\text{el}} = \frac{\pi^2 a^3}{E^*}\left(\frac{2}{15}p_0^2 + \frac{2}{3}p_0 p_0' + p_0'^2\right),$$

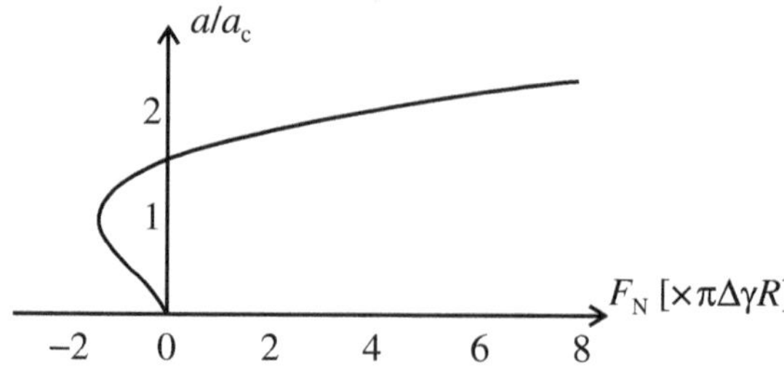

Figure 10.1 Contact radius as a function of the normal force in the JKR model.

and a term

$$\delta_{\mathrm{adh}} = \frac{\pi a}{E^*} p_0'$$

is added to the penetration depth $\delta_0 = (\pi a / 2E^*) p_0$, as proven in [156, section 5.5]. The value of p_0' is defined by the condition that the total energy $U_{\mathrm{el}} + U_{\mathrm{surf}}$ has a minimum with respect to a:

$$p_0' = -\sqrt{\frac{2\Delta\gamma E^*}{\pi a}}.$$

By integrating the pressure $p(r)$ over the contact area, we can express the contact radius a as a function of the resulting normal force $F_{\mathrm{N}} = \int 2\pi r p(r)\,\mathrm{d}r$:

$$a^3 = \frac{3R}{4E^*}\left(F_{\mathrm{N}} + 3\Delta\gamma\pi R + \sqrt{6\Delta\gamma\pi R F_{\mathrm{N}} + (3\Delta\gamma\pi R)^2}\right). \tag{10.1}$$

The relation (10.1) is plotted in Fig. 10.1. Note that the equilibrium of the contact becomes unstable when the force F_{N} reaches the critical value

$$F_{\mathrm{c}} = -\frac{3\pi}{2}\Delta\gamma R. \tag{10.2}$$

In this situation, corresponding to a finite critical radius

$$a_{\mathrm{c}} = \left(\frac{9\pi\,\Delta\gamma\,R^2}{8E^*}\right)^{1/3},$$

the two surfaces separate. The opposite of F_{c} can be considered as the *adhesion force* F_{adh} of the contact.

In Fig. 10.2 the deformed shape of the spheres is shown. The spheres meet at 90° at the edges, where the pressure tends (theoretically) to infinity.

10.2 The Derjaguin–Muller–Toporov model

According to the Derjaguin, Muller and Toporov (DMT) model [71] two elastic spheres pressed together by a normal force F_{N} are supposed to follow the Hertz

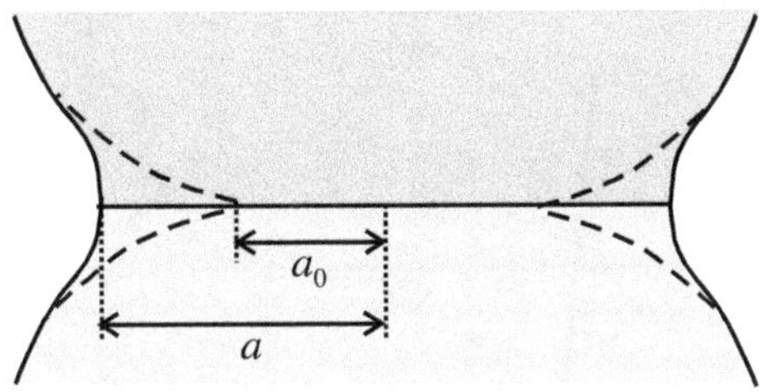

Figure 10.2 Deformation profiles for two elastic spheres in contact without adhesion (dashed curves) and with adhesion (solid curves).

theory in their area of contact, whereas a long-range attractive force acts outside this area. The contact radius in this case is given by

$$a^3 = \frac{3R}{4E^*}(F_N - F_c),$$

where the pull-off force

$$F_c = -2\pi \Delta\gamma R.$$

The value of F_c is assumed to be equal to the force required to separate two *rigid* spheres with the same radii interacting via a Lennard–Jones (LJ) potential [34]:

$$F(z) = \frac{8\Delta\gamma\pi R}{3}\left[\frac{1}{4}\left(\frac{z_0}{z}\right)^8 - \left(\frac{z_0}{z}\right)^2\right]. \tag{10.3}$$

The expression (10.3) has indeed a minimum $F = F_c$ at $z = z_0$.

The conditions of applicability of either the DMT or the JKR model were established by Tabor [328], who introduced a parameter to quantify the ratio between the elastic deformation at the point of separation and the range of surface forces:

$$\mu_T = \left(\frac{R(\Delta\gamma)^2}{E^{*2}z_0^3}\right)^{1/3}.$$

For small stiff spheres ($\mu_T \ll 1$) the elastic deformation is negligible and the DMT model is applicable. For large compliant spheres ($\mu_T \gg 1$) the JKR model is a better approximation.

10.3 The Maugis–Dugdale model

The DMT model and the JKR model can be seen as limit cases of the theory developed by Maugis and Dugdale [207]. Here, a constant adhesive force (per unit area) σ_{adh} is supposed to act over a distance h_0 between the contacting spheres. The parameter σ_{adh} corresponds, arbitrarily, to the minimum force in the LJ potential

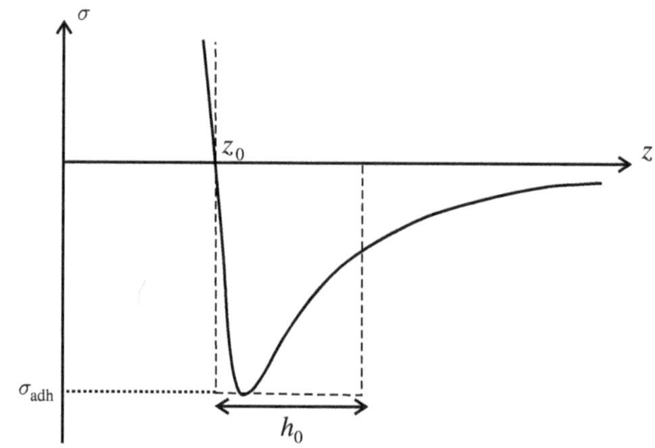

Figure 10.3 Comparison between a Lennard–Jones force profile (continuous curve) and the corresponding Dugdale profile (dashed lines).

and the parameter h_0 is determined by matching the work of adhesion to that of LJ: $\sigma_{adh} h_0 = \Delta\gamma$ (Fig. 10.3). This leads to the relation $h_0 = 0.971 z_0$, where z_0 is the equilibrium distance in LJ. The adhesive force extends up to a radius $c > a$, where a is the radius of intimate contact of the two spheres, as determined by the Hertzian relation (4.16). The radius c is implicitly obtained from the relation

$$\frac{2\sigma_{adh}}{\pi \Delta\gamma R}\left(\left(c^2 - 2a^2\right) \arccos\frac{a}{c} + a\sqrt{c^2 - a^2}\right) + \frac{16\sigma_{adh}^2}{\pi \Delta\gamma E^*}\left(\sqrt{c^2 - a^2} \arccos\frac{a}{c} - c + a\right) = 1.$$

The normal force

$$F_N = \frac{4E^* a^3}{3R} - 2\sigma_{adh}\left(c^2 \arccos\frac{a}{c} + a\sqrt{c^2 - a^2}\right)$$

and the total compression is

$$\delta = \delta_0 - \frac{2\sigma_{adh}}{E^*}\sqrt{c^2 - a^2}.$$

Introducing the *Maugis parameter*

$$\lambda = \sigma_{adh}\left(\frac{9R}{2\pi \Delta\gamma E^{*2}}\right)^{1/3},$$

the DMT and the JKR relations are recovered when $\lambda \ll 1$ or $\lambda \gg 1$ respectively. This is not suprising, since $\lambda = 1.16\mu_T$. An empirical fit of the MD model, which is valid for any value of the parameter λ, has been proposed by Carpick *et al.* [47].

10.4 The Persson theory with adhesion

The contact mechanics theory for rough surfaces introduced in Section 8.3 can be extended to adhesive contacts. As a preliminary step for determining the contact area at different magnifications in the presence of adhesion, the concepts of effective surface tension and detachment stress must first be introduced.

Effective surface tension

Suppose that a rubber block is squeezed against a rough hard substrate by a pressure p. If the contact is complete, we can define an *effective surface tension* $\gamma_{\rm eff}$ satisfying the relation

$$U_{\rm el} + U_{\rm adh} = -\gamma_{\rm eff} A_0,$$

where $U_{\rm el}$ is the elastic energy stored in the asperity contact regions, $U_{\rm adh}$ is the adhesion energy associated with the bonding between the two solids and A_0 is the apparent contact area. The adhesion energy can be written as

$$U_{\rm adh} = -\Delta\gamma A_{\rm con},$$

where $\Delta\gamma$ is the change of surface tension upon contact, introduced in Section 10.1, and $A_{\rm con}$ is the total contact area (which is larger than A_0). The expressions (8.20) and (8.21) for $U_{\rm el}$ and $A_{\rm con}$ allow us to determine $\gamma_{\rm eff}$ for any surface with a given roughness power spectrum $S(k)$.

Experimentally, the effective surface tension $\gamma_{\rm eff}$ can be estimated from the pull-off force $F_{\rm adh}$ required to detach a sphere squeezed against a rough surface. This force is expected to be the same as in the JKR theory, with $\Delta\gamma$ replaced by $\gamma_{\rm eff}$ [248]:

$$F_{\rm adh} = \frac{3\pi}{2}\gamma_{\rm eff} R.$$

A relation between $\gamma_{\rm eff}$ and the rms roughness $h_{\rm rms}$ in good agreement with the Persson theory has been measured by Peressadko *et al.* using rubber balls [240].

More generally, we can define the effective surface tension $\gamma_{\rm eff}(\zeta)$ at the magnification ζ with the relation

$$U_{\rm el}(\zeta) + U_{\rm adh} = -\gamma_{\rm eff}(\zeta) A_{\rm con}(\zeta),$$

where $U_{\rm el}(\zeta)$ is the elastic energy stored in the asperity contact regions and $A_{\rm con}(\zeta)$ is the real contact area at the magnification ζ. A possible relation between $\gamma_{\rm eff}$ and ζ, estimated numerically, is shown in Fig. 10.4. At low magnification $\gamma_{\rm eff}$ is smaller than $\Delta\gamma$ due to the contribution of the roughness-induced elastic deformation energy [248]. As ζ increases, the limit value $\Delta\gamma$ is approached.

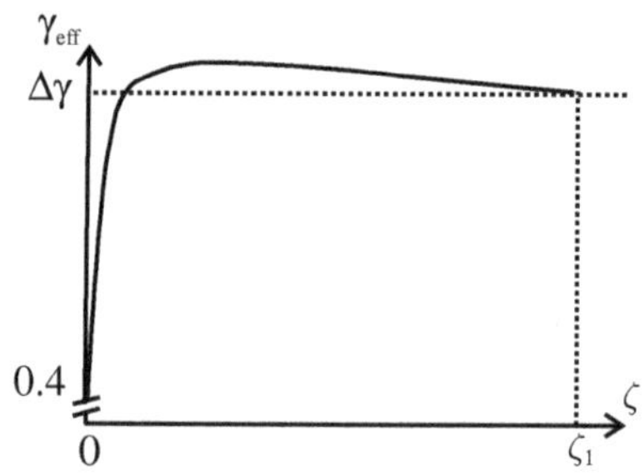

Figure 10.4 Dependence of the effective surface tension γ_{eff} on the magnification ζ. Parameter values: $k_0 = 0.17/h_{\text{rms}}$, $k_1 = 100$, $D_{\text{f}} = 2.2$, $p = 0.05E^*$, $\Delta\gamma = 2.5k_0E^*$. Adapted from [248] with permission from The European Physical Journal (EPJ).

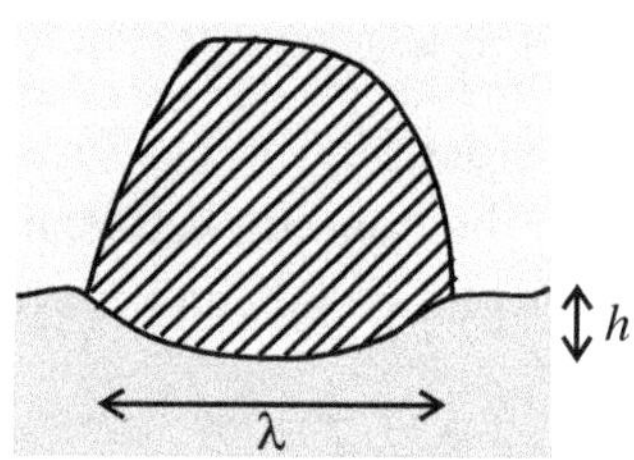

Figure 10.5 Contact formed by a rubber surface completely squeezed against a rough hard substrate. The dashed region corresponds to high stress.

Detachment stress

In the presence of adhesion the first boundary condition for the diffusion equation (8.12) becomes

$$P(-\sigma_{\text{a}}(\zeta), \zeta) = 0, \tag{10.4}$$

where σ_{a} is the largest possible tensile stress at a given magnification ζ. The detachment stress $\sigma_{\text{a}}(\zeta)$ can be related to the effective surface tension γ_{eff} as follows. If the rubber fills a cavity of height h and width λ, as in Fig. 10.5, the strain $\varepsilon \sim h/\lambda$ corresponds to the elastic energy $U_{\text{el}} \sim \lambda^3 E(h/\lambda)^2$. This quantity is balanced by the adhesion energy $U_{\text{adh}} \sim \gamma_{\text{eff}}\lambda^2$. Thus, $\sigma_{\text{a}}(\zeta) \sim E(h/\lambda) \sim \sqrt{\gamma_{\text{eff}}(\zeta)E/\lambda}$. A precise relation is obtained using the Griffith criterion, Eq. (13.7):

$$\sigma_{\text{a}}(\zeta) = \sqrt{\frac{2\gamma_{\text{eff}}(\zeta)E^*k}{\pi^2}}. \tag{10.5}$$

Area of contact

The Persson equation (8.12) can also be solved with the boundary condition (10.4). As a result, one gets the following relation between the normal force F_{N} and the contact area [250]:

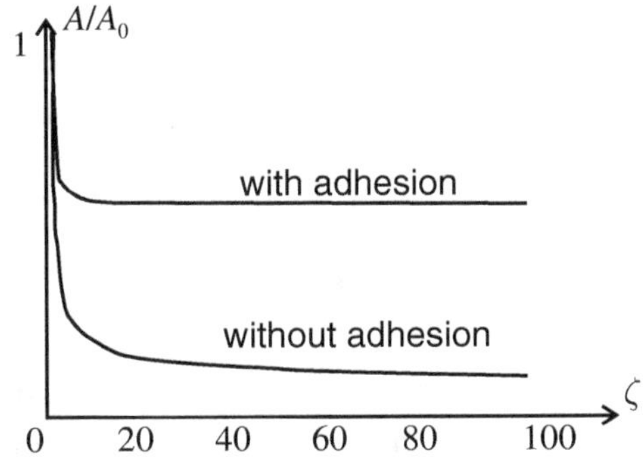

Figure 10.6 Dependence of the projected area of contact on the magnification ζ with and without adhesion. The parameter values are the same as in Fig. 10.4. Adapted from [248] with permission from The European Physical Journal (EPJ).

$$\frac{dF_N}{d\zeta} = -\sigma_a(\zeta)\frac{dA}{d\zeta}. \tag{10.6}$$

If σ_a did not depend on the magnification, Eq. (10.6) would imply that the normal force is the sum of the applied load and the adhesion load, as in the DMT model (Section 10.2):

$$F_N \approx (p + \sigma_a)A_0. \tag{10.7}$$

However, since σ_a usually depends on ζ, the relation (10.7) has no general applicability.

The dependence of the contact area on ζ can be estimated numerically with an iterative procedure, as described in [248]. The result is shown in Fig. 10.6. In the presence of adhesion the apparent contact area equals the real contact area (i.e. the rubber is in complete contact with the substrate) already at a small magnification $\zeta \approx 10$. However, this is the case only if the fractal dimension $D_f \sim 2$. For larger values of D_f the area A decreases continuously with increasing ζ.

Adhesion and rubber friction

For rough surfaces, the contribution of adhesion to rubber friction is usually negligible. This can be seen by comparing the detachment stress σ_a at a given length scale λ to the pressure p (which is usually $\sim$ 1 MPa). If a tire tread rubber is squeezed against a road surface, the effective elastic modulus $E^* \sim 1$ MPa and the interfacial surface tension $\gamma \sim 1$ meV/A^2 [263], so that, according to Eq. (10.5), adhesion is important only on length scales $\lambda < 10$ nm. These values are well below the typical cut-off length scales, which, as mentioned in Section 8.1, are in the order of 1 μm.

10.5 Adhesion in biological systems

An elastic object can adhere to a rigid rough substrate only if the strain energy of the deformed material is larger than the interfacial binding energy. This means that the Young's modulus E of the solid must be low enough. A possible way to reduce E, which is exploited in many biological adhesive systems, is to form foam-like or fiber-like structures. Foam-like structures are adopted in the adhesion pads of cicadas whereas hierarchical fiber structures are observed in other animals. Even if the Young's modulus of the keratin-like proteins forming these systems is in principle quite large (of the order of 1 GPa, i.e. three orders of magnitude more than in rubber), the effective modulus resulting from the non-compact shape is quite small [249], which makes adhesion possible even on very rough surfaces.

11

Thermal and electric effects

Heat transfer has a key importance in friction and wear phenomena. For instance, when a rubber tire slides on an asphalt road, the friction coefficient depends significantly on the temperature increase in the contacting asperities. If such an increase is ignored, the friction can be overestimated, especially at high velocities. In this chapter, we will first discuss the glass transition and introduce the concept of flash temperature in viscoelastic materials. After that, we will move to the contact of elastic materials and discuss simple expressions for the heat transfer and the electric conductivity at the interface. We will also show that, in both cases, friction has a negligible influence on the contact resistance.

11.1 Thermal effects in polymers

The existence of rubbery and glassy regions in polymers (Section 9.2) has a simple physical explanation. At low frequencies the molecular chains forming the polymer have enough time to flip betwen different configurations and rearrange, while this is not possible at high frequencies. This causes a soft response in the first case, and significant hardening in the second one. The flipping process is thermally activated and a strong temperature dependence of the viscoelastic modulus $E(t)$ is expected. Indeed, the characteristic relaxation time τ changes dramatically around a so-called *glass transition temperature* T_g. Below T_g, τ is quite long and the material behaves like a glass. Above T_g, τ is very short and the polymer presents a rubbery behavior. All polymers show this general trend, but the extent of each regime, and the detailed response within each regime, depend on the molecular structure of the material. The glass transition temperature is 70–100 °C for most polymers. Heavily cross-linked polymers (so-called *elastomers*) are the most likely to behave as ideal rubber.

The time dependence and the temperature dependence of the viscoelastic modulus have a remarkable analogy. In fact, it is experimentally well established that

$$E(t, T) = E(a_T t, T_0),$$

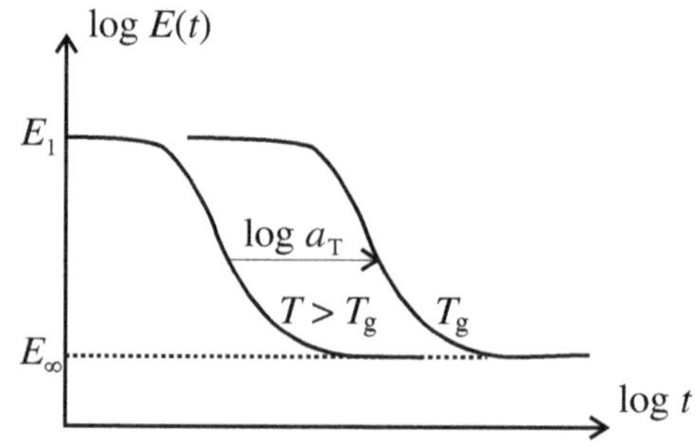

Figure 11.1 Temperature dependence of the viscoelastic modulus.

where T_0 is a reference temperature (conventionally set equal to the glass transition temperature T_g) and the *shift factor* a_T is a function of the temperature T. The dependence $a_T(T)$ is well approximated by the *Williams–Landel–Ferry* (WLF) *equation* [342]:

$$\log a_T = \frac{C_1(T - T_0)}{C_2 + T - T_0}. \tag{11.1}$$

In this empirical relation the constants $C_1 \approx -17.44$ and $C_2 \approx 51.6$ K are approximately the same for all amorphous polymers. In this way the curves $E(t)$ recorded at different temperatures can be overlapped, on a logarithmic scale, onto a single *master curve* by simply shifting them by the distance a_T, as shown in Fig. 11.1. Similarly, for the frequency dependence,

$$E(\omega, T) = E(\omega/a_T, T_0).$$

Note that in practice, a temperature increase of 10 °C may shift any curve representing a function of $E(\omega)$ (and specifically the loss function) to higher frequencies by one decade.

The WLF equation can be recovered by assuming that the viscosity η of the polymer follows the semi-empirical *Doolittle equation* $\eta = A\exp(Bv_0/v_f)$, where v_0 is the volume of closest packing of the molecules and the 'free' volume v_f accessible to a molecule is supposed to depend linearly on the temperature [77].

11.2 Flash temperature

Consider again a rubber block sliding on a rough hard substrate, as in Section 9.4. The viscoelastic energy dissipation heats the rubber locally in the same regions where the energy is dissipated. The resulting temperature increase ΔT (so-called *flash temperature*) is larger in the smaller asperity contact regions. If the sliding velocity $v > 1$ mm/s the flash temperature is usually not negligible since, in this case, the heat has not enough time to flow away from the contact.

On a wavy surface, with periodicity λ_0, we can refer to Fig. 9.8 and observe that the heat dissipated in the unit volume of an asperity with linear size l, pressed by a normal force F_N, is $Q \approx \mu_k F_N l / l^3 = \mu_k p$, where p is the average pressure in the contact area. According to Hertz theory, $p \approx k_0 h_{rms} |E(\omega_0, T)|$, where $k_0 = 2\pi/\lambda_0$ and h_{rms} is the rms roughness amplitude. If v is high enough, the heat diffusion becomes negligible. In this case a temperature increase $\Delta T \approx Q/\rho c_V$ is expected, where ρ is the mass density and c_V is the specific heat capacity of the material. Taking into account the relation (9.13):

$$\Delta T \approx (k_0 h_{rms})^2 \frac{\text{Im}\, E(\omega_0, T)}{\rho c_V}. \tag{11.2}$$

Since in most applications the perturbing frequencies ω_0 are below the frequency ω_{max} at which the loss function has a maximum (Section 9.2) and the temperature increase ΔT shifts the viscoelastic spectrum to higher frequencies (Section 11.1), Eq. (11.2) implies that the flash temperature will reduce the friction force.

If the block is sliding on an arbitrarily rough surface, an implicit expression for the flash temperature at a depth $\lambda = 2\pi/k$ inside the block can be derived in the framework of Persson theory [251]:

$$T_k = T_0 + \frac{1}{\pi} \iint \frac{4k^2}{q^2 + 4k^2} \frac{4k'}{q^2 + 4k'^2} \left(1 - e^{-\alpha k^2 R/v}\right) dq\, dk', \tag{11.3}$$

where T_0 is the background temperature, α is the thermal diffusivity,[1] and R is the average radius of the 'macroasperity' contact regions (Section 8.3). As a result, the temperature of a tire tread rubber on an asphalt road is expected to increase by 80 °C at a depth of 10 μm, but only by few degrees at a depth of few mm.

The friction coefficient μ_k can still be expressed by Eq. (9.14), where the flash temperature T_k, as defined by Eq. (11.3), enters the definition of the effective elastic modulus E^*. The velocity dependence of μ_k, including the flash temperature, is compared to the dependence $\mu_k(v)$ without thermal effects in Fig. 9.9. Neglecting the flash temperature, μ_k increases monotonically with v. Taking the flash temperature into account, μ_k reaches a maximum at $v \approx 1$ cm/s. Beyond this value, the shift of the viscoelastic module $E(\omega)$ makes the rubber less viscous and results in less energy dissipation.

Nowadays, the distribution of flash temperature on the macroasperity contacts of an automotive tire can be easily visualized with infrared cameras; see Fig. 11.2. Note that, even if the background temperature T_0 does vary slowly with time, this quantity is very important in rubber friction. In this context, we remark that the full

[1] The thermal diffusivity depends on the thermal conductivity κ as $\alpha = \kappa/\rho c_V$.

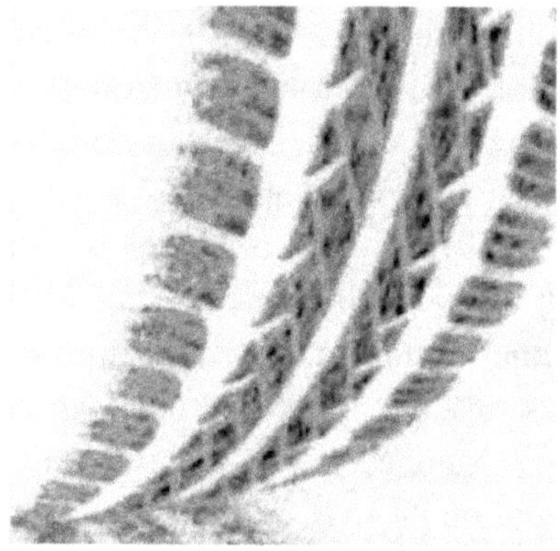

Figure 11.2 Infrared photograph of a tyre rolling on an asphalt road. The temperature of the rubber is higher in the darkest spots, arising from macro-asperity contacts with the road. Reproduced from [251] with permission from IOP Publishing.

tire temperature is usually reached on a time scale of half an hour from the start of driving.

11.3 Heat transfer between rough surfaces

Consider the contact between two rough elastic blocks with thickness d_1 and d_2 (Fig. 11.3) and suppose that the temperature at the outer surfaces is kept fixed at T_1 and T_2, respectively, with $T_2 > T_1$. Introducing the average distance δ between the macroasperity contact regions (Section 8.3), we can write the heat transfer at the interface as

$$q = \alpha \Delta T',$$

where $\Delta T'$ is the temperature variation through the distance δ and α is the *heat transfer coefficient*.[2] Since the heat transfer in each block is

$$q = \pm \frac{\kappa_i}{d_i} \Delta T_i,$$

where $\Delta T_i = T_i' - T_i$ is the temperature variation with respect to the outer surface and κ_i is the thermal conductivity,

$$q = \frac{\Delta T_0}{d_1/\kappa_1 + d_2/\kappa_2 + 1/\alpha}.$$

It follows immediately from dimensional considerations that

$$\alpha \sim \frac{\kappa p}{E^* \delta_0}, \tag{11.4}$$

[2] Not to be confused with the thermal diffusivity!

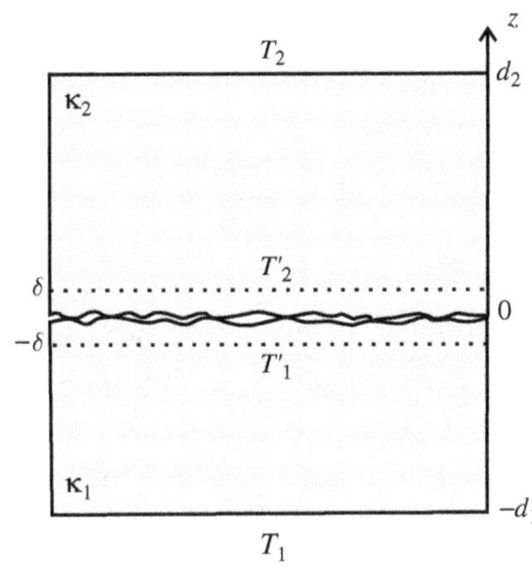

Figure 11.3 Heat transfer between two rough elastic surfaces in contact.

where p is the squeezing pressure and δ_0 is the length parameter introduced in Section 8.4. In the case of rubber in contact with a road surface it is estimated that $\alpha \sim 10$ W/m^{2}K [266]. This value is one order of magnitude smaller than the heat transfer resulting from forced convection.

In general, the contact resistance can always be neglected if $d_i \gg \kappa/\alpha$ or, equivalently, if $d_i \gg \delta_0 E^*/p$. This is not the case in MEMS, where the hard and very flat surfaces and the high-resistivity materials characterizing these devices result in contact heat transfer coefficients of the order of 10^4–10^5 W/m^{2}K. For low loads we can use Eq. (8.23) and conclude that

$$\alpha \sim \frac{\kappa}{E^*} k_{\mathrm{N}},$$

where $k_{\mathrm{N}} = \mathrm{d}p/\mathrm{d}\delta$ is the normal stiffness of the contact. For self-affine fractal surfaces Eq. (8.25) implies that α is determined by the longest wavelength roughness components of the surfaces. Thus, the real contact area or, equivalently, the friction force has no influence on the contact resistance (provided that the Hurst coefficient H is not too small).

11.4 Electric contact resistance

Measuring the electrical contact conductance G is an efficient method for estimating the contact area between two metal surfaces. For a macroscopic circular contact with radius a:

$$G = 2a/\rho, \tag{11.5}$$

where ρ is the resistivity of the material. However, Eq. (11.5) cannot be applied on the nanoscale, where the mean free path l of the electrons is larger than the linear size of the contact. In this case, Sharvin noticed that the problem is analogous to the effusion of gas molecules through a small orifice in the Knudsen regime [306], so that

$$G = \frac{3\pi a^2}{4\rho l}.$$

This means that in the *Sharvin regime* the conductance is proportional to the contact area. However, if the contact is only a few atoms wide, quantization effects appear and the conductance is expressed by a multiple of the quantum unit $2e^2/h$, where h is Planck's constant. This result has been verified by nanoindentation experiments on atomic-scale gold contacts, where the normal force was measured simultaneously with the conductance [347, 295].

The problem of the electrical contact resistance between rough surfaces is completely analogous to that of the thermal contact resistance. Thus, the electric current crossing a rough interface between two elastic solids depends on the electric potential drop ΔV as

$$J = \alpha_{\mathrm{el}} \Delta V,$$

where

$$\alpha_{\mathrm{el}} \sim \frac{p\kappa_{\mathrm{el}}}{E^* \delta_0}, \tag{11.6}$$

and κ_{el} is the electrical conductivity (compare Eq. (11.4)). Also in this case the influence of friction on the contact resistance is expected to be negligible. However, we should keep in mind that the variation of electric conductivity between metals and bad conductors can embrace more than 20 orders of magnitude.[3] This makes the electric contact resistance very sensitive to contamination or oxide layers at the contacting interface. Still, if the layers are broken in several tiny spots, as often occurs in practice, the resistance is almost the same as without layers [7], and Eq. (11.6) remains applicable.

[3] Typical values of κ_{el} are $\sim 10^7$ S/m in metals and $\sim 10^{-14}$ S/m in silicon oxide and rubber.

12

Plastic contacts

In this chapter we consider the transition from elastic to plastic behavior (the *yield* point). This transition implies that the material undergoes irreversible shape changes in response to external forces. A simple example is a piece of metal permanently bent into a new shape. Several physical mechanisms can cause plastic deformation. Plasticity in metals is usually associated with the motion of dislocations, while in brittle materials it is caused predominantly by slip at microcracks. After introducing the most important criteria for yielding, the concept of plastic flow and the definition of hardness, we will consider various examples of indentation, sliding and rolling involving plastically deformed objects. These processes are severely affected by the friction at the contact interfaces, which is also discussed in the chapter. We will also mention the importance of plasticity in geotechnics, where it determines the safety of a structure founded on a soil. In this context, a peculiar role is played by the angle of internal friction of the materials.

12.1 Plasticity

A typical stress–strain curve for a material in simple tension is shown in Fig. 12.1. The initial part of the curve is a straight line with a slope equal to the Young's modulus E of the material. The linear relationship between σ and ε ends at a certain point, corresponding to the *yield strength* Y. At this point plastic deformation occurs. The value of Y depends on the manufacturing process and on the purity of the material. For metals, it is typically in the range of 10–100 MPa. If the material is stressed further in the plastic range and the load is released, the recovery is elastic, with the same value of E as in the first loading. This key assumption was carefully verified by Tabor in a series of measurements on soft metals using spherical and conical indenters [327, 321]. A subsequent loading of the material results in an increased value of the yield strength, as seen in Fig. 12.1. This effect is known as *work hardening* or *strain hardening*. A different result is obtained if

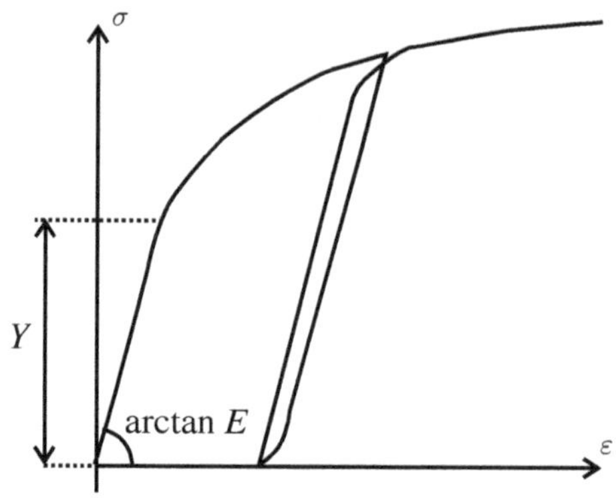

Figure 12.1 Stress–strain curve in a metal.

the material is reloaded in simple compression. In this case the new yield strength is lower than Y (*Bauschinger effect*) [15]. Experiments with pure shear, such as the torsion of a tube, result in similar behavior. In this case, the yield strength is usually denoted by k.

Physical interpretation

Plasticity in metals is ultimately due to the relative motion (*slip*) along specific crystallographic planes, which is caused by the shear stress. *Slip planes* are usually parallel to the planes of closest packing of atoms, where the resistance to slip has a minimum. Within each plane there are preferred *slip directions* corresponding to the atomic rows with the largest density of atoms. The shear strength required to produce a slip can be estimated assuming that the slip occurs by uniform displacement of adjacent atomic planes. If the shear stress varies sinusoidally with the lateral displacement (Fig. 12.2), it is not difficult to prove that the maximum shear stress

$$\tau_{\max} = \frac{Gb}{2\pi a}, \tag{12.1}$$

where a is the interplanar spacing, b is the interatomic spacing in a crystal plane and G is the shear modulus of the material. In a first approximation $\tau_{\max}$ should be of the order of 10 GPa. However, this is in contrast to the experimental results, where the values of $\tau_{\max}$ are one to three orders of magnitude lower. This discrepancy is simply explained by the lattice defects (e.g. dislocations) which are always present in real crystals.

Dislocations are possibly the most important defects in solids. An *edge dislocation* is generated by an extra half-plane inserted in a crystal lattice as shown in

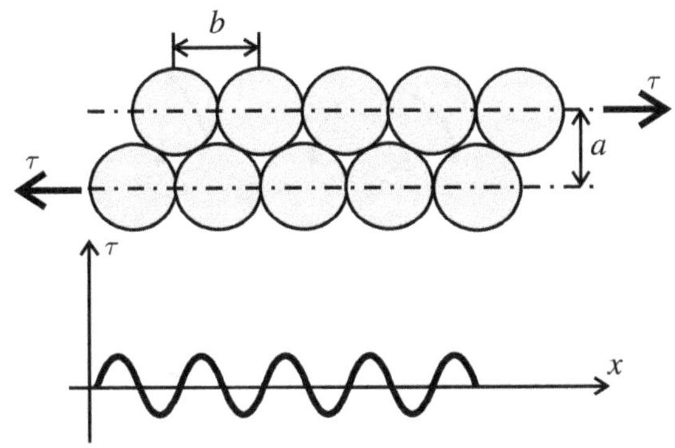

Figure 12.2 Slip mechanism on the atomic scale.

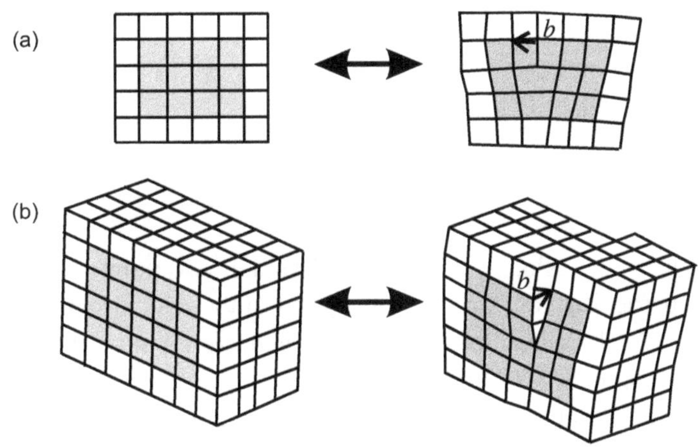

Figure 12.3 (a) An edge dislocation and (b) a shear dislocation in a cubic crystal.

Fig. 12.3(a). A *screw dislocation* is obtained by cutting the lattice along a half-plane and shifting the two resulting faces parallel to the cut (Fig. 12.3(b)). After passing around a dislocation line, the displacement vector **u** is incremented by a finite quantity **b** (*Burgers vector*), which is equal to one of the lattice periods:

$$\oint \mathrm{d}\mathbf{u} = -\mathbf{b}.$$

When a dislocation moves in a slip plane containing it and **b**, each atom moves much less than a lattice constant, which explains why the stress required to move the dislocation is much smaller than the theoretical shear stress (12.1).

Even if dislocations can be produced at the surface of a crystal, for instance using a nanoindenter, most of them are generated in the bulk. The most important process is the *Frank–Read mechanism* schematically shown in Fig. 12.4 [94]. In this

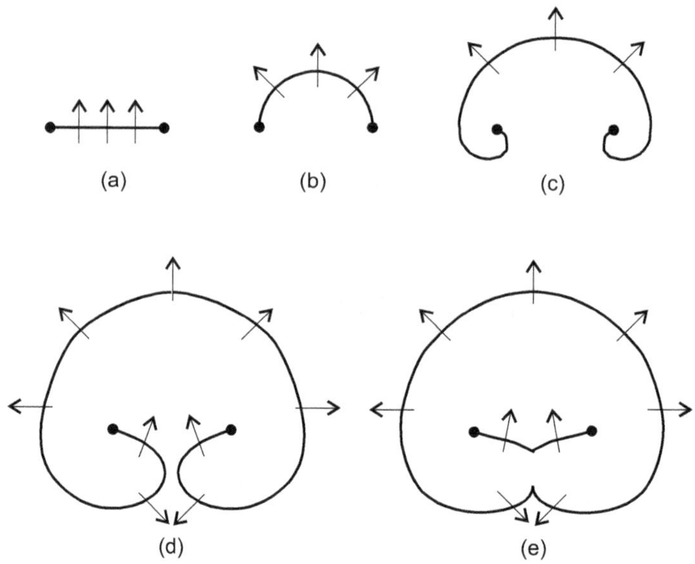

Figure 12.4 The Frank–Read mechanism.

process a segment of dislocation firmly anchored at two points (e.g. two defects) responds to a perpendicular force by bowing out. If the critical semicircular configuration in Fig. 12.4(b) is reached, the dislocation becomes unstable and the process continues as shown in Fig. 12.4(c). When the two segments come into contact, they annihilate each other and a dislocation loop propagates under the action of the shear stress. At the same time a new segment appears and the whole process is repeated. The final result is a sequence of nested dislocation loops. The phenomenon of work hardening can be also explained by the repulsive action of the stress field produced by dislocations which get stuck as the plastic flow proceeds.

12.2 Criteria of yielding

To predict the yielding of ductile materials under stress two criteria are commonly adopted. Note that in both cases the material is supposed to be isotropic and the Bauschinger effect is disregarded. According to the *Tresca criterion* yield is achieved when the maximum shear stress $\tau_{\max}$ in simple tension reaches the value $Y/2$ or, equivalently, when

$$\max\left(|\sigma_i - \sigma_j|\right) = Y, \tag{12.2}$$

where σ_i ($i = 1, 2, 3$) are the principal normal stresses of the material. Equation (12.2) defines a hexagon in the *deviatoric plane* $\sigma_1 + \sigma_2 + \sigma_3 = 0$ shown in

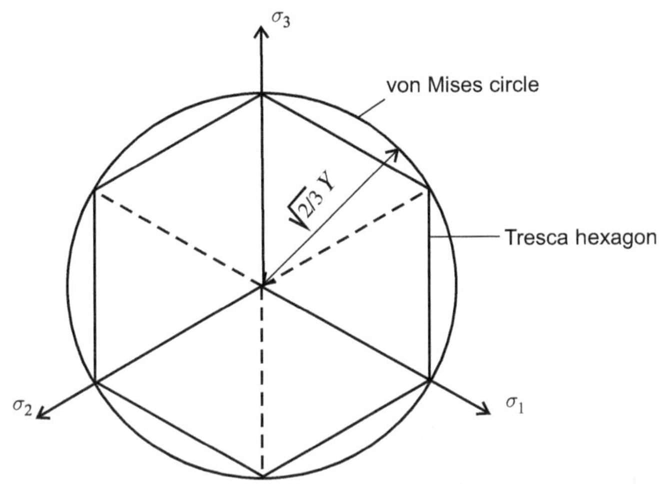

Figure 12.5 Deviatoric yield loci of Tresca and von Mises.

Fig. 12.5. In the *von Mises criterion* [339] the hexagon is replaced by the circle circumscribing it, so that

$$\sqrt{\frac{(\sigma_1-\sigma_2)^2+(\sigma_2-\sigma_3)^2+(\sigma_3-\sigma_1)^2}{2}} = Y. \tag{12.3}$$

The radius of this circle is $\sqrt{2/3}Y$. Interestingly, while the von Mises criterion was intended to be an approximation of the Tresca criterion, it is in a better agreement with the experiments, although the quantitative differences between the two criteria are very small. More important is the fact that the left hand side of Eq. (12.3), i.e. the so-called *von Mises stress*, is proportional to the strain energy associated with the shear of the deformed material, which corresponds to the first term on the right hand side of Eq. (3.9). This observation gives a more physical basis to the von Mises criterion. In pure shear the yield strength is $k = Y/2$ in the Tresca criterion and $k = Y/\sqrt{3}$ in the von Mises criterion, as seen from Eq. (12.2) and Eq. (12.3) with $\sigma_1 = -\sigma_2 = k$.

From the results of Section 4.3 it follows that the contact between two spheres and the axial contact between two cylinders enter the plastic regime, according to the Tresca criterion, when the maximum pressure $p_0 \approx 1.61Y$ and $1.67Y$ respectively. For two ellipsoidal surfaces in contact, intermediate values of the critical pressure are expected. If a blunt wedge or a blunt cone is pressed against a flat surface, yield initiates at the apex of the indenter when the characteristic pressure p_0 (see Section 4.4) reaches the value $0.5Y$.

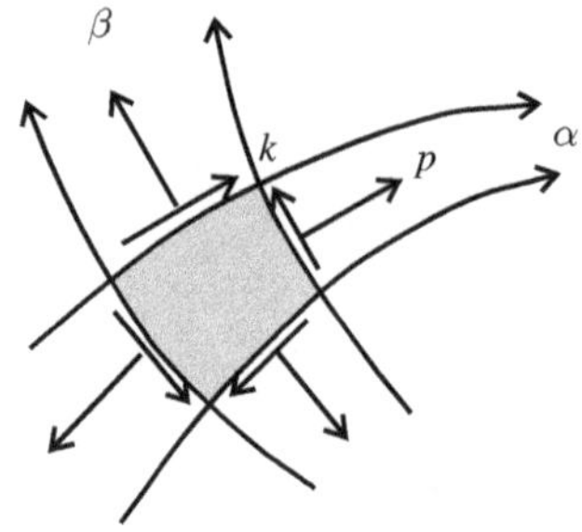

Figure 12.6 Stress components on an element bound by slip lines.

12.3 Plastic flow

Suppose now that the plastic deformation is so severe that the elastic deformation can be neglected. In this case the material can be considered incompressible and in a state of *plastic flow* with a constant shear stress $k = Y/2$ (following the Tresca criterion).

In the case of plane deformation the flow occurs with a variable hydrostatic pressure $p = (\sigma_1 + \sigma_2)/2$ and it is possible to introduce a curvilinear net of α and β *slip lines*, which are mutually perpendicular and oriented along the directions of maximum shear stress (Fig. 12.6). At a given position in the material [138, section VI.3]:

$$\sigma_x = p - k\sin 2\varphi, \qquad \sigma_y = p + k\sin 2\varphi, \qquad \tau_{xy} = k\cos 2\varphi, \tag{12.4}$$

where the angle φ defines the orientation of the α slip lines. Along the slip lines, it can be shown that [135]

$$p \pm 2k\varphi = \text{const.} \tag{12.5}$$

In this way the variation of p in a region of plastic flow can be determined from the values of p on a free surface limiting the region.

12.4 Plastic indentation

Consider a rigid wedge with half-angle α indenting a perfectly plastic half-space as in Fig. 12.7 [139]. The plastic flow occurs in the gray regions in the figure. Since the half-space is incompressible, the areas of the two striped triangles must be equal. If the friction is negligible the wedge faces do not sustain any shear stress and the slip lines meet them at 45° (see Section 3.2). The same occurs at the free surfaces. Using Eqs. (12.5) it can be concluded that the pressure on the half-space is uniform and equal to

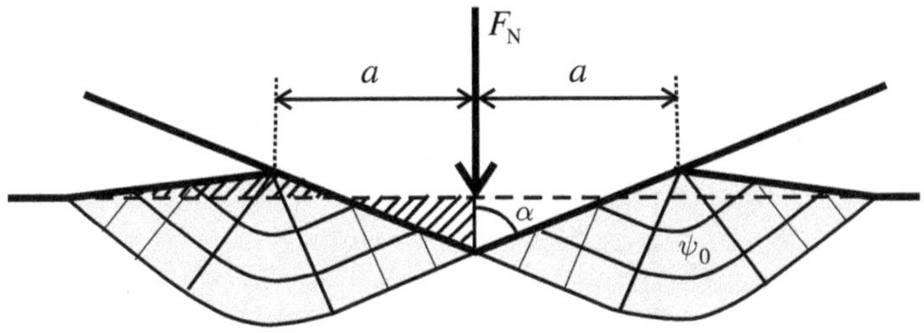

Figure 12.7 Rigid wedge indenting a perfectly plastic half-space without interfacial friction.

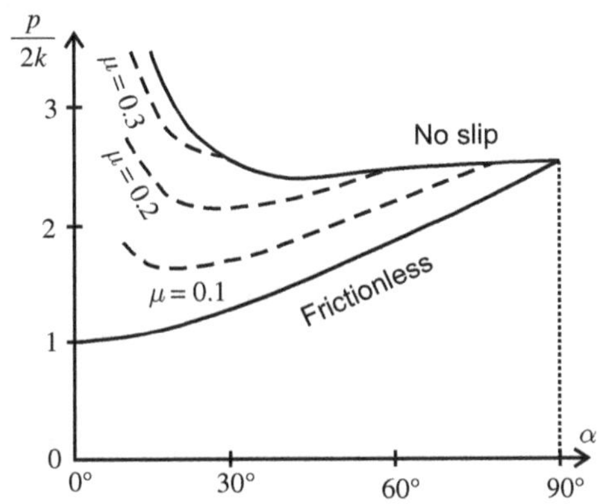

Figure 12.8 Pressure on the half-space as a function of the half-angle of the wedge. Adapted from [129] with permission from Elsevier.

$$p = 2k(1 + \psi_0), \tag{12.6}$$

where ψ_0 is the angle of turning of the slip lines. This angle is determined from geometry considerations leading to the relation

$$\cos(2\alpha - \psi_0) = \frac{\cos \psi_0}{1 + \sin \psi_0}.$$

The dependence $p(\alpha)$ is represented by the lower curve in Fig. 12.8.

In the presence of friction at the contact area, the slip lines meet the wedge faces at an angle $\lambda < 45°$. This angle is related to the coefficient of friction μ by the expression

$$\cos 2\lambda = \mu(1 + 2\psi + \sin 2\lambda),$$

where the fan angle ψ is again determined by the geometry of the problem. The indentation pressure becomes

$$p = k(1 + \sin 2\lambda + 2\psi)(1 + \mu/\tan\alpha),$$

and its dependence on α is also shown in Fig. 12.8 for different values of μ.

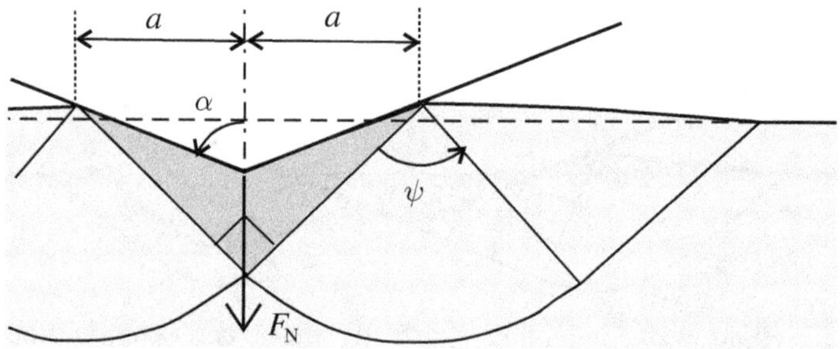

Figure 12.9 Blunt rigid wedge indenting a perfectly plastic half-space in the presence of high friction. The 'false nose' is highlighted in dark gray. Adapted from [159] with permission from Elsevier.

However, if the friction is so high that λ becomes zero, the slip at the wedge faces disappears and the surface of the indented material will adhere to the wedge. If the wedge is blunt ($\alpha > 45°$), as the slip lines at the apex cannot meet at an angle $< 90°$, a $90°$ 'false nose' of undeformed material will adhere to the faces of the wedge (Fig. 12.9) and move rigidly with it [159].

The case of a rigid cone indenting a plastic half-space can be solved with the method of the slip lines [188], even if the problem is not 2D, provided that the half-angle $\alpha > 52.5°$. In this case the pressure is not uniform, but rises up to a maximum at the apex. The slip line field does not differ significantly from that of a wedge, although the slip lines and the deformed surface are no longer straight. The average pressure $\overline{p}$ acting on the half-space is slightly larger than for a wedge.

In the case of a cylinder indenting a plastic half-space $\overline{p} \approx 5.69k$ [82]. If $\mu > 0.14$ a cone of undeformed material sticks to the cylinder, and $\overline{p}$ increases slightly. If the indenter has a spherical shape, $\overline{p}$ is of the order of $6k$ and almost independent of the penetration depth [289].

12.5 Compression and traction of a plastic wedge

Consider now a rigid plane pushed by a normal force F_N against a wedge undergoing plastic deformation (Fig. 12.10). In the absence of friction the pressure at the interface is uniform and given again by Eq. (12.6). However, the fan angle ψ_0 is related now to the half-angle α of the wedge by the expression [138, section 8.3]

$$\tan\alpha = \frac{(1+\sin\psi_0)^2}{\cos\psi_0(2+\sin\psi_0)}. \tag{12.7}$$

Equation (12.7) has a real solution only if $\alpha > 26.6°$. In the limit case the displaced surfaces become vertical and the wedge apex undergoes a simple compression.

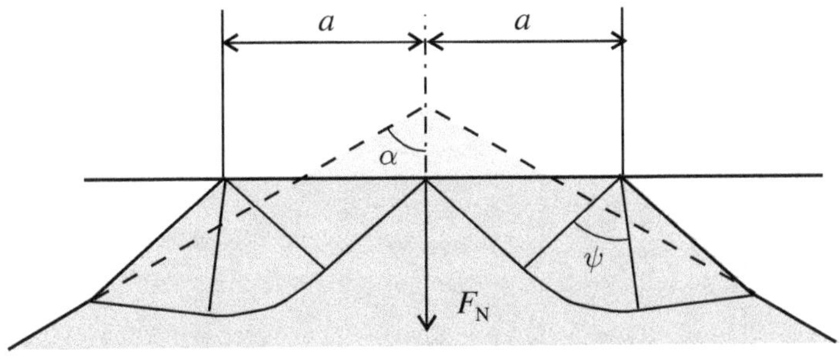

Figure 12.10 Perfectly plastic wedge crushed by a rigid half-space.

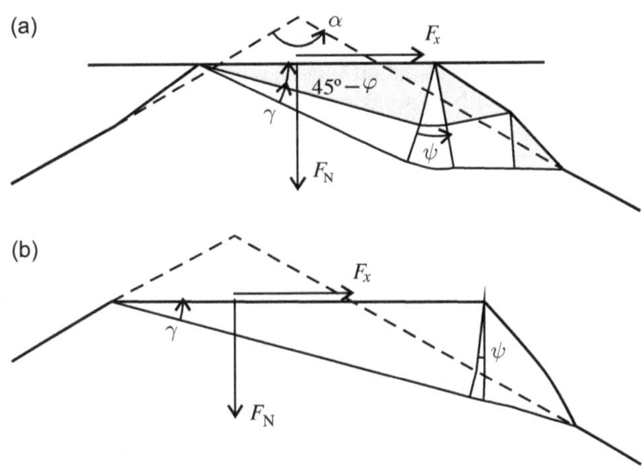

Figure 12.11 Deformation of the plastic wedge in the presence of an increasing tangential force F_x: (a) first stage and (b) second stage. Adapted from [155] with permission from Elsevier.

If a tangential force F_x is gradually superimposed on F_N, we can distinguish two stages [155]. In the first stage the three triangles highlighted in Fig. 12.11(a) move as rigid objects. The slip lines meet the plane at the angles $45° \pm \varphi$, where, for a given ratio F_x/F_N, φ is defined by the relation

$$\frac{\sin\varphi\cos\varphi}{(1+\psi_0)-(\varphi+\sin^2\varphi)} = \frac{F_x}{F_N},$$

and ψ_0 is determined from Eq. (12.7). The contact pressure is

$$p = k(1 + 2\psi + \cos 2\varphi) \tag{12.8}$$

where the fan angle $\psi = \psi_0 - \varphi$. The angle φ increases with F_x from 0° to 45°, and at this point the first stage ends. In the second stage (Fig. 12.11(b)) the pressure is

$$p = k(1 + 2\psi), \tag{12.9}$$

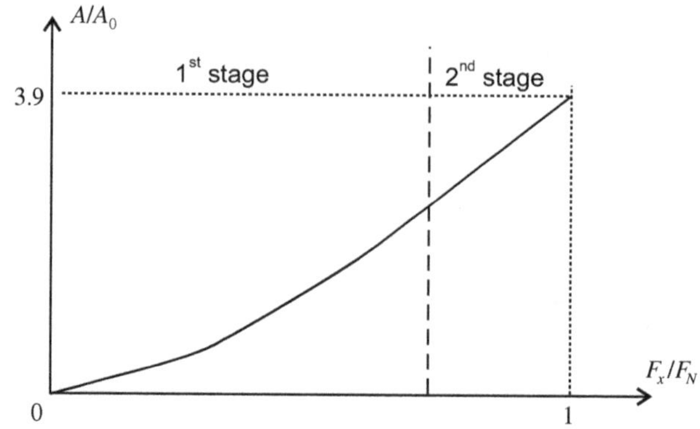

Figure 12.12 Growth of the contact area of a plastic wedge with half-angle $\alpha = 60°$ under the action of a constant load F_N and an increasing tangential force F_x. Adapted from [155] with permission from Elsevier.

where $\psi = (F_\mathrm{N}/F_x - 1)/2$. When $F_x = F_\mathrm{N}$ the angle $\psi = 0$ and the wedge starts sliding.

From the previous discussion it follows that the contact area A increases even if F_N remains constant. The dependence of A on F_x is given by $A = A_0(p_0/p)$, where the relation $p(F_x)$ in the two stages is described by Eqs. (12.8) and (12.9) and the initial value of $p = p_0$ is given by Eq. (12.6). This process is known as *junction growth* and its result is illustrated in Fig. 12.12. Note that this growth can be interrupted by premature sliding caused by contaminants at the interface. In Section 14.5 we will see how a junction can also grow due to thermally activated plastic flow even without a tangential force.

12.6 Hardness

The *indentation hardness* of a material is the ratio between the normal force F_N acting on a rigid tool indenting the material and the area A of the impression left after the indenter has been completely retracted:

$$H = F_\mathrm{N}/A.$$

Depending on the shape of the indenter, different definitions of hardness have been proposed. In a *Brinell test* the indenter is a sphere of radius R, and the hardness is defined as

$$H_\mathrm{B} = \frac{F_\mathrm{N}}{2\pi R^2\left(1 - \sqrt{1-(a/R)^2}\right)},$$

where a is the radius of the impression. In a *Vickers test* the sphere is replaced by a square pyramid with half-angle $\alpha = 68°$. This value corresponds to the ratio

$a/R = 0.375$ in the Brinell test, which is in the middle of the range where the test was originally supposed to be most reliable. Expressing the impression area as a function of α and the diagonal length d, it is not difficult to see that the hardness measured in a Vickers test is

$$H_V = \frac{2F_N \sin\alpha}{d^2}.$$

For nanoindentation experiments (see Section 12.8) the *Berkovich test* is more common [21]. In this case, the indenter is a three-sided pyramid. This shape is easier to grind to a sharp point. The half-angle α of the Berkovich indenter is 65.35° in order to get the same projected area-to-depth ratio as a Vickers indenter.

Note that the hardness is not related to the elastic properties of a material. Rubber is elastically soft but plastically hard while metals behave the other way around. This can be seen by comparing the indentation hardness of steel (a few GPa) or rubber (a few hundred MPa) with the values of the Young's modulus in Section 3.3.

The Bowden–Tabor model

In the 1940s Bowden and Tabor speculated that friction results from shearing the cold-welded junctions formed between two solids [32]. In this case, assuming that all the junctions are in a state of incipient plastic flow, the contact area A would be equal to F_N/H, where H is the hardness of the material, and the friction force F_{fric} would be equal to YA, where Y is the yield strength. According to the discussion in section 12.4, this would lead to the relation (2.1) with a coefficient of friction $\mu = Y/H \approx 0.33$, which is indeed a typical value for two metal surfaces in contact (see Table 2.1). Nevertheless, finite-element simulations on typical surface profiles have shown that only very few junctions are expected to be in a state of incipient plastic flow [144]. Furthermore, friction is found to vary with the material properties of the contacting surfaces. Thus, in spite of its popularity, the Bowden–Tabor model cannot be considered as a 'proof' of Amontons' law.

12.7 Plowing

The plowing of a plastic half-space by a rigid wedge with half-angle α can be also studied using the slip line theory [156, section 7.6]. If the normal force F_N is constant and the tangential force F_x is gradually increased, the wedge penetrates deeper and deeper while sliding till $F_x = F_N \tan\alpha$. At this point the wedge starts to ride up. As the plowing goes on, the wedge apex reaches the free surface and enters a state of steady sliding during which a plastic wave is pushed along the surface (Fig. 12.13). The width of the contact strip formed between wedge and surface, as determined from slip line theory, is

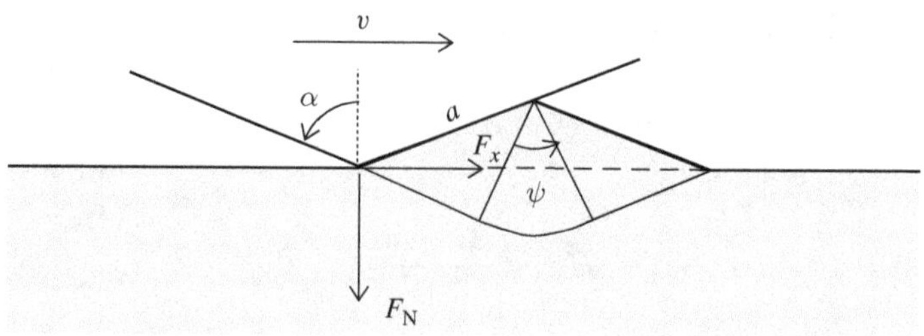

Figure 12.13 Rigid wedge plowing a perfectly plastic half-space.

$$a = \frac{F_N}{2k(1 + \psi)\sin\alpha},$$

where k is the yield strength in pure shear and $\psi = 2\alpha - 90°$. In the presence of friction the indentation depth is much deeper and the wedge builds up a larger hill ahead. The case of a 3D indenter is much more complex since the material is also displaced sideways.

12.8 Elastic–plastic indentation

If the size of the plastic zone is comparable to the region which remains elastically deformed, the problem can usually be solved only with finite element methods. Due to their simple geometry, indentations made with a rigid sphere or axisymmetric rigid indenters are an exception, and analytical models have succeeded in reproducing experimental results. Here we will briefly discuss the models introduced by Johnson and by Oliver and Pharr.

The cavity model

Consider a rigid tool indenting a deformable material. If the average contact pressure $\overline{p}$ is between Y and approximately $3Y$, the plastically deformed region is surrounded by elastic material. The upper limit comes from the results of Section 12.4 (remembering that $k = Y/2$) and marks the transition to fully plastic uncontained flow. This limit defines the hardness H of the material. The deformation in the elastic–plastic regime can be estimated numerically and, in general, it is found that the plastic flow leads to a flattening of the pressure distribution. This can be seen in Fig. 12.14 for the contact between a rigid sphere and an elastic–plastic half-space [132]. Furthermore, the subsurface displacements are approximately radial. This has also been observed experimentally when measuring the strain produced by blunt indenters [301].

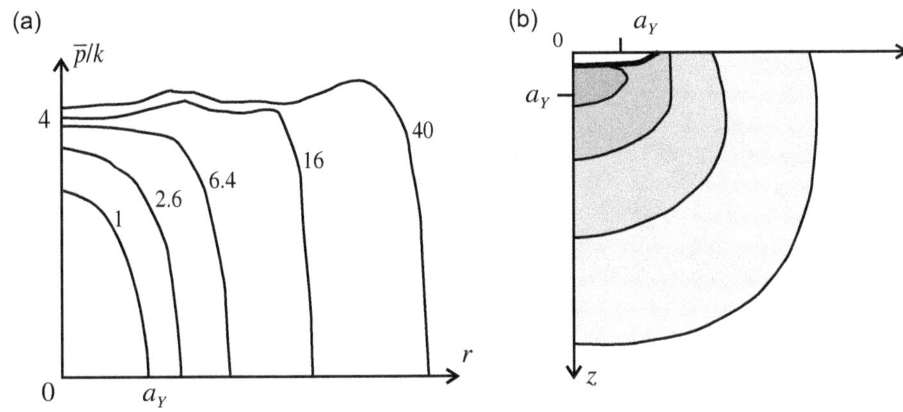

Figure 12.14 (a) Pressure distribution in the contact between a rigid sphere and an elastic–plastic half-space, for values of the ratio $F_N/Y = 1.0$, 2.6, 6.4, 16 and 40 (a_Y is the contact radius at the point of first yield). (b) Corresponding plastic zones. Adapted from [132] with permission from John Wiley and Sons.

Based on the previous result, Johnson proposed the following 'cavity model' [156, section 6.3]. In this model the contact area is supposed to be enclosed in a hemispherical core of radius a. Inside the core the hydrostatic pressure p is constant. Outside the core we distinguish between a plastic zone with radius c and an elastic zone beyond it (Fig. 12.15). Stresses and displacements have radial symmetry and, at the interface between core and plastic zone, the pressure p in the core equals the radial component σ_r of the stress immediately outside it. The radial displacement of this interface must accommodate the volume of material. As a result, the radius c can be estimated from the contact radius a and the half-angle α of the indenter, which is supposed to be conical, as

$$c = a\left(\frac{E}{6Y(1-\nu)\tan\alpha} + \frac{2(1-2\nu)}{3(1-\nu)}\right)^{1/3}, \tag{12.10}$$

and the core pressure takes the value

$$p = 2Y\left(\frac{1}{3} + \ln\frac{c}{a}\right).$$

If the material is incompressible ($\nu = 1/2$) the expression (12.10) is considerably simplified:

$$c = a\left(\frac{E}{3Y\tan\alpha}\right)^{1/3}.$$

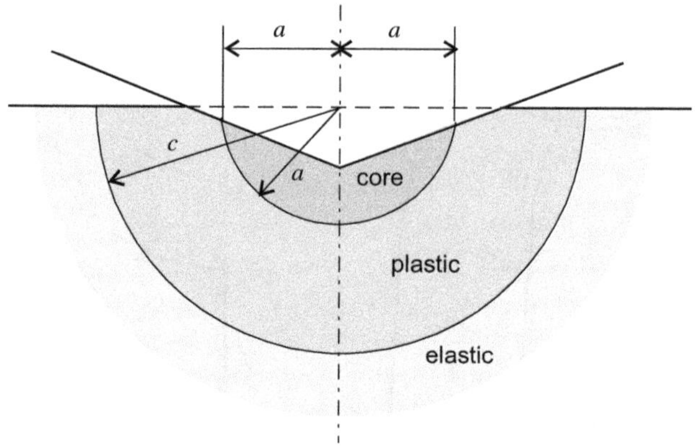

Figure 12.15 Johnson's cavity model for an elastic–plastic half-space indented by a rigid cone.

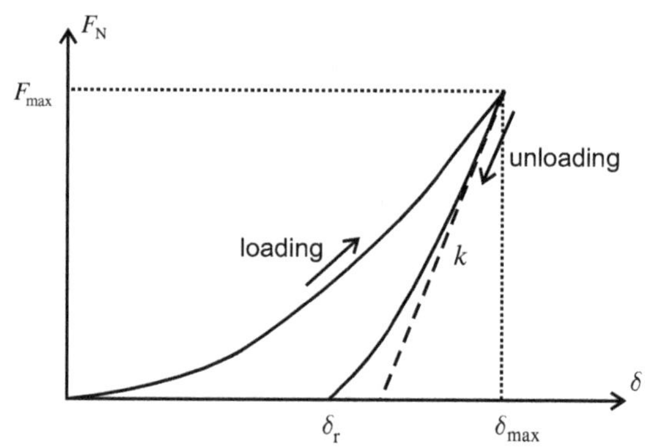

Figure 12.16 Loading cycle of a sharp indenter on a ductile material.

When the fully plastic state is reached, $E/Y \tan\alpha \approx 40$ so that $c \approx 2.3a$. This result justifies a practical rule requiring a spatial distance of at least five times the indenter diameter between consecutive indentation tests.

The Oliver–Pharr method

Suppose now that a sharp tip is indented into an elastic–plastic half-space and subsequently retracted from it. The typical response of the material in the cycle is shown in Fig. 12.16. The precise relation between the normal force F_N and the penetration depth δ in the loading phase is not relevant for the rest of the analysis. Upon unloading, the dependence on $F_N(\delta)$ is initially linear and a slope $k = (dF_N/d\delta)_{max}$ can be defined. A residual depth δ_r is observed at zero load (Fig. 12.17). The area enclosed by the loading and unloading curves corresponds to the energy which is dissipated plastically in the indentation process.

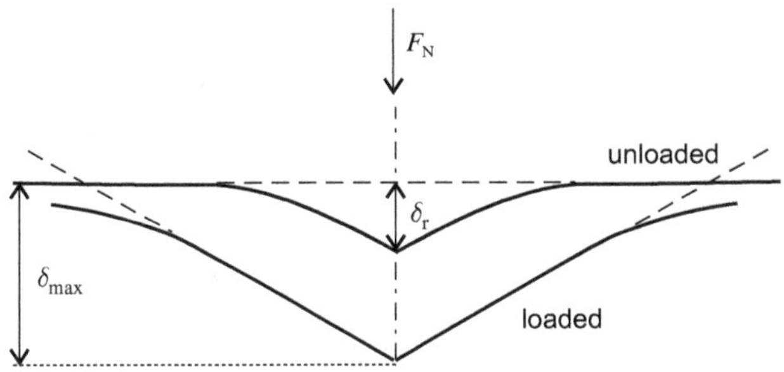

Figure 12.17 Contact depth as defined by the unloading process.

The unloading curve in Fig. 12.16 can be used to determine the hardness H of the substrate and the effective elastic modulus E^* of the contacting materials. In the widely accepted method introduced by Oliver and Pharr [237]

$$\delta_r = C\frac{F_{max}}{k}, \tag{12.11}$$

where F_{max} is the maximum loading force and the constant C depends on the tip geometry.[1] For a conical tip $C \approx 0.72$. The projected contact area is

$$A = \pi(\delta_{max} - \delta_r)^2 \tan^2\alpha,$$

where α is the half-angle of the indenter. The ratio A/F_{max} is equal to H whereas

$$E^* \approx \frac{k}{2}\sqrt{\frac{\pi}{A}}. \tag{12.12}$$

However, a serious limitation of this model is neglecting the pile-up of material at the edge of the contact area.

As a result of the loading cycle, the area around the impression ends up in a state of radial compression and circumferential tension, which can be quantified by finite element methods. Residual stresses are also left if a surface is hit by round shaped hard particles (*shot peening*). This process can considerably increase the resistance to fatigue of the material, which makes shot peening very important for aerospace applications.

12.9 Rolling on plastically deformed bodies

Consider an elastic cylinder which repeatedly rolls on an elastic–plastic half-space and suppose that the elastic limit is exceeded already in the first run. In this case it

[1] Equation (12.11), and the precise value of C, can be derived from the Sneddon theory for elastic contacts mentioned in Section 4.3.

may happen that the residual stress builds up to such values that subsequent runs result in entirely elastic deformation. This phenomenon is known as *shakedown*. According to *Melan's theorem* [215] an elastic–plastic contact will shake down if a time-independent distribution of residual stress can be found such that, if the distribution is superimposed on the elastic stress due to the load, the yield point is not reached. If the cylinder is rolling freely, Melan's theorem implies that the ratio between the shakedown limit and the elastic limit load is 1.66 [156, section 9.2]. However, this ratio is reduced in the presence of friction. The case of an elastic sphere rolling on an elastic–plastic half-space is much more complicated. Neglecting the pressure reduction caused by the formation of a groove, the aforementioned ratio becomes 4.7.

The case of a rigid cylinder rolling on a perfectly plastic half-space has been investigated by Mandel using the slip line theory [203]. A remarkable result is that the surface of the half-space is displaced backwards by the cylinder. An opposite behavior is observed on an elastic–plastic half-space, where the material is displaced forwards.

12.10 Rough plastic contacts

It is also instructive to study the plastic contact between a periodic surface and a rigid half-space. This problem has been solved for serrated surfaces (Fig. 12.18) [54]. For low values of the average pressure $\overline{p}$, the asperities are independent and the deformation occurs as described in Section 12.5. The situation changes when the deformation fields of adjacent asperities start to overlap. In the case in the figure (half-angle $\alpha = 65°$) this happens when the ratio between the contact length l and the periodicity λ is 0.36. The deformation ends when $\overline{p}$ reaches the limit value for plastic indentation of a wedge ($5.14k$ in the present case, see Section 12.5).

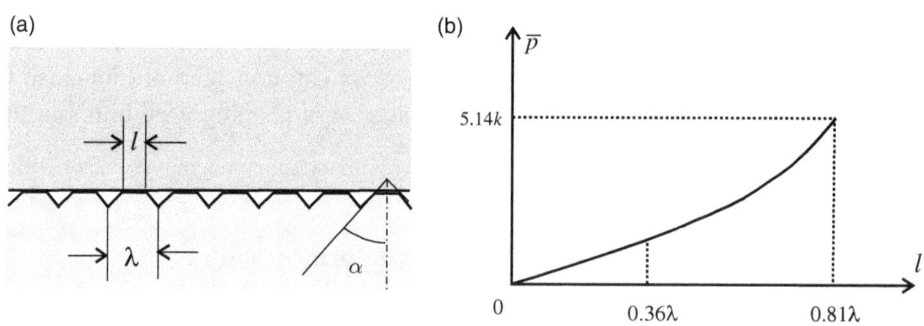

Figure 12.18 (a) Serrated plastic surface squeezed by a rigid half-space. (b) Average pressure $\overline{p}$ as a function of the contact length. Adapted from [54] with permission from Elsevier.

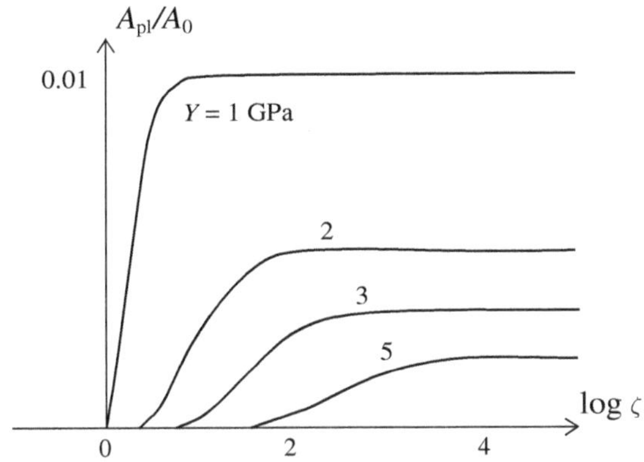

Figure 12.19 Normalized plastic contact area vs. logarithm of the magnification for different values of the yield strength ($E = 10^{11}$ Pa, $\nu = 0.3$, $\overline{p} = 10$ MPa). Adapted from [250] with permission from Elsevier.

The Persson theory can also be applied to elastic–plastic solids by replacing the boundary condition (8.15) with $P(Y, \zeta) = 0$, where Y is the yield strength of the material [250]. With increasing magnification ζ the contact area 'flows' into no-contact where the stress $\sigma < 0$, and into plastic contact where $\sigma > Y$. In Fig. 12.19 the ratio between the plastic contact area A_{pl} and the apparent contact area A_0 is plotted as a function of ζ for different values of Y. The asymptotic limit of this ratio is $\overline{p}/Y$.

12.11 Plasticity of geomaterials

Soils are composed by small solid particles ('grains') ranging from 1 μm to a few mm in size, and their shear deformation is accompanied by sliding of the grains one over another. Furthermore, the voids between particles can be filled by water and air. In *frictional materials*, such as gravels, sands and silts, the motion of the grains is opposed by dry friction. In *cohesive solids*, such as clays, the grains are protected by thick water films and they are not worn out when stress is applied. If a soil is confined laterally and compressed, the void fraction decreases irreversibly in a so-called *consolidation* process. If the load is subsequently released, the corresponding stress–strain curve resembles that of metals.

However, in contrast to metals, the shear strength τ_{max} of a soil depends significantly on the normal stress σ. The failure surface can be described by the *Mohr–Coulomb* (MC) *criterion*:

$$\tau_{max} = \sigma \tan \varphi + c, \tag{12.13}$$

where φ is the *angle of internal friction* (typically around 25–35°) and the parameter c is the *cohesion* of the soil. The angle φ corresponds to the slope of

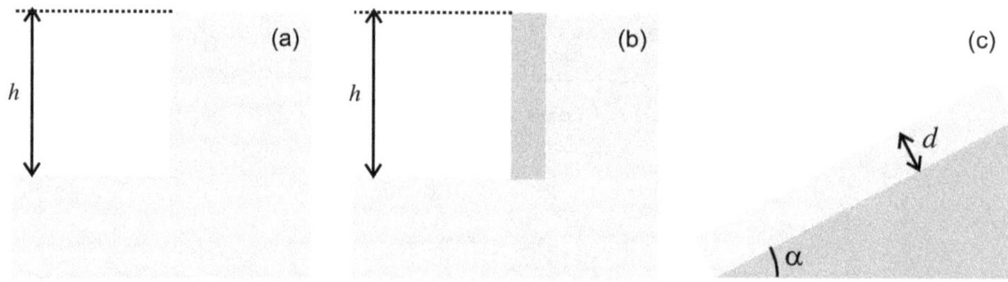

Figure 12.20 The stabilities (a) of a vertical bank, (b) of a retaining wall and (c) of an infinite slope depend on the angle of internal friction φ and can be assessed using the MC criterion.

the mound formed when the material is piled onto a horizontal plane. The stress-independent cohesion parameter is determined by cementation between sand grains and electrostatic attraction between clay particles.

In dry frictional materials $c = 0$, whereas saturated clays are described by the Tresca criterion, which is recovered from (12.13) with $\varphi = 0$ and $c = k$. Note that in terms of the principal stresses σ_i, the MC criterion takes the form

$$(\sigma_1 - \sigma_3) + (\sigma_1 + \sigma_3) \sin\varphi = 2c \cos\varphi. \tag{12.14}$$

The MC criterion has important applications in geotechnical engineering and a few formulas are worth being mentioned.

1. A vertical bank of height h and density ρ (Fig. 12.20a) will not collapse provided [79]
$$h < \frac{2c}{\rho g} \frac{\cos\varphi}{1 - \sin\varphi}.$$
2. In order to determine the thrust on a retaining wall of height h (Fig. 12.20(b)) one has to distinguish two cases depending on whether the wall is free to move outward (*active state*) or into the soil (*passive state*). According to *Rankine's formulas*, the lateral force acting on the wall is
$$F = \frac{1}{2}\rho g h^2 \tan^2\left(\frac{\pi}{4} \mp \frac{\varphi}{2}\right) - 2ch \tan\left(\frac{\pi}{4} \mp \frac{\varphi}{2}\right),$$
where the upper and lower signs refer to the active and passive state respectively.
3. An infinite slope of thickness d tilted by an angle α as in Fig. 12.20(c) is stable if the *factor of safety*
$$\frac{c}{\rho g d \cos\alpha} + \frac{\tan\varphi}{\tan\alpha} > 1. \tag{12.15}$$

For the derivation of these formulas and further results the reader is referred to the textbooks on this subject [62].

13

Fracture

Under the action of stress a solid object can separate into two or more pieces. The failure of the object is called *fracture* if the load is monotonic, and *fatigue* if the load is cyclic. The classical theory of fracture mechanics in elastic materials assumes the pre-existence of cracks and develops criteria for the catastrophic growth of these cracks. Crack propagation is a difficult topic, and, in spite of significant research efforts, a clear picture of this phenomenon is still missing. Crack propagation is also important for understanding friction in viscoelastic materials. In fact, when a rubber block slides on a smooth substrate, most energy dissipation is due to the opening crack propagating at the interface.

13.1 Fracture modes

In an isotropic homogenous material three fundamental fracture modes can be distinguished, depending on the orientation of the loading direction with respect to the crack (Fig. 13.1). In the *opening mode* a tensile stress acts normally to the plane of the crack. In the *sliding mode* a shear stress acts parallel to the plane of the crack and perpendicular to the crack front. In the *tearing mode* a shear stress acts parallel to the plane of the crack and to the crack front. In practice, the three modes can also be combined.

As shown by Westergaard [341] the components of the stress tensor at a distance r from a crack tip take the form

$$\sigma_{ij} = \frac{K}{\sqrt{2\pi r}} f_{ij}(\theta), \tag{13.1}$$

where the *stress intensity factor* K and the dimensionless functions f_{ij} depend on the crack mode, the crack geometry and the loading conditions (the angle θ is referred to the crack direction). The relation (13.1) breaks down in a *process zone*

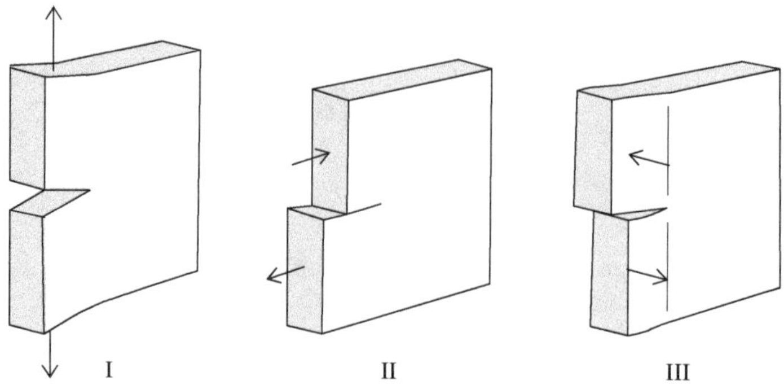

Figure 13.1 The three fracture modes.

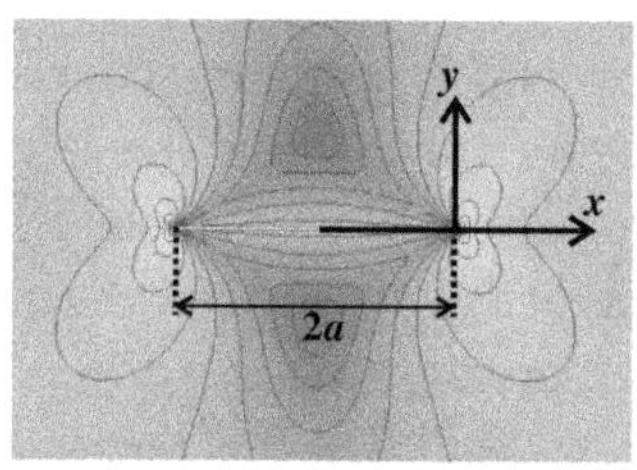

Figure 13.2 Stress distribution (logarithmic scale) around a linear crack in the opening mode.

very close to the tip, where plastic deformation occurs. The size of this zone can be very small (comparable to the unit cell of the crystal) if the material is brittle.

For the opening mode of a crack of length $2a$ in an infinite plate (Fig. 13.2) the stress intensity factor is

$$K_{\mathrm{I}} = \sigma\sqrt{\pi a}, \tag{13.2}$$

where σ is the (uniform) stress at infinity. The stress components near the crack tip are

$$\sigma_{x,y} = \frac{K_{\mathrm{I}}}{\sqrt{2\pi r}} \cos\frac{\theta}{2}\left(1 \mp \sin\frac{\theta}{2}\sin\frac{3\theta}{2}\right), \tag{13.3}$$

where the polar coordinates are centered around the points $x = \pm a$. The displacement field is described by the equations

$$u_x = \frac{K_{\mathrm{I}}}{2G}\sqrt{\frac{r}{2\pi}} \cos\frac{\theta}{2}\left(f(\nu) - \cos\theta\right),$$
$$u_y = \frac{K_{\mathrm{I}}}{2G}\sqrt{\frac{r}{2\pi}} \sin\frac{\theta}{2}\left(f(\nu) - \cos\theta\right), \tag{13.4}$$

where ν is Poisson's ratio, $f(\nu) = 3 - 4\nu$ for plane strain, $f(\nu) = (3 - \nu)/(1 + \nu)$ for plane stress, and G is the shear modulus of the material. If the crack is subjected to an in-plane stress τ at infinity:

$$K_{\mathrm{II}} = \tau\sqrt{\pi a}.$$

A similar expression applies to the factor K_{III} in the presence of an out-of-plane shear stress τ.

A simple relation holds also for the opening mode of a penny-shaped crack of radius a:

$$K_{\mathrm{I}} = 2\sigma\sqrt{a/\pi},$$

where σ is the stress at infinity. Several formulas for stress intensity factors can be found in dedicated handbooks [308, 225]. Alternatively, the factors K can be estimated using finite element methods.

13.2 The Griffith criterion

When the stress intensity factor K reaches a critical value (the so-called *fracture toughness*) K_{c}, a crack becomes unstable and starts to propagate. In the case of Fig. 13.2 the elastic strain energy (per unit width) released by the opening crack, as estimated from (3.9) using (13.3) and (13.4), is[1]

$$\Gamma \equiv \frac{\mathrm{d}U_{\mathrm{el}}}{\mathrm{d}a} = \frac{\pi a \sigma^2}{8G}(f(\nu) + 1), \tag{13.5}$$

or, using Eq. (3.11),

$$\Gamma = K_{\mathrm{I}}^2/E$$

for plane stress, and

$$\Gamma = K_{\mathrm{I}}^2(1 - \nu^2)/E$$

for plane strain.

If the propagation is very slow, which is the case at the instability onset, the energy release rate Γ is equal to the surface energy 2γ per unit length.[2] In this way one gets the following relations for the critical stress, corresponding to the famous *Griffith criterion* [123]:

$$\sigma_{\mathrm{c}} = \sqrt{\frac{2E'\gamma}{\pi a}}, \tag{13.6}$$

[1] We use the symbol Γ instead the more common G to avoid confusion with the shear modulus.

[2] The factor 2 comes from the fact that two surfaces are created during crack propagation.

where $E' = E$ for plane stress and

$$E' = E/(1 - \nu^2) \tag{13.7}$$

for plane strain. Note that in both cases σ_c depends on the crack length.

From a comparison with Eq. (13.2) it is possible to relate the fracture toughness K_{Ic} in the opening mode to the surface tension γ:

$$K_{\mathrm{Ic}} = \sqrt{2E'\gamma}.$$

In contrast to σ_c, K_{Ic} does not depend on the crack length. The energy release rates can be also estimated for the sliding mode, where

$$K_{\mathrm{IIc}} = \sqrt{\frac{16G\gamma}{f(\nu) + 1}},$$

and for the tearing mode:

$$K_{\mathrm{IIIc}} = \sqrt{4G\gamma}.$$

The Griffith criterion is in good agreement with experiments on brittle materials, but not on ductile materials, where plastic deformation is not negligible. This effect can be taken into account by adding a term γ_p (i.e. the plastic dissipation per unit area of crack growth) to the surface tension γ in Eq. (13.6) [146]. The radius of the plastic *process zone* around the crack tip, $r_p(\theta)$, can be estimated by using the principal stresses σ_1 and σ_2 in the von Mises criterion (12.3). For the opening mode one gets

$$r_p = \frac{1}{4\pi}\left(\frac{K_{\mathrm{I}}}{Y}\right)^2 \left(\frac{3}{2}\sin^2\theta + g(\nu)(1 + \cos\theta)\right), \tag{13.8}$$

where $g(\nu) = 1$ for plane stress and $g(\nu) = (1 - 2\nu^2)$ for plane strain. The relation (13.8) is plotted in Fig. 13.3 for both cases. Note that the process zone is much larger for plane stress.

13.3 Dynamic fracture

If the driving stress increases beyond σ_c, a crack in a brittle solid can suddenly start to propagate at a rate comparable to the velocity of sound. For a crack propagating in an infinite plate with a steady velocity v, the dynamic stress intensity factor can be written as [96]

$$K(v) = \frac{1 - v/c_R}{\sqrt{1 - v/c_l}} K_0(a),$$

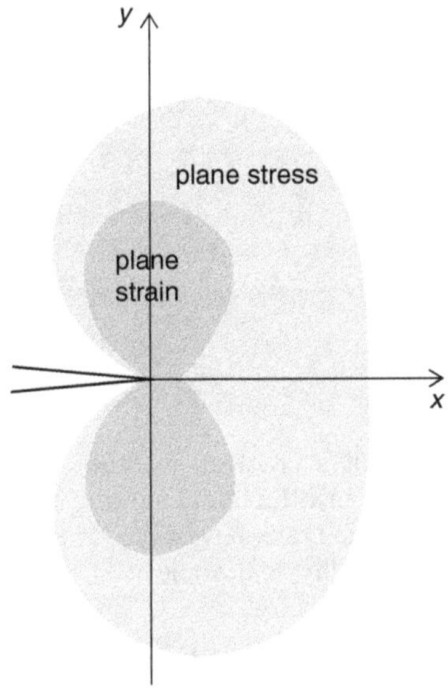

Figure 13.3 Process zones for the opening mode of a crack under plane stress and plane strain ($\nu = 1/3$).

where $c_{\rm l}$ is the longitudinal wave speed, $c_{\rm R}$ is the Rayleigh wave speed (Section 3.5) and $K_0(a)$ is the static stress intensity factor introduced in Section 13.2. The energy release rate depends on v as

$$\Gamma(v) \approx \frac{1-\nu^2}{E} K_0^2(a) \left(1 - \frac{v}{c_{\rm R}}\right). \qquad (13.9)$$

If the crack propagates in an infinitely long strip of width $2b$ under non-steady conditions [204]:

$$\Gamma(v) \approx W \left(1 - \frac{b\dot{v}/c_{\rm l}^2}{(1 - v^2/c_{\rm R}^2)^2}\right), \qquad (13.10)$$

where W is the energy (per unit width) stored in the unit length of the strip far ahead of the crack tip. Note that the energy release rate in Eq. (13.9) depends on the tip position, whereas in Eq. (13.10) it depends on the tip acceleration $\dot{v}$. Equation (13.10) can also be written as $F = m_{\rm eff}\dot{v}$, where the 'force' F is proportional to the difference $W - \Gamma(v)$ and the effective mass

$$m_{\rm eff} \propto \frac{b}{c_{\rm l}^2(1 - v^2/c_{\rm R}^2)^2}.$$

In order to observe crack propagation on standard brittle materials such as soda-lime glasses a spatial resolution of the order of 1 μm and an acquisition rate beyond 10^6 frames/s are required, which is not possible with current imaging techniques.

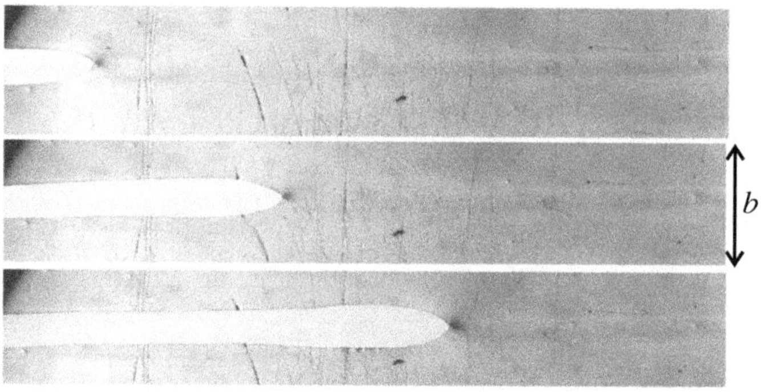

Figure 13.4 Time evolution of a crack in a finite strip of soft polyacrylamide gel. Note the transition from an effectively infinite medium accompanied by a change of the tip shape when the crack length $l \sim b$. Reproduced from [113] with permission from the American Physical Society.

To circumvent this problem Goldman *et al.* used a soft aqueous gel whose shear wave speed is about three orders of magnitude less than in glass (Fig. 13.4). An excellent agreement with both Eqs. (13.9) and (13.10) was found. Furthermore, the use of gel avoided the formation of branching instabilities, which are commonly observed in brittle materials beyond a critical velocity $v_c \approx 0.4c_R$ [28].

13.4 Fracture in rubber-like materials

In rubber materials it has been observed experimentally that the energy release rate depends on the tip velocity v and the temperature T as

$$\Gamma(v, T) = \Gamma_0[1 + f(v, T)],$$

where $f \to 0$ as $v \to 0$, and Γ_0 is a threshold value below which no fracture occurs [264]. Note that, in this case, Γ includes a contribution of viscoelastic energy dissipation which may occur far away from the crack tip in addition to the energy required to break the bonds at the tip. The values of Γ_0 are usually of the order of few tens of J/m^2. For simple hydrocarbon elastomers, if the increase of temperature in the contact area is negligible, $f(v, T) = f(a_T v)$, where a_T is the factor in the WLF equation (Section 11.1).

In the context of the Persson theory it can be proven that, for a given velocity v, the tip radius a_{tip} is implicitly determined from the relation [255]

$$a_{tip} = a_0 \left(1 - \frac{2}{\pi} E_0 \int_0^{2\pi v/a_{tip}} \frac{\sqrt{1 - (\omega a_{tip}/2\pi v)^2}}{\omega} \mathrm{Im} \frac{1}{E(\omega)} \mathrm{d}\omega \right)^{-1},$$

where $E(\omega)$ is the viscoelastic modulus of the material, $E_0 = E(0)$ and a_0 is the tip radius for a very slowly propagating crack. Once the relation $a_{tip}(v)$ is known, the energy release rate is obtained as

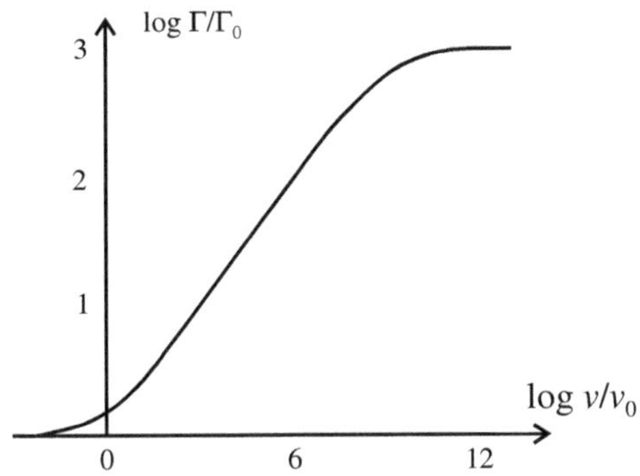

Figure 13.5 Energy release rate vs. crack velocity in a viscoelastic material (for styrene–butadiene at room temperature $v_0 \sim 1$ nm/s and $\Gamma_0 \approx 30$ J/m^2). Adapted from [264] with permission from IOP Publishing.

$$\Gamma(v) = \Gamma_0 a_{\text{tip}}(v)/a_0.$$

A possible dependence of $\Gamma(v)$, estimated numerically, is shown in Fig. 13.5 [264]. Note that Γ increases by three orders of magnitude at high crack tip velocities.

The sliding (or rolling) of a rubber block on a smooth substrate can be seen as the combination of a closing crack at the leading edge of the contact region and an opening crack at the trailing edge. Associated with the crack propagation, a frictional stress $\tau_{\text{fric}} = \Gamma(v)/l$, where l is the linear size of the contact area, can be observed. In the case of styrene–butadiene rubber filled with carbon black experiments by Lorenz *et al.* have shown that the energy dissipation at the opening crack gives the main contribution to the frictional shear stress in the area of real contact [192].

14

Stick–slip

The alternation of stick and slip phases in the sliding of two surfaces past each other is one of the most intriguing aspects of tribology. In this chapter we will discuss general models for stick–slip and relate this effect to the constitutive relations defining the kinetic friction force. An important concept, which significantly modifies the velocity dependence of the friction, is the so-called 'contact ageing'. Contact ageing may be due to plastic flow, capillary condensation or the interdiffusion of polymer chains. On large scales it is also responsible for the generation of seismic waves, as first recognized by Dieterich and Ruina.

14.1 Stick–slip

Consider a rigid block of mass m, thickness d and width L lying on a flat surface and connected to a spring with stiffness k, as shown in Fig. 14.1. If the spring is pulled with a constant velocity v the block will not move till the time $t_c = F_s/kv$, where F_s is the static friction force. When $t = t_c$ the block will suddenly start sliding, but the motion will be slowed down by the kinetic friction force F_k. When the block is sliding the coordinate x of its center of mass varies with time as

$$x(t) = vt - \frac{F_k}{k} - A\sin(\omega_0 t + \alpha),$$

where $\omega_0 = \sqrt{k/m}$ is the resonance frequency of the system. The amplitude A and the phase shift α can be derived from the conditions $x = 0$ and $\dot{x} = 0$ at $t = t_c$:

$$A = \sqrt{\frac{v^2}{\omega_0^2} + \frac{(F_s - F_k)^2}{k^2}}, \qquad \alpha = \arctan\frac{\omega_0(F_s - F_k)}{kv} - \frac{\omega_0 F_s}{kv}.$$

It is not difficult to determine the time t_{slip} after which the body comes to rest:

$$t_{slip} = \frac{2}{\omega_0}\arctan\left(\frac{\omega(F_s - F_k)}{kv}\right).$$

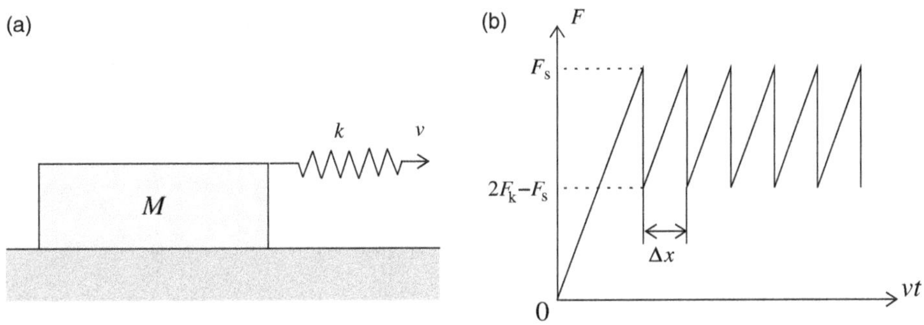

Figure 14.1 (a) A block of mass m pulled by a spring over a flat substrate. (b) Time dependence of the spring force in the quasi-static limit $v \to 0$.

After this time, the block will stick again for a time

$$t_{\text{stick}} = \frac{2(F_s - F_k)}{kv}$$

and so on.

If v is very low the previous expressions are simplified. In this case $t_{\text{stick}} \gg t_{\text{slip}} \approx \pi/\omega_0$ (i.e. half of the period of the free harmonic oscillator) and the spring force oscillates between the values F_s and $2F_k - F_s$ with a characteristic sawtooth profile, as shown in Fig. 14.1(b). The slip length Δx is approximately equal to vt_{stick} or

$$\Delta x \approx 2(F_s - F_k)/k. \tag{14.1}$$

Note that the same behavior would be observed if the block were elastic and its upper surface were pulled at constant speed [245, section 9.2]. In this case the spring constant k is replaced by the lateral stiffness Gd of the block, where G is the shear modulus of the material, and the slip time

$$t_{\text{slip}} \approx 2\pi L/c_s, \tag{14.2}$$

where $c_s = \sqrt{G/\rho}$ is the shear velocity of sound.

14.2 Contact ageing

From the simple model discussed in Section 14.1 we expect that the stick–slip occurs indepedently of the value of the spring constant k. However, it is experimentally well established that the stick–slip disappears if k or the driving velocity v are high enough. A typical 'kinetic phase diagram' is shown in Fig. 14.2(a). The stick–slip motion corresponds to the gray area, while steady sliding is observed in the rest of the (k, v)-plane. A diagram with this form can be reproduced assuming

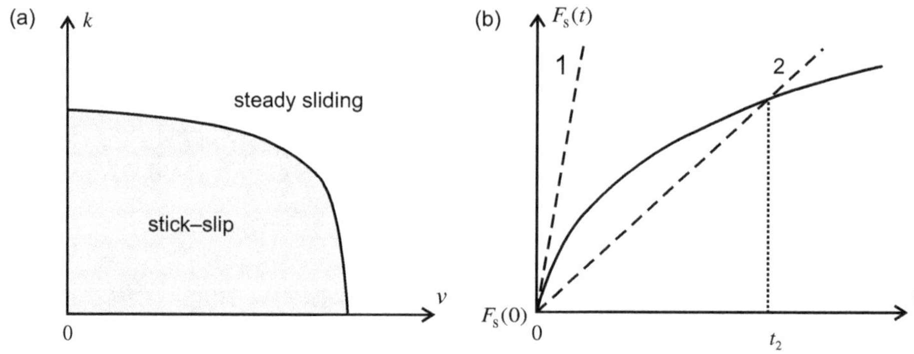

Figure 14.2 (a) A typical kinetic phase diagram. (b) Time variation of the static friction force (solid line). If the corresponding variation of the spring force is described by the dashed line 1, steady sliding will be observed. Otherwise (dashed line 2) stick–slip is expected.

that the static friction force F_s increases with the time t of stationary contact. If the initial rate of increase of F_s is lower than k (curve 1 in Fig. 14.2(b)), steady sliding will be observed. Otherwise, stick–slip must occur.

The increase of the static friction with time can have different origins. The most common mechanisms are the formation of capillary bridges, the increase of contact area due to thermally activated plastic flow, chain interdiffusion for polymers or solids covered by grafted monolayer films, and stress relaxation at the interface. Capillary bridges are discussed in Section 24.2. The other mechanisms are briefly described below.

14.3 Lubricated friction

We will consider two types of adsorption system. In the first case the interaction between the lubricant molecules and the substrate is supposed to be so weak that the adsorbate layer can fluidize at the onset of sliding (Fig. 14.3(a)). This is a common situation when saturated hydrocarbons are adsorbed on metal or mica surfaces. In the second case the interaction is much stronger and fluidization occurs. This happens in fatty acid–metal oxide systems, where sliding occurs between the tails of the grafted hydrocarbon chains (see Fig. 14.3(b) and also Section 24.1). In this case the static friction force increases with the time of stationary contact because of interdiffusion and other relaxation processes.[1] In both cases the friction force is conveniently described introducing a *state variable* $\theta(t)$, which evolves with time in a specific way.

[1] We assume that the normal force is not too large, see Section 24.1.

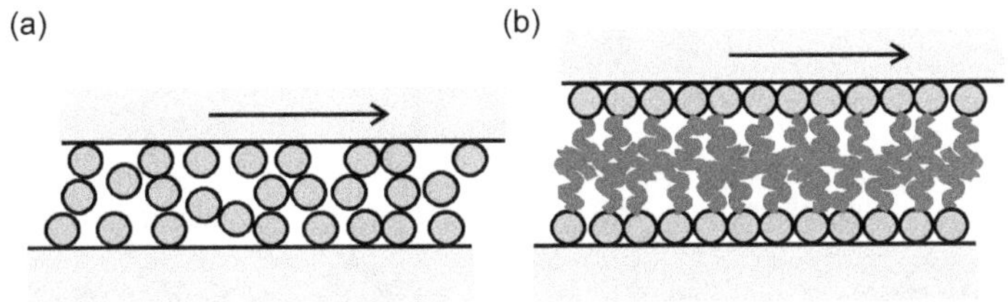

Figure 14.3 Schematic representation of lubricating layers with (a) low and (b) high corrugation of the interaction potential.

Small corrugation

Consider a block of mass m sliding under boundary lubricated conditions. If the lateral corrugation of the interaction potential is very low, the friction force can be assumed to depend on the state variable θ as

$$F = F_k + (F_s - F_k) \cdot \theta(t) + m\gamma\dot{x}, \tag{14.3}$$

where θ evolves according to the equation [44]

$$\dot{\theta} = \frac{\theta(1-\theta)}{\tau} - \frac{\theta\dot{x}}{D}. \tag{14.4}$$

In Eq. (14.4) τ is the characteristic time over which the film freezes, and D is the characteristic length over which the melting transition takes place.

In practice, θ describes the degree of melting of the lubrication layer. The lubricant is in the fluid state if $\theta = 0$ and in the solid state if $\theta = 1$. If the block is moving and is suddenly stopped at $t = 0$, the friction force will increase with time as

$$F(t) = F_k + (F_s - F_k)\frac{\theta_0}{\theta_0 + (1-\theta_0)e^{-t/\tau}},$$

Under these assumptions, it can be seen by linear stability analysis that the stick–slip takes place in the gray area of the (k, v) plane shown in Fig. 14.4(a). Above a critical velocity $v_c(k)$ stick–slip is not possible and the block slides continuously. The dependence of the critical velocity v_c on the spring constant k is described by the equation [244]

$$\left(1 - \frac{v_c}{v_0}\right)\left(\gamma\tau + 1 - \frac{v_c}{v_0}\right) = \frac{k}{k_0}, \tag{14.5}$$

where $v_0 \equiv D/\tau$ and

$$k_0 \equiv \frac{m}{\tau^2}\left(\frac{\tau(F_s - F_k)}{mD\gamma} - 1\right).$$

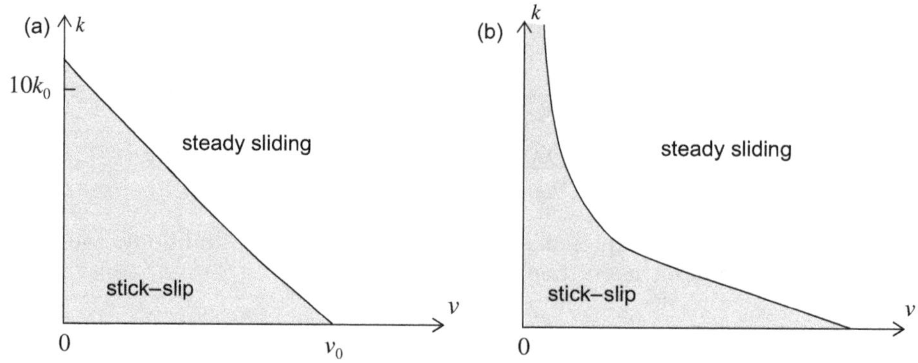

Figure 14.4 Kinetic phase diagram in the (k, v) plane for boundary lubrication (a) with small corrugation and (b) with large corrugation. The gray regions correspond to the values where stick–slip is observed.

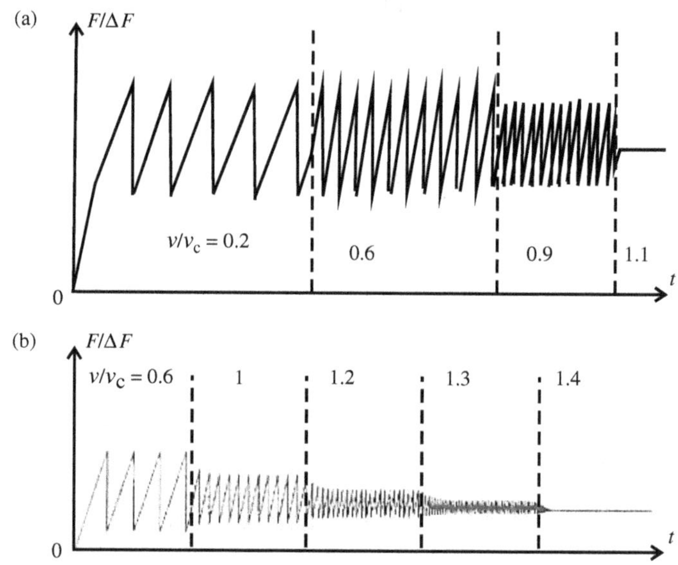

Figure 14.5 Time dependence of the spring force on a lubricated block when the support velocity v is increased stepwise on a surface potential with (a) small corrugation and (b) large corrugation. Adapted from [245] with permission from Springer.

If the driving velocity v increases, the friction peaks and the amplitude of the stick–slip oscillations is slightly reduced, as shown in Fig. 14.5(a). At the critical velocity v_c, the stick–slip phenomenon disappears all of a sudden. Although the model above, introduced by Carlson and Batista [44], is rather phenomenological, it contains all the ingredients of 'rate and state' theories, which are quite successful in describing macroscopic friction phenomena.

Large corrugation

If the lateral corrugation of the adsorbate–substrate interaction potential is large, a good assumption for the variation of the friction force is [245, section 12.2]

$$F = F_\mathrm{k} + (F_\mathrm{s} - F_\mathrm{k})\left(1 - \mathrm{e}^{-\theta/\tau}\right) + m\gamma\dot{x},$$

where the contact age variable θ satisfies the equation

$$\dot{\theta} = 1 - \frac{\dot{x}\theta}{D}. \tag{14.6}$$

Note that for stationary contact ($\dot{x} = 0$) the variable θ coincides with the time t of stationary contact and the friction force increases asymptotically from F_k to F_s. A possible variation of the friction force with the driving velocity is shown in Fig. 14.5(b). Here the average friction does not change significantly at increasing values of v, whereas the friction peaks diminish till they vanish without abrupt variations.

The boundary line in the (k, v) plane, which separates stick–slip motion from steady sliding, is now given by

$$\frac{k}{k_0} = \left(1 + \frac{v_\mathrm{c}}{\gamma D}\right)\left(\frac{\Delta F_0}{m\tau v_\mathrm{c}}\mathrm{e}^{-D/v_\mathrm{c}\tau} - \frac{\gamma v_\mathrm{c}}{D}\right) \tag{14.7}$$

(Fig. 14.4(b)). Depending on the value of the spring constant k, the transition can be continuous, as in Fig. 14.5(b), or discontinuous, as in the case of small corrugation. If the separation between the chains is large enough, stick–slip may also disappear from the whole (k, v) plane and the boundary lubrication film may be considered to be in a liquid-like state.

Comparison with the experiments

Experimental results in a certain agreement with the models above have been obtained with the surface force apparatus (Section 17.3). Figure 14.6(a) shows the transition from stick–slip to sliding observed with a thick hydrocarbon film. Here, the corrugation of the adsorbate–substrate interaction is very weak. In this case the amplitude of the stick–slip spikes is nearly independent of the driving velocity up to the critical value v_c at which the transition occurs. When grafted chain molecules are used, as in Fig. 14.6(b), the corrugation is much larger and the amplitude of the spikes is found to decrease continuously when v increases up to a different critical value.

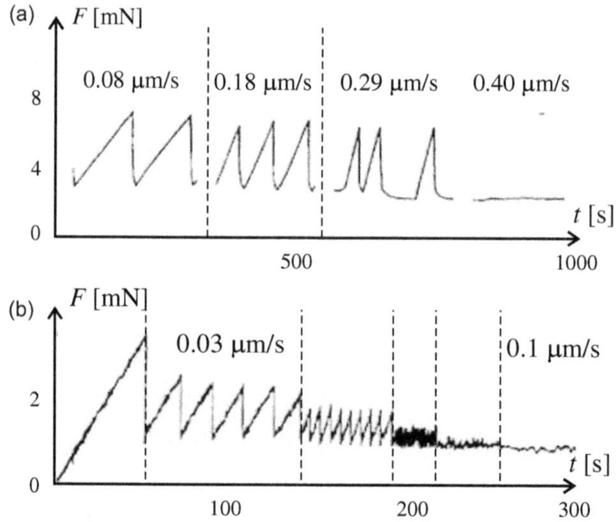

Figure 14.6 Spring force F as a function of time (a) for hexadecane and (b) for mica surfaces coated with DMPE molecules, as measured in a surface force apparatus. Adapted from [348] with permission from Elsevier.

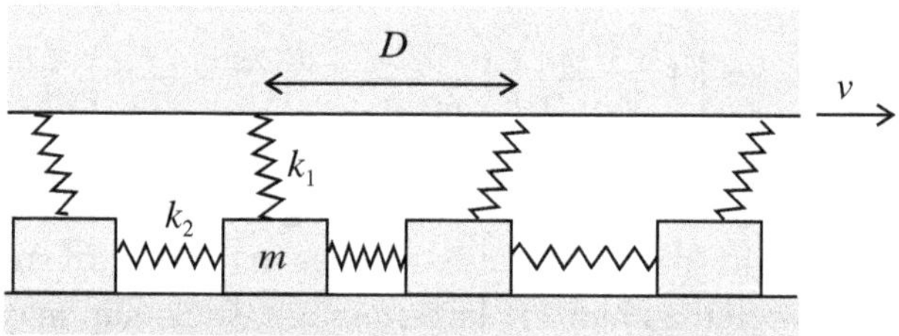

Figure 14.7 The Burridge–Knopoff (BK) model.

14.4 The Burridge–Knopoff model

Consider the model in Fig. 14.7 [39]. Each of the N_0 small blocks in contact with the substrate is connected to the (sliding) large block by a spring k_1 and to the neighboring blocks by two springs of stiffness k_2. The potential energy stored in the springs connected to the block i is

$$U = \frac{1}{2}k_1(x - u_i)^2 + \frac{1}{2}k_2(u_{i+1} - u_i)^2 + \frac{1}{2}k_2(u_{i-1} - u_i)^2,$$

where x is the position of the surface of the big block connecting the block i and u_i is the displacements of the small block. The mean field force acting on the block i is

$$F = k_1(x - u_i) + k_2(u_{i+1} + u_{i-1} - 2u_i).$$

In order to initiate sliding, a critical displacement $u_i^{(c)}$ must be reached, at which F equals the static friction F_s. It is not difficult to see that the energy barrier $\Delta E = U(u_i^{(c)}) - U(u_i)$ preventing the motion of the small block depends on F, and vanishes for $F = F_s$, as

$$\Delta E = U_0 \left(1 - (F/F_s)^2\right),$$

where

$$U_0 = \frac{F_s^2}{2(k_1 + 2k_2)}.$$

The probability p that the block remains pinned changes with the time t according to the *master equation*

$$\frac{\mathrm{d}p}{\mathrm{d}t} = -f_0 \exp\left(-\frac{\Delta E}{k_B T}\right) p(t). \tag{14.8}$$

According to Kramers' theory for thermally activated processes, the prefactor f_0 depends on the resonance frequency $\omega_0 = \sqrt{k_1/m} \sim c_s/D$, where D is the average displacement of the blocks, and on the damping coefficient γ as[2] $f_0 \sim \omega_0^2/(2\pi\gamma)$. In the next chapter[3] we will show that, in most cases, $\gamma \sim \omega_0$, so that $f_0 \sim \omega_0/2\pi$.

If $U_0 \ll k_B T$ and $t \ll 1/f_0$ it can be proven that the number N of small blocks remaining in their original positions at time t is [245, section 11.3]

$$N(t) \approx N_0 \left(1 - \frac{k_B T}{2U_0} \ln f_0 t\right).$$

When a small block jumps, the force on the big block changes by $-F_s/N_0$. Hence $F(t) \approx F(0) - F_s(\Delta N/N_0)$, where $\Delta N = N - N_0$. Observing that $F(0)$ is the kinetic friction force F_k and assuming that $F_k \approx F_s/2$ (see section 16.3) we conclude that F increases logarithmically with the sliding velocity as

$$F \approx F_k + \frac{k_B T}{4U_0} F_s \ln\left(\frac{v}{v_0}\right), \tag{14.9}$$

where

$$v_0 = \frac{F_s f_0 k_B T}{2U_0 k_1}.$$

The distribution of slip events can be studied numerically. As a result, it is found that, when $k_2 \sim k_1$, the events are very localized. However, this is not the case when $k_2 \gg k_1$. The time variations of the friction force and of the fraction of moving blocks expected in this case are shown in Fig. 14.8. When a block rapidly

[2] More precisely, $f_0 = \sqrt{\kappa_{\min}\kappa_{\max}}/2\pi\gamma$, where $\kappa_{\min}$ ($\kappa_{\max}$) is the curvature in the vicinity of the energy minimum (maximum).

[3] See Eq. (15.28).

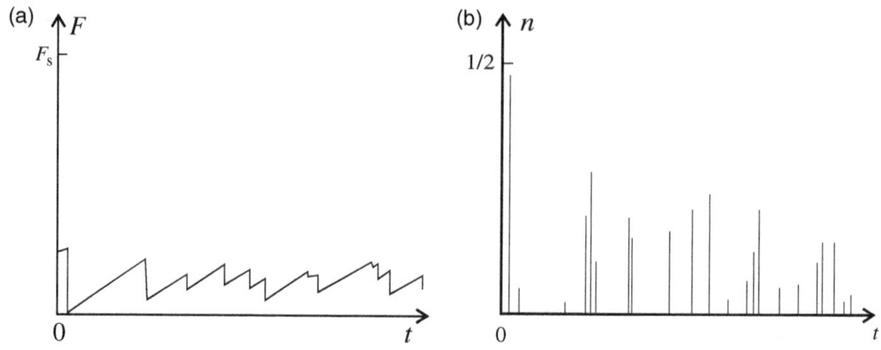

Figure 14.8 (a) Time variation of the friction force and (b) the fraction of moving blocks in the BK model when $k_2 \gg k_1$.

slips forwards, it can pull neighboring blocks over their barriers, giving rise to a wide distribution of avalanche sizes (provided that v is low enough).

14.5 Plastic flow

A logarithmic increase of the contact area with time can be also observed if the local pressure in the contact exceeds the plastic yield strength.

Creep in metals

The plastic flow in a solid can be seen as the result of 'melting' and 'refreezing' of small stress blocks. As in the Burridge–Knopoff model thermal activation plays a key role and the stress σ and the strain rate $\dot{\varepsilon}$ at a temperature T can be related by an equation which is analogous to (14.9) [247]:

$$\sigma = \frac{Y}{2}\left(1 + \frac{k_B T}{U_0} \ln \frac{\dot{\varepsilon}}{\dot{\varepsilon}_0}\right), \tag{14.10}$$

where Y is the macroscopic yield strength (during uniaxial tension) and U_0 is the pinning energy of dislocations (for metals, $U_0 \sim 1$ eV). In Eq. (14.10) the characteristic strain rate

$$\dot{\varepsilon}_0 = \frac{4 f_0 (1+\nu) Y k_B T}{3 E U_0},$$

and the attempt frequency f_0 depends on the width D of a typical stress block and on the velocity of sound c as $f_0 \approx c/2\pi D$.

Suppose now a cylinder is squeezed against a rigid substrate by applying a normal force F_N. If the yield threshold is reached, the residual stress relaxes slowly.

Since the volume is constant during plastic flow, it is not difficult to prove that the strain rate $\dot{\varepsilon}$ and the stress σ depend on the contact area A and its time variation $\dot{A}$ as

$$\dot{\varepsilon} = -\frac{\dot{A}}{A}, \qquad \sigma = -\frac{F_{\rm N}}{A}.$$

Introducing the parameter $\xi = \Delta A(t)/A_0$ and the creep law (14.10) we obtain the differential equation

$$\frac{k_{\rm B}T}{U_0} \ln \frac{\dot{\xi}/\dot{\varepsilon}_0}{1+\xi} = -\frac{\xi}{1+\xi}. \tag{14.11}$$

If $k_{\rm B}T \ll U_0$, Eq. (14.11) can be approximated as

$$\frac{k_{\rm B}T}{U_0} \ln \frac{\dot{\xi}}{\dot{\varepsilon}_0} = -\xi$$

and, as a result, the contact area is found to increase logarithmically with the time t of stationary contact as

$$\frac{\Delta A(t)}{A_0} = \frac{k_{\rm B}T}{U_0} \ln\left(1 + \frac{t}{\tau}\right), \tag{14.12}$$

where the characteristic time $\tau = k_{\rm B}T/U_0\dot{\varepsilon}_0$. For metals at room temperature $\dot{\varepsilon} \sim 10^8\ {\rm s}^{-1}$ so that $\tau \sim 10^{-10}$ s and, in practice, $\Delta A \propto \ln(t/\tau)$.

Note that, even if the mechanism of thermal activation described above involves dislocations, a logarithmic increase of the contact area with time is a general although not completely understood effect. A beautiful demonstration of contact ageing on an amorphous material is given in Fig. 14.9, where the contact formed by rough acrylic plastic is visualized using optical techniques. Indentation experiments on ice [31] have also shown that the contact area increases logarithmically with the time of contact.

Sliding friction

We can combine the previous results with those in Section 14.3 and assume that, if a block is sliding, the contact area changes with time as

$$A = A_0\left(1 + B_\perp \ln \frac{\theta}{\theta_0}\right),$$

where $B_\perp = k_{\rm B}T/U_0$ and the state variable θ is defined by Eq. (14.6). If the shear stress depends on the sliding velocity v according to Eq. (14.9), the friction force can be written as

$$F \approx F_{\rm k}\left(1 + B_\perp \ln \frac{\theta}{\theta_0} + B_\parallel \ln \frac{\dot{x}}{v_0}\right), \tag{14.13}$$

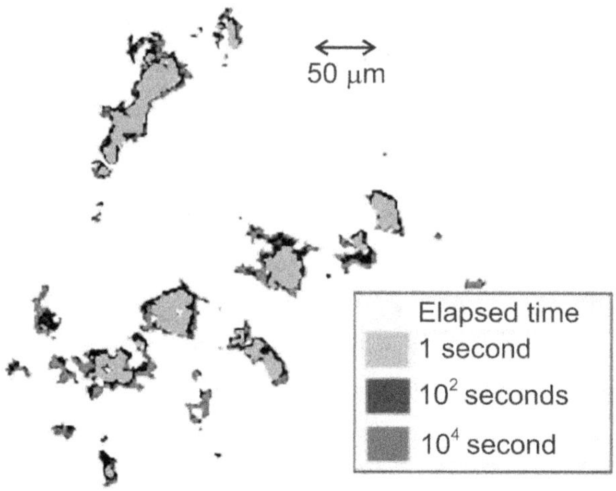

Figure 14.9 Time variation of the contact area between rough acrylic plastic and glass. Adapted from [74] with permission from Springer.

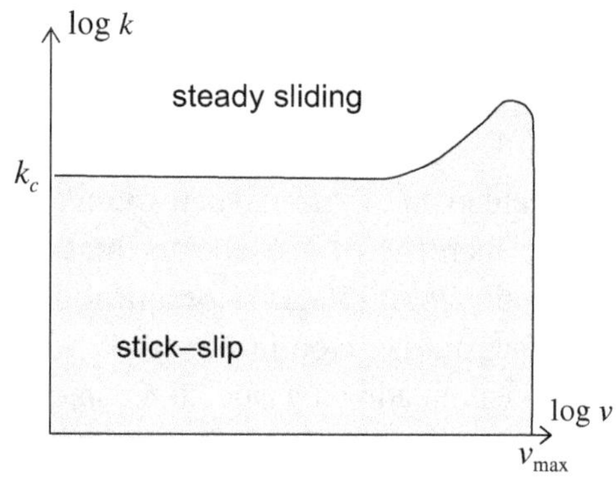

Figure 14.10 Kinetic phase diagram for plastic flow in a solid. Adapted from [245] with permission from Springer.

where

$$B_{\parallel} = \frac{F_{\mathrm{s}}}{F_{\mathrm{k}}} \frac{k_{\mathrm{B}} T}{4U_0}.$$

In this case, the kinetic phase diagram has the form in Fig. 14.10 [245, section 13.3]. At low values of v the critical stiffness separating the regions of stick–slip and steady sliding is

$$k_{\mathrm{c}} = \frac{F_{\mathrm{k}}}{D}(B_{\perp} - B_{\parallel}). \tag{14.14}$$

The nearly vertical boundary corresponds to the value

$$v_{\max} = \frac{F_k}{m\gamma}(B_\perp - B_\parallel)$$

of v at which the kinetic friction force of steady sliding changes from a velocity weakening to a velocity strengthening regime. These results are in a certain agreement with experiments by Baumberger *et al.* on Bristol board surfaces, where wear effects are negligible and the results are highly reproducible [14].

14.6 Earthquakes

The models discussed in the previous sections have important applications in the interpretation of earthquakes. For instance, we can use Eq. (14.2) with the velocity of the shear waves in the Earth's crust, $c_s \sim 1$ km/s, to estimate that a large earthquake involving a rupture lenght $L \sim 100$ km should have a time duration of about one minute. Equation (14.13) is consistent with the phenomenological *Dieterich–Ruina* (DR) *law* [73, 297] according to which the friction coefficient depends on a state variable θ as

$$\mu = \mu_0 + a \ln \frac{v}{v_0} + b \ln \frac{\theta v_0}{L}, \tag{14.15}$$

where L is a characteristic sliding distance, a and b are empirical parameters, and μ_0 and v_0 are reference values. The frictional strengthening observed with the time of contact is well reproduced if θ satisfies an 'ageing equation' of the form

$$\frac{d\theta}{dt} = 1 - \frac{\theta v}{L}.$$

The meaning of the quantities a, b and L can be easily seen if the sliding velocity is suddenly increased, as shown in Fig. 14.11.

From the results in Section 14.5 we expect that the sliding is stable and no earthquakes nucleate if $a > b$. Vice versa, if $a < b$ nucleation is possible as the result

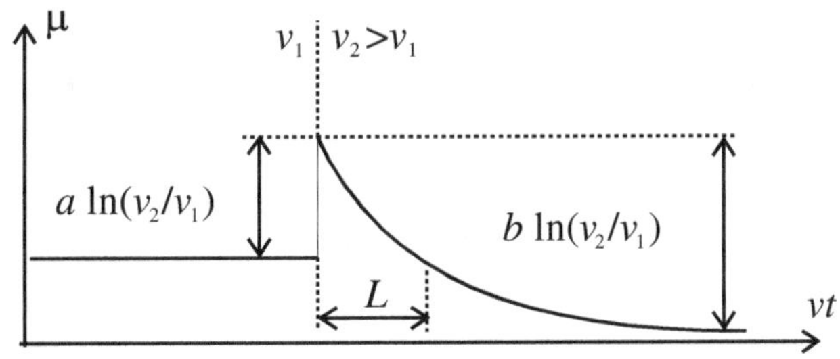

Figure 14.11 Response of the friction coefficient to a sudden variation of the sliding velocity in the Dieterich–Ruina model.

of a velocity perturbation the critical value of which depends on the normal stress σ. Experimentally, it is found that the difference $a - b$ varies with the temperature. In the case of granite the difference is positive below 300 °C, meaning that earthquakes cannot nucleate in the upper part of the Earth's crust, although they can propagate in that region, where they are strongly damped.

Based on the previous discussions, one would also expect that v and θ evolve in a cyclic way. However, this in no way means that the earthquake occurrence is periodic. Indeed, faults are often segmented with jogs and steps and every earthquake perturbs the stress field at the site of future earthquakes. As a result, the timing of earthquakes can be advanced or delayed, making these processes very difficult to predict.

Another important quantity is the energy released during the earthquake. This is usually expressed by the *seismic moment*

$$M = GAd,$$

where G ($\sim$ 30 GPa) is the shear modulus of the rocks, A is the area of the fracture, and d is the average displacement along the fracture. According to the famous *Gutenberg–Richter law* [126], the probability of an earthquake with 'magnitude' M is inversely proportional to M:

$$p(M) \propto 1/M.$$

This empirical law can be recovered with the Burridge–Knopoff model assuming a velocity weakening friction force [45].

The DR model has been extended to 3D by Rice, who also introduced a characteristic *creep length* l_c [288]. The quantity l_c corresponds to the minimum linear size l of a fault area which must slip simultaneously in order for a stick–slip instability to develop. Since the volume element is connected to the surrounding solid by an effective elastic spring $k_{\text{eff}} \sim \rho c^2 l$, the condition $k_{\text{eff}} = k_c$, with k_c defined by Eq. (14.14), implies that

$$l_c = \frac{D\rho c^2}{\tau_0(B_\perp - B_\parallel)}.$$

For wet granite at ~10 km below the Earth's surface, the creep length l_c is the order of 1 m.

Even if it is now well established that earthquakes result from stick–slip instabilities, since they usually occur by sudden slip along a pre-existing fault in the Earth's crust, it is quite remarkable that this conclusion was reached only in the 1960s [33].

Part III
Nanotribology

15

Atomic-scale stick–slip

Our review on nanotribology starts with a description of the Prandtl–Tomlinson model, which reproduces the stick–slip motion of a particle elastically driven on a crystal surface. Combined with Kramer's theory for thermally activated processes, this model also predicts the temperature and velocity dependence of atomic-scale friction observed experimentally. Several analytical expressions will be presented to illustrate these concepts. Stick–slip can be suppressed if the normal force is reduced statically or by means of mechanical excitations. The length of the jumps across the surface lattice is also influenced by the normal force, as well as by the damping coefficient describing the coupling with phonon and electron excitations in the bulk. Stick–slip is also observed if the particle is pulled with a constant lateral force rather than using a spring. The motion of a chain of equal particles on a crystal surface is interpreted by the Frenkel–Kontorova model, which is also introduced in this chapter.

15.1 The Prandtl–Tomlinson model

The basic features of atomic-scale stick–slip are captured by a model which was introduced by Prandtl in 1928 [277] to describe plastic deformation in crystals. The model is often attributed also to Tomlinson [332] and, for this reason, we will follow the common acceptance and call it *Prandtl–Tomlinson* (PT) *model*.

Consider a nano-asperity ('tip') of mass m pulled on a corrugated substrate potential $U_{\text{int}}(x)$ by a spring of stiffness k, which is driven by a solid support at constant velocity v along the x direction. The equation of motion of the tip is

$$m\ddot{x} + m\gamma\dot{x} + k(x - vt) + U'_{\text{int}}(x) = 0, \tag{15.1}$$

where the damping coefficient γ describes the coupling with phonon and (for metal surfaces) electron–hole excitations in the substrate,[1] as described in Section 15.8. At finite temperature T a noise term $\xi(t)$ is added to the right hand side of Eq. (15.1). The force $\xi(t)$ is responsible for the Brownian motion of the tip according to the fluctuation–dissipation theorem:[2]

$$\langle \xi(t)\xi(t') \rangle = 2m\gamma k_B T \delta(t - t'),$$

where k_B is Boltzmann's constant. Note that the delta function is justified by the fact that the characteristic frequencies of the phonon and electron excitations are much shorter than the hopping rate of the tip between the minima of the interaction potential (see below).

For sake of simplicity we will consider a sinusoidal potential U_{int} with amplitude U_0 and periodicity a such that the total potential reads

$$U(x,t) = -U_0 \cos \frac{2\pi x}{a} + \frac{1}{2}k(x - vt)^2. \tag{15.2}$$

In this case, it is convenient to introduce a dimensionless parameter

$$\eta = \frac{4\pi^2 U_0}{ka^2}. \tag{15.3}$$

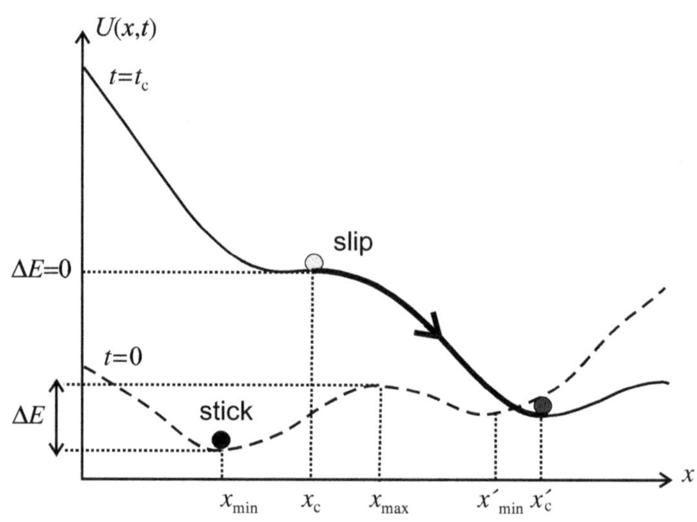

Figure 15.1 The equilibrium of a particle driven by a spring in a sinusoidal potential is broken at a critical time t_c when the particle suddenly hops from the minimum x_c to the next one, x'_c.

[1] A corresponding term, with a different damping coefficient, may be introduced to describe the coupling with the spring support.

[2] The coupling of the point mass to the spring makes the application of this theorem somehow questionable.

A typical experimental situation, corresponding to the value $\eta = 10$, is shown in Fig. 15.1. We will show below that the motion is continuous if $\eta < 1$, whereas stick–slip is observed if $\eta > 1$. More generally, if the potential $U_{\rm int}$ is not sinusoidal, the stick–slip condition is $-U_{\rm int}^{\prime\prime({\rm max})}/k > 1$, where $U_{\rm int}^{\prime\prime({\rm max})}$ is the maximum value of the second derivative of $U_{\rm int}$ with respect to x.

If v is low enough, the tip position $x_{\rm min}$ is determined by the equilibrium condition $\partial U/\partial x = 0$, which, in our case, becomes

$$\eta \sin\frac{2\pi x}{a} + \frac{2\pi x}{a}(x - vt) = 0.$$

The tip moves slowly when the spring starts to be pulled along the direction x, but becomes faster and faster when the critical position $x_{\rm c}$ defined by $\partial^2 U/\partial x^2 = 0$ is approached and the equilibrium of the tip becomes unstable. The critical position is given by

$$x_{\rm c} = \frac{a}{2\pi}\arccos\left(-\frac{1}{\eta}\right), \tag{15.4}$$

and is reached at the time

$$t_{\rm c} = \frac{a}{2\pi v} f(\eta),$$

where

$$f(\eta) = \sqrt{\eta^2 - 1} + \arccos(-1/\eta). \tag{15.5}$$

The corresponding spring force

$$F_{\rm s} = \frac{ka}{2\pi}\sqrt{\eta^2 - 1} \tag{15.6}$$

can be seen as the *static friction* force acting on the tip. Note that the value of $F_{\rm s}$ defined by Eq. (15.6) is lower than the maximum force $F_{\rm max} = (ka/2\pi)\eta$. The maximum is reached slightly before hopping when the tip, which is rapidly accelerated, overcomes the velocity v of the spring support [312].

When $t = t_{\rm c}$ the tip slips into a new minimum position $x_{\rm min} = x'_{\rm c}$. If the damping coefficient γ is high enough, the tip will 'land' into the first minimum of the potential $U(x, t_{\rm c})$ after the 'take-off' position $x_{\rm min}$. Longer jumps observed in underdamped conditions are discussed in Section 15.4. In the limit $\eta \gg 1$ the landing position is $x'_{\rm c} = 5a/4$ whereas, if $\eta \to 1$, $x'_{\rm c}$ coincides with $x_{\rm c} = a/2$. At each jump an energy amount $\Delta U = U(x_{\rm c}) - U(x'_{\rm c})$ is released from the contact area. The (average) *kinetic friction* force can be simply estimated from the energy drop ΔU as [97]

$$F_{\rm k} = \Delta U/a, \tag{15.7}$$

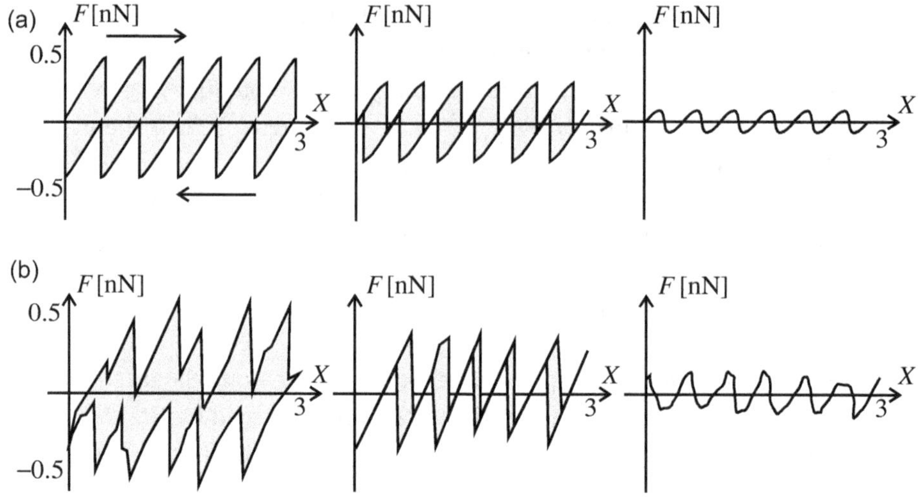

Figure 15.2 (a) Theoretical friction loops in the Prandtl–Tomlinson model when (left to right) $\eta = 5$, $\eta = 3$ and $\eta = 1$. (b) Experimental friction loops on NaCl(001) (Section 18.3) corresponding to normal force values of (left to right) 4.5, 3.3 and −0.5 nN. Adapted from [312] with permission from the American Physical Society.

whereas the variation of the spring force in the jump is given by $\Delta F = k(x_c' - x_c)$. After the jump, the stick–slip process is repeated again and again. If, at a certain point, the motion of the slider is suddenly inverted, the spring force F, plotted as a function of the support position X, defines a characteristic 'friction loop', as seen in Fig. 15.2(a). If η is reduced, ΔF will eventually become larger than F_s, meaning that the tip overcomes the support during the jump and F becomes negative for a short period. If $\eta < 1$ the equation $\partial^2 U/\partial x^2 = 0$ has no real solutions, and the motion becomes continuous. In this case, the kinetic friction force $F_k = 0$ (in the quasi-static limit that we are considering) and we may say that a state of *static superlubricity* has been reached.

The relations above are considerably simplified if $\eta \gg 1$. In this case it is not difficult to prove that the tip is essentially pinned, most of the time, and the spring force $F = k(vt - x)$ increases linearly with t as

$$F(t) \approx \frac{\eta}{1+\eta} kvt \approx kvt, \tag{15.8}$$

till the first jump occurs when $t_c = a\eta/2\pi v$.

Critical positions and kinetic friction

The values of the critical positions x_c and x_c' and of the kinetic friction force F_k can be approximated by series expansions [112]. If $\eta \to 1$:

$$x_{\mathrm{c}} = \frac{a}{2\pi}\left(\pi - \sqrt{2(\eta - 1)} + \frac{5}{6\sqrt{2}}(\eta - 1)^{3/2} - \cdots\right), \tag{15.9}$$

$$x'_{\mathrm{c}} = \frac{a}{2\pi}\left(\pi + 2\sqrt{(\eta - 1)} - \frac{19}{15\sqrt{2}}(\eta - 1)^{3/2} + \cdots\right). \tag{15.10}$$

The energy drop ΔU can be estimated by substituting the expressions (15.9) and (15.10) into (15.2), with $t = t_{\mathrm{c}}$, and expanding again. Dividing by the periodicity a, one gets for the kinetic friction force

$$F_{\mathrm{k}} = \frac{ka}{2\pi}\left(\frac{9}{4\pi}(\eta - 1)^2 - \frac{9}{5\pi}(\eta - 1)^3 + \cdots\right), \tag{15.11}$$

whereas, in a first approximation, the spring force variation in the jump is

$$\Delta F \approx 3\sqrt{2(\eta - 1)}\frac{ka}{2\pi}.$$

If $\eta \gg 1$ the take-off and landing positions of the tip are approximately given by

$$x_{\mathrm{c}} = \frac{a}{2\pi}\left(\frac{\pi}{2} + \frac{1}{\eta} + \frac{1}{6\eta^3} + \frac{3}{40\eta^5} + \cdots\right),$$

$$x'_{\mathrm{c}} = \frac{a}{2\pi}\left(\frac{5\pi}{2} - 2\sqrt{\frac{\pi}{\eta}} + \frac{1}{\eta} - \frac{\pi^{3/2}}{3\eta^{3/2}} + \cdots\right).$$

Using again (15.2), with $t = t_{\mathrm{c}}$, to estimate the energy drop ΔU, we obtain for the kinetic friction force:

$$F_{\mathrm{k}} = \frac{ka}{2\pi}\left(\eta - \pi + \frac{4}{3}\sqrt{\frac{\pi}{\eta}} - \frac{1}{2\eta} + \cdots\right), \tag{15.12}$$

whereas, in a first approximation, $\Delta F \approx ka$ independently of η. The relations (15.11) and (15.12) are plotted in Fig. 15.3(a).

15.2 Energy barrier

In Section 15.3 we will also need analytical expressions for the energy barrier ΔE preventing the tip jump in the PT model. This barrier is defined as

$$\Delta E = U(x_{\mathrm{max}}) - U(x_{\mathrm{min}}), \tag{15.13}$$

where x_{max} is the position of the first maximum after the position x_{min} where the tip is pinned (see Fig. 15.1). As long as the corrugation of the interaction potential U_{int} is large enough, the dependence of ΔE on the spring force F is given by the power law (*ramped creep*)

$$\Delta E \propto (\mathrm{const.} - F)^{3/2}.$$

However, the linear approximation (*linear creep*)

$$\Delta E \propto (\text{const.} - F)$$

is also in good agreement with experimental results at sufficiently low speed. In both cases, the coefficients of proportionality can be estimated by series expansions [112].

Ramped creep

If U_{int} has a sinusoidal shape the general expression for the time variation of ΔE is

$$\Delta E(t) \approx \frac{8U_0}{\eta(\eta^2-1)^{1/4}}\left(\frac{\pi v}{a}\right)^{3/2}(t_c - t)^{3/2}. \tag{15.14}$$

If $\eta \gg 1$ the dependence $\Delta E(F)$ follows the same power law:

$$\Delta E(F) \approx 2\sqrt{2}\left(1-\frac{F}{F_s}\right)^{3/2} U_0. \tag{15.15}$$

In Fig. 15.4 the approximated relation (15.15), with $\eta = 10$, is plotted with the exact dependence $\Delta E(F)$. Such a topological variation of the energy landscape, which is known as *fold catastrophe*, has been reported in other driven systems, e.g. superconducting quantum interference devices, mechanically deformed glasses, and stretched proteins [334].

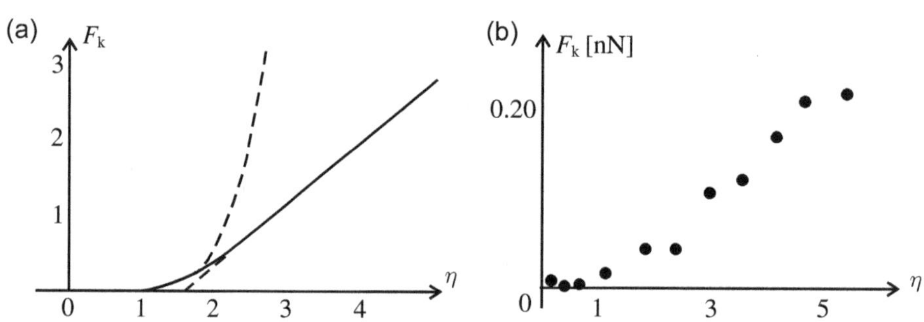

Figure 15.3 (a) Kinetic friction force F_k in the Prandtl–Tomlinson model as a function of the parameter η: the exact solution (continous curve) may be compared with the approximations (15.11) and (15.12) derived in the text when $\eta \to 1$ and $\eta \gg 1$ (dashed curves). (b) The experimental results for NaCl(001). Adapted from [312] with permission from the American Physical Society.

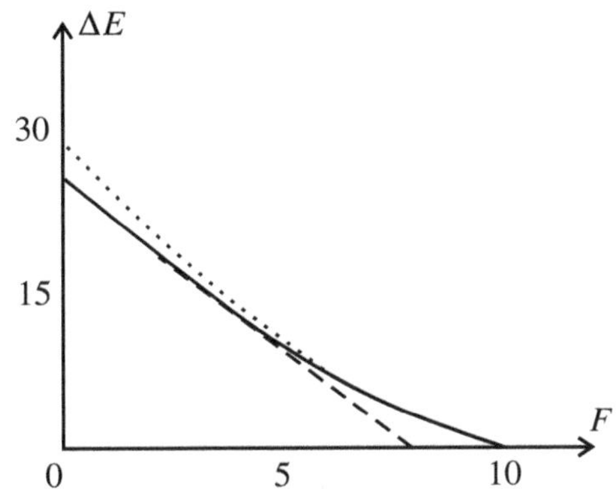

Figure 15.4 Ramped approximation (dotted curve) and linear approximation (dashed line) for the dependence $\Delta E(F)$ of the energy barrier if $\eta = 10$.

Linear creep

In the linear creep regime the choice of values of t or F around which the energy barrier can be approximated is completely arbitrary. Here we choose the symmetric configuration, where the energy barriers for forward and backward jumps are equal. If $\eta \gg 1$ it can be proven that

$$\Delta E \approx \left(2 + \frac{1}{2\eta}\right) U_0 - \frac{a}{2} F. \tag{15.16}$$

The relation (15.16) is represented by the dashed line in Fig. 15.4. Note that the constant slope of the $\Delta E(F)$ line is equal to $a/2$.

15.3 Thermal effects

At a finite temperature T, the spring force $F_s(T)$ inducing a jump is lower than F_s at 0 K. The actual values of $F_s(T)$ are not unique, and the most probable value taken by this force can be determined from the Kramers' theory already used in Section 14.4. In order to do that we introduce the probability p that the tip remains pinned. Ignoring reverse jumps, p changes with the time t according to the master equation (14.8). Assuming, as in Section 14.4, that $\gamma \sim \omega_0$, the attempt rate $f_0 \sim \omega_0/2\pi$, where $\omega_0 \approx \sqrt{\eta k/m}$ is the natural frequency of vibration of the point mass m in the variable potential well $U_{\text{int}}(x, t)$. If $\eta \gg 1$ the rate f_0 does not change significantly till the critical point is approached and f_0 rapidly drops to zero [302]. Since, according to Eq. (15.8), the force variation rate $\mathrm{d}F/\mathrm{d}t \approx kv$, we can express the time derivative of p as a function of F. Using the expression (15.16) for the energy barrier $\Delta E(F)$, the condition that the probability variation has a maximum, $\mathrm{d}^2 p/\mathrm{d}F^2 = 0$, yields

$$F_s(v, T) \approx \frac{2F_s}{\pi} + \frac{2k_B T}{a} \ln \frac{v}{v_0}, \tag{15.17}$$

where

$$v_0 = \frac{2 f_0 k_B T}{ka}.$$

If the energy barrier is approximated by the power law (15.15), one gets in a similar way:

$$F_s(v, T) \approx F_s \left[1 - \left(\frac{k_B T}{2\sqrt{2} U_0} \right)^{2/3} \left(\ln \frac{v_0}{v} \right)^{2/3} \right], \tag{15.18}$$

where

$$v_0 = \frac{\pi \sqrt{2}}{2} \frac{f_0 k_B T}{ka}.$$

Similar expressions, with F_s replaced by F_k and approximated by Eq. (15.12), are expected for the kinetic friction force $F_k(v, T)$.

At very low velocities or high temperatures the tip can repeatedly jump back and forth across the energy barrier. In this case it can be proven that the friction force is proportional to the sliding velocity [171]:

$$F_s(v, T) \approx \alpha(T) v, \qquad \alpha(T) \propto \frac{k}{2\pi f_0} \frac{2U_0}{k_B T} \exp\left(\frac{2U_0}{k_B T} \right). \tag{15.19}$$

Note that the 'equilibrium damping' coefficient $\alpha(T)$ in Eq. (15.19) is independent of γ. In practice, the regime predicted by Eq. (15.19) can be entered only if the driving velocity

$$v \ll v_c \equiv 2\pi f_0 a \exp\left(-\frac{2U_0}{k_B T} \right). \tag{15.20}$$

With typical values of $f_0 \sim 100$ kHz, $U_0 \sim 0.25$ eV, $k \sim 1$ N/m and $a \sim 0.5$ nm (see Section 18.4) such a 'thermolubric' regime would be observed, at room temperature, only for unrealistic values of $v \ll 1$ pm/s. However, the threshold value for the onset of thermolubricity increases rapidly with temperature. According to Eq. (15.20) $v_c = 1$ nm/s at $T = 200$ °C and $v_c = 10$ nm/s at $T = 300$ °C.

The distribution of the actual values of the static friction force at a given temperature can also be estimated from the expression [302]

$$P(\Delta E) = \frac{3(\Delta E/(k_B T))^{1/3}}{2(v/v_0)} \exp\left(-\frac{\Delta E}{k_B T} - \frac{e^{-\Delta E/(k_B T)}}{(v/v_0)} \right), \tag{15.21}$$

for the energy barrier, where

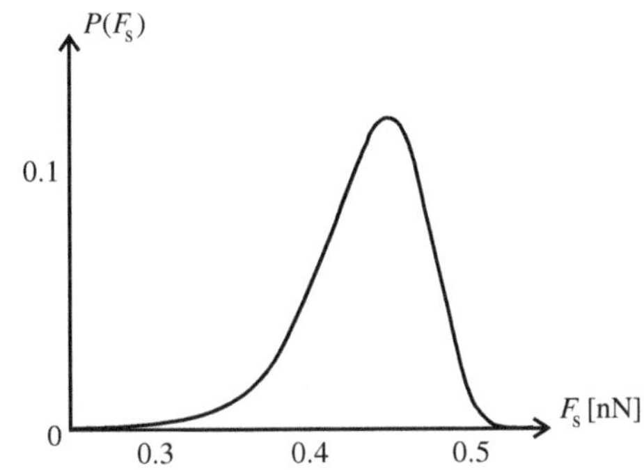

Figure 15.5 Distribution of the friction force in the PT model at room temperature, according to Eq. (15.21).

$$v_0 = \frac{k_B T \sqrt{\eta^2 - 1}}{2\gamma m a}.$$

Inserting the relation (15.15) for the dependence of $\Delta E(F)$ (with $F = F_s$ and $F_s = F_s(0)$), the distribution $P(F_s)$ turns out to be left-skewed, as seen in Fig. 15.5.

15.4 Long jumps

Depending on the value of the parameter η, other equilibrium positions can be observed beyond x'_c [214]. A jump across κ periodicities into the position $x = x'^{(\kappa)}_c$ is indeed possible if the conditions $\partial U/\partial x = 0$ and $\partial^2 U/\partial^2 x > 0$ are satisfied in at least κ points. It is not difficult to see that a new minimum appears when the parameter η exceeds one of the values η_k defined by $f(\eta_k) = \kappa\pi$. Substituting the expression (15.5) for the function $f(\eta)$, one gets $\eta_1 = 1$, $\eta_2 = 4.603$, and $\eta_3 = 7.790$ for the conditions of appearance of one, two and three minima respectively. In the same way as in Section 15.1 it can be proven that, if $\eta \gg 1$,

$$x'^{(\kappa)}_c = \frac{a}{2\pi}\left(\frac{\pi}{2} + 2\kappa\pi - 2\sqrt{\frac{\kappa\pi}{\eta}} + \frac{1}{\eta} - \frac{(\kappa\pi)^{3/2}}{3\eta^{3/2}} + \cdots\right),$$

whereas the kinetic friction force F_k is given by averaging the expression (15.12), with π replaced by $\kappa\pi$, over the distribution of κ-jumps. Still, the problem of determining which of the available minima will be actually occupied by the tip goes beyond the quasi-static model discussed so far.

The answer to this question depends on the damping coefficient γ in the equation of motion (15.1). Numerical solutions show that the η-γ plane can be divided into different regions, as shown in Fig. 15.6. At zero temperature a jump across κ periodicities occurs (for a fixed value of η) only if $\gamma_{c,\kappa}(\eta) < \gamma < \gamma_{c,\kappa-1}(\eta)$ (when

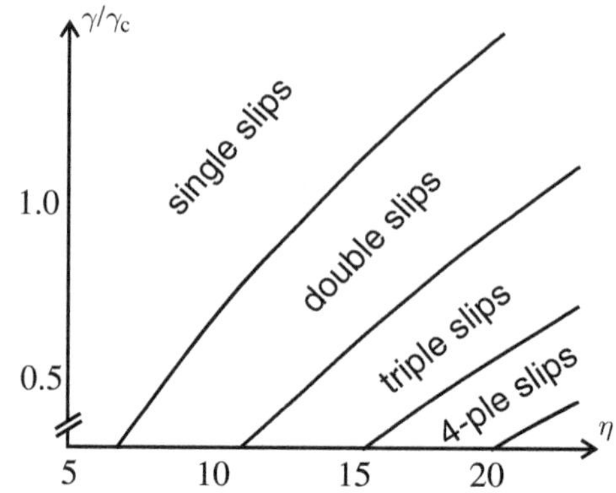

Figure 15.6 Depending on the values of the damping coefficient γ in Eq. (15.1) long jumps of different lengths are possible. Adapted from [112] with permission from the American Physical Society.

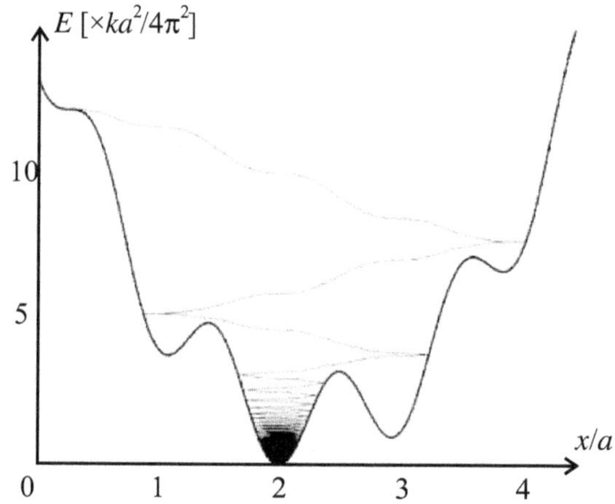

Figure 15.7 Potential profile $U(x)$ corresponding to a critical position (thick curve) and variation of the total energy E (thin curve) in the slip phase. Parameter values: $\gamma = 0.1\gamma_c$ with $\eta = 12.8$. Adapted from [112] with permission from the American Physical Society.

$\kappa > 1$) or $\gamma > \gamma_{c,1}(\eta)$ (when $\kappa = 1$), provided that $\gamma \gtrsim 1$. If $\gamma \lesssim 0.3\gamma_c$ the curves $\gamma_{c,\kappa}(\eta)$ start to bifurcate and a chaotic regime is established, in which the landing position can significantly change with little variations of η and γ. At finite temperature the curves separating the regions corresponding to jumps with different values of κ are smeared out and shifted towards larger values of η [112].

To get better insight into the 'landing' process at zero temperature, the total energy $E = (1/2)m\dot{x}^2 + U(x)$ can be plotted as a function of the actual point mass position when an underdamped long jump occurs, as done in Fig. 15.7. The particle bounces back and forth until it is conveyed by the potential profile $U(x)$ into the second minimum beyond the take-off position, where the particle stops. Had the damping coefficient been γ slightly larger, the tip would have ended up in another minimum.

15.5 Dynamic superlubricity

As seen in Section 15.1, the average kinetic friction F_k (at zero temperature and in the quasi-static limit) becomes zero if the parameter $\eta \leq 1$. However, low values of η are difficult to control. Here, we will show how a state of ultralow (average) friction can be also achieved for larger and in principle arbitrary values of η if the contact region is mechanically excited while sliding.

Suppose first that the tip oscillates perpendicularly to the plane of sliding at a given frequency ω. In this case we can assume that the amplitude U_0 of the interaction potential $U_{\rm int}$ varies with time as $U_0(1+\alpha)\sin\omega t$, where α is the relative amplitude of the oscillations. If ω is much larger than the 'washboard' frequency $2\pi(v/a)$, $U_{\rm int}$ may remain close to the minimum value $U_0(1-\alpha)$ long enough to observe a thermally activated jump. In this case, one can see that the dependence of F_k in Fig. 15.3(a) is retained if η is replaced by the effective parameter [313]

$$\eta_{\rm eff} = \eta(1-\alpha).$$

This means that the superlubric state will be reached for any value of η, provided that $\alpha > 1 - 1/\eta$.

To simulate transverse vibrations we add an oscillating term to the support coordinate in the PT model, so that the elastic energy stored in the driving spring reads

$$U_{\rm el}(x,t) = \frac{k}{2}(vt + \beta a \sin\omega t - x)^2. \tag{15.22}$$

In this case, if the driving frequency $\omega \gg 2\pi v/a$, it can be proven that the average friction force is approximately given by [88]

$$F_k(\eta,\beta) \approx \frac{ka}{2\pi}\left[\eta - \pi(1+2\beta) + \frac{4}{3}\sqrt{\frac{\pi}{\eta}}\left((1+\beta)^{3/2} - \beta^{3/2}\right)\right]. \tag{15.23}$$

The dependence of F_k on the oscillation amplitude β is plotted in Fig. 15.8 (filled circles). Introducing the noise term $\xi(t)$ in the equation of motion, the $F_k(\beta)$ curve is lowered due to thermally activated jumps, as shown by the empty circles in the figure.

15.6 Constant driving force

Suppose now that the point mass m in the Prandtl–Tomlinson model is pulled by a constant force F rather than by an elastic spring driven with constant velocity.

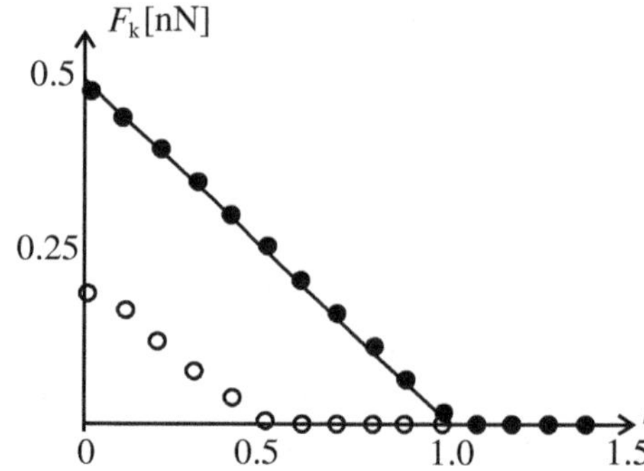

Figure 15.8 Average friction force F_k at increasing lateral oscillation amplitude. The results at $T = 0$ K (filled circles) are well reproduced by the continuous line defined by Eq. (15.23). The empty circles correspond to $T = 300$ K. Adapted from [294] with permission from AIP Publishing.

We first consider the underdamped limit $\gamma \ll \omega_0$, where $\omega_0 = (2\pi/a)\sqrt{U_0/m}$. If $F < F_{c1}$, where

$$F_{c1} = \frac{4\gamma}{\pi}\sqrt{mU_0}, \tag{15.24}$$

the particle is locked in the minimum of the total potential. The threshold value (15.24) is determined by equating the work done by the force F over a period a to the energy loss over the same period, which is obtained by integrating the power dissipation $m\gamma\dot{x}^2$ [36]. If $F > F_s$, where $F_s = 2\pi U_0/a$, the particle is in a state of steady sliding. For intermediate values of the driving force, $F_{c1} < F < F_s$, the particle can be either locked or running, depending on the initial conditions. In this sense the system is said to be *bistable*. However, this conclusion strictly holds only if the temperature $T = 0$. Thermal fluctuations can indeed 'kick' the particle out of the locked or the running state. If the fluctuations are infinitesimal, a new threshold force

$$F_{c2} \approx 3.4\frac{\gamma}{\omega_0}F_s$$

separates the two regimes [35, p. 295]. For larger values of γ, the relation between the critical forces F_{c1}, F_{c2} and the damping coefficient is shown in Fig. 15.9.

Consider now the stick–slip motion of the particle at finite temperature assuming that γ is large enough to avoid long jumps. According to Kramers' theory, if the energy barrier $\Delta E \approx 2U_0 - Fa/2 \gg k_BT$, the jump rates are given by [169]

$$w_\pm = f_0 \exp\left(-\frac{2U_0 - Fa/2}{k_BT}\right),$$

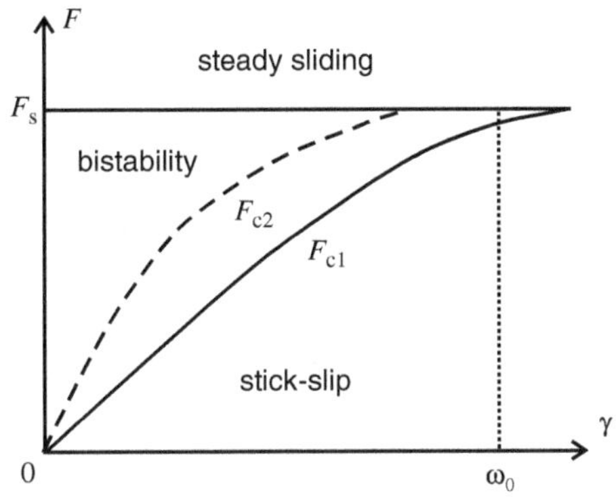

Figure 15.9 Phase diagram in the (F, γ) plane for a particle in an inclined sinusoidal potential. Reproduced from [35] with permission from Springer.

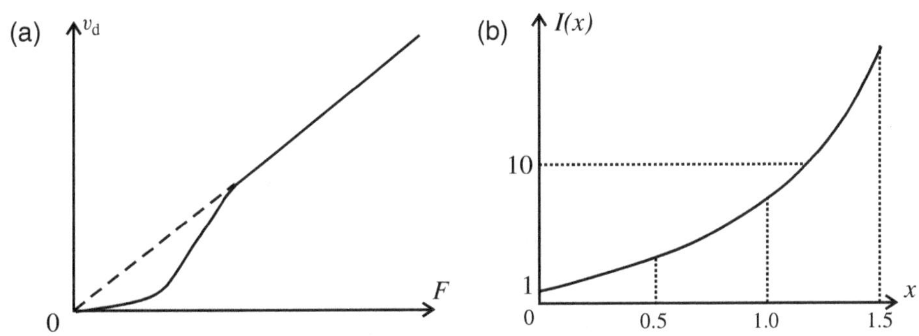

Figure 15.10 (a) Relation between drift velocity and driving force corresponding to Eq. (15.25). (b) The Risken–Vollmer integral (15.27).

where $f_0 \approx 2\pi U_0/ma^2\gamma$. As a result, the particle will advance with a drift velocity

$$v_d = a(w_+ - w_-) = af_0 e^{-2U_0/(k_B T)} \left(e^{Fa/(2k_B T)} - e^{-Fa/(2k_B T)} \right). \tag{15.25}$$

The relation $v_d(F)$ is represented by the continuous curve in Fig. 15.10(a). For very low values of F, the response of the system is linear:

$$v_d \approx \frac{f_0 a^2}{k_B T} e^{-2U_0/(k_B T)} F.$$

This is also the case (with a different slope) if $F \gg 4U_0/a$:

$$v_d \approx \frac{F}{m\gamma}. \tag{15.26}$$

The actual velocity of the particle, neglecting the thermal contribution from the random force $\xi(t)$, oscillates around the value (15.26) with a frequency $\omega = 2\pi v_d/a$ and an amplitude

$$\Delta v = \frac{2\pi U_0}{a\omega\sqrt{m^2\omega^2 + \gamma^2}}.$$

A general solution accounting for long jumps was obtained by Risken and Vollmer using Kramers' theory again [292]. If the equation of motion is reformulated as a Fokker–Planck equation for the distribution function of the point coordinate and velocity, the drift velocity and the driving force, in the limit of low damping, are found to be related by the equation

$$v_d = \frac{F}{m\gamma I(U_0/(k_B T))},$$

where the integral $I(x)$ is defined as

$$I(x) = \frac{1}{4\pi^{3/2}} \frac{\int_0^{2\pi} e^{x(1+\cos s)}\, ds}{\displaystyle\int_0^{\infty} \frac{e^{-u}}{\int_0^{2\pi} \sqrt{u + x(1+\cos s)}\, ds}\, du}. \tag{15.27}$$

The dependence (15.27) is plotted in Fig. 15.10(b).

15.7 The Frenkel–Kontorova model

The *Frenkel–Kontorova* (FK) *model* was first introduced to describe dislocations in solids [95] and subsequently applied in different contexts.[3] In surface science, the FK model is often used to interpret the physical behavior of adsorbed monolayers, especially in connection to competing incommensurate periodicities.

In the FK model a chain of particles of mass m connected by elastic springs interacts with a periodic potential, which mimics the structure of a crystal surface, as in the PT model (Fig. 15.11). The total potential thus reads

$$U_{tot} = \sum_n \left(-U_0 \cos\frac{2\pi x_n}{a} + \frac{1}{2}k(x_{n+1} - x_n - b)^2 \right),$$

where b is the equilibrium distance between two particles in the chain. Static friction can be probed by adding an external force F_{ext} adiabatically increasing till sliding initiates.

The ratio a/b is very important. For any *irrational* value of a/b there is a critical value η_c of the parameter η defined by (15.3) such that the static friction force F_s

[3] A detailed review of this model is given in the book by Braun and Kivshar [35].

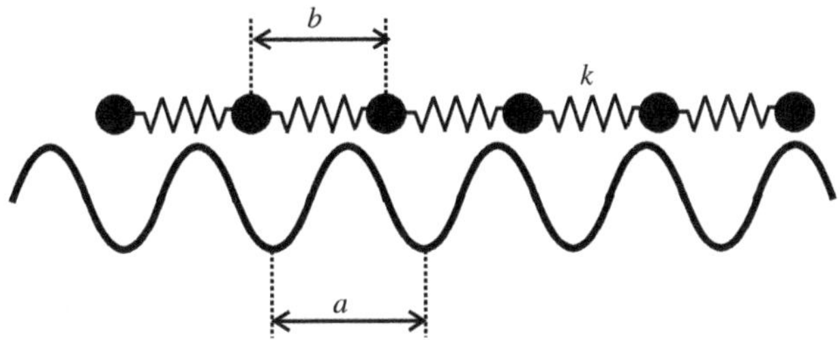

Figure 15.11 In the FK model, a chain of particles connected by springs moves over a periodic potential. The quantities a and b are the periodicity of the potential and the equilibrium distance of the springs respectively.

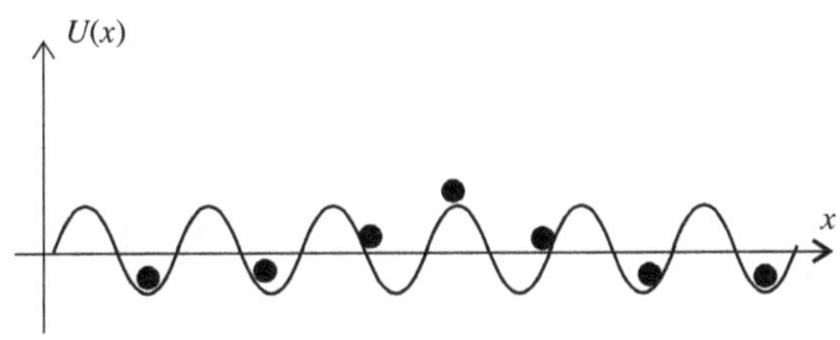

Figure 15.12 Adding an extra atom (and spring) in the FK model results in a so-called kink.

vanishes when $\eta < \eta_c$. The transition to a finite F_s is accompanied by a second-order phase transition to an incommensurate structure (*Aubry transition*) [268]. The parameter η_c takes a maximum value ($= 1$) when the ratio a/b is equal to the *golden mean* $(1 + \sqrt{5})/2 \approx 1.618$ [35]. If the average distance between consecutive particles is close to a and η is low, the position of the particle n in the chain is given by

$$x_n = na + \frac{a\xi_n}{2\pi},$$

where ξ_n is a small number. Considering n as a continuous variable, the motion of the chain can be described by the *sine-Gordon equation* [35]

$$\frac{\partial^2 \xi}{\partial t^2} = \frac{\partial^2 \xi}{\partial x^2} + \sin \xi,$$

which has important applications also in the theory of Josephson junctions and coupled pendula.

An interesting situation is observed when the chain has one particle more (or fewer) than the number of minima in the substrate potential. In this case, assuming periodic boundary conditions, a *kink* (or antikink) appears, as shown in Fig. 15.12. The motion of kinks in the presence of an external force has a key role in determining the frictional response of the chain. A kink can indeed be interpreted as an elementary excitation (*soliton*) with an effective mass $m_{\text{eff}} = (2m/\pi^2)\sqrt{\eta}$ and an energy

$$E = m_{\text{eff}}c^2 + \frac{1}{2}m_{\text{eff}}v^2,$$

where $c = a\sqrt{k/m}$ is the velocity of sound in the free chain. The activation energy for kink motion is called the *Peierls–Nabarro barrier* and is usually much smaller than the amplitude U_0 of the substrate potential. This means that kinks move much easier than the particles on the substrate (see also Section 12.1).

Combining lateral and normal motion

We will now extend the FK model to an interesting problem. A molecular chain, approximated by a series of units connected by equivalent springs (with stiffness k and equilibrium length b) interacting with the substrate via a sinusoidal potential of amplitude U_0 and period a, is pulled up from one of its ends by a weaker spring perpendicularly to the substrate surface (Fig. 15.13(a)). The interaction potential between each unit and the substrate can be written as

$$U_{\text{int}} = U_0(z)f(x, y) + U_1(z),$$

where the first term on the right hand side accounts for the corrugation of the potential energy parallel to the surface, and $U_1(z)$ describes the distance dependence of the (x, y)-averaged potential perpendicular to the surface.

This problem has been studied theoretically in the case of a polyfluorene chain on a Au(111) substrate [165], since this system was investigated by AFM at low temperature (Section 19.4). In this case, the interaction between the molecular units and the substrate is well described by realistic expressions for U_0, U_1 and f introduced by Steele [316]. As a result, the units are found to be sequentially detached from the surface with periodicity b. The normal force suddenly changes every time a unit is detached, as shown in Fig. 15.13(b). A small continuous modulation related to the periodicity of the substrate along the sliding direction is also observed, as seen in Fig. 15.13(c). This modulation is not periodic but shows extrema or shoulders which gradually shift with respect to the main variation and reflect the misfit between the chain structure (b-periodicity) and the atomic lattice of the Au(111) surface.

15.8 Electronic and phononic friction

To conclude the chapter, we briefly discuss the physical origins of the damping coefficient γ in Eq. (15.1). This quantity can be estimated using various techniques

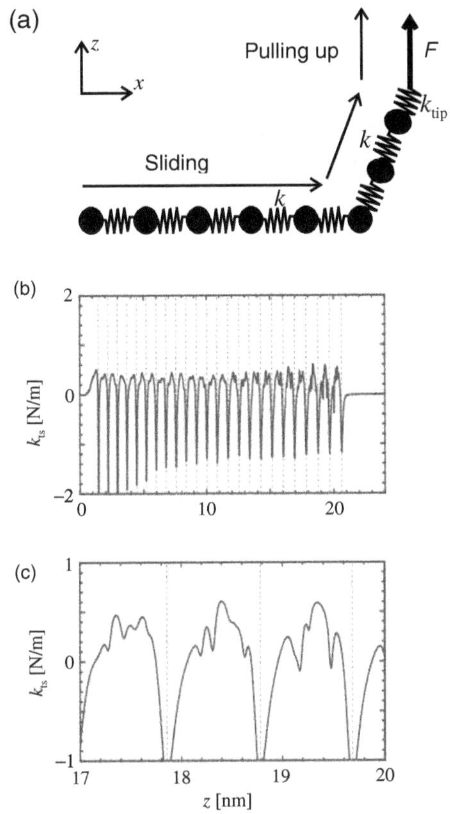

Figure 15.13 (a) Schematic of the interactions experienced during the pulling up of a molecular chain from a periodic substrate. (b) Simulated normal force gradient for a realistic choice of parameter values. (c) The zoom shows a small variation due to the sliding of the molecular chain on the substrate, which demonstrates practically frictionless motion due to the incommensurability of the substrate periodicity and the molecular spacing. Adapted from [165] with permission from the National Academy of Sciences, USA.

such as infrared spectroscopy, inelastic helium scattering, quartz crystal microbalance, and surface resistivity measurements.

Phononic friction

Consider an atom or a small molecule of mass m sliding on a substrate with a characteristic frequency ω_0 associated with parallel or perpendicular oscillations of the adsorbate. If we interpret an oscillation as a collision between adsorbate and substrate, it is possible to quantify the order of magnitude of the phononic contribution to γ. During the collision time $\tau \sim 1/\omega_0$ the adsorbate transmits an energy $(m/m_{\text{eff}})\varepsilon$ to the substrate, where ε is the oscillation energy and the effective mass m_{eff} of the substrate is easily determined as follows. The displacement field extends up to a distance $\sim c_s\tau$, where c_s is the transverse velocity of sound in the substrate. The corresponding volume is $\sim (c_s\tau)^3$ so that $m_{\text{eff}} \sim \rho(c_s\tau)^3$, where ρ is the substrate density. Since the collisions occur with a frequency $f = \omega_0/2\pi$, the energy transfer per unit time is

$$\frac{\mathrm{d}\varepsilon}{\mathrm{d}t} \approx \frac{\omega_0}{2\pi}\frac{m}{m_{\text{eff}}}\varepsilon,$$

which correspond to an energy decay with damping constant[4]

$$\gamma_{\text{phon}} \sim \frac{m\omega_0^4}{\rho c_s^3}. \tag{15.28}$$

Smaller values are obtained for a large molecule sliding on the surface. Note that Eq. (15.28) is expected to remain valid also for liquid adsorbates at low coverage, but not for solid adsorbate layers.

Electronic friction

Consider now a gas of atoms or small molecules adsorbed on a thin metal film. In case of covalent bonding (*chemisorption*) a resonance state appears close to the Fermi level ε_F of the metal. If Γ is the resonance width and ρ_a is the corresponding density of states, the contribution to electronic friction is [242]

$$\gamma_{\text{el}}^{(\text{cov})} = 2\frac{m_e}{m}\omega_F\Gamma\rho_a(\varepsilon_F)\left\langle\sin^2(\theta)\right\rangle,$$

where $\omega_F = \varepsilon_F/\hbar$ and the average depends on the symmetry of the adsorbate orbital. In case of van der Waals bonding (*physisorption*), the following relation has been derived by considering the metal as a semi-infinite jellium[5] and the molecule–substrate interaction in the dipole approximation [261]:

$$\gamma_{\text{el}}^{(\text{vdW})} = \frac{e^2}{\hbar a_0}\frac{(k_F\alpha)^2}{(k_F z)^{10}}\frac{m_e}{m}\frac{\omega_F}{\omega_p}k_F a_0 I(z), \tag{15.29}$$

where a_0 is the Bohr radius, α is the static electric polarizability, ω_p is the plasma frequency, k_F is the Fermi wave vector and I is a function of the distance z between the adsorbate and the jellium edge. Since I is constant except at very low distances ($\sim$ 0.1 nm) the vdW contribution decreases as z^{-10}. In contrast, the Pauli repulsion appearing when the electron clouds of adsorbate and substrate start to overlap

[4] According to a more accurate estimation the right hand side of Eq. (15.28) is multiplied by a factor $(3/8\pi)$ [256].

[5] In the jellium model the crystal structure is ignored and the electrons are supposed to interact with a positive charge uniformly distributed in the metal.

results in an exponential decay with the distance z. Equation (15.29) can be applied to light noble gases and saturated hydrocarbons on metals.

Note that, experimentally, the electronic damping can be simply estimated from the increase $\Delta\rho$ in the resistivity of a thin film as

$$\gamma_{\mathrm{el}} = \frac{n^2 e^2 d}{m n_{\mathrm{a}}} \Delta\rho,$$

where d is the film thickness, n is the density of conduction electrons and n_{a} is the number of adsorbates per unit area. For chemisorbed systems $\gamma_{\mathrm{el}} \sim 10^{10}$–$10^{12}\ \mathrm{s}^{-1}$, while $\gamma_{\mathrm{el}} \sim 10^{8}$–$10^{9}\ \mathrm{s}^{-1}$ for physisorbed systems in the low-coverage limit. These values are lowered if the coverage increases.

16

Atomic-scale stick–slip in two dimensions

In this chapter we will discuss how the magnitude of the static and kinetic friction on a particle elastically driven on a crystal surface is influenced by the pulling direction. The problem is more complex if, instead of a point mass, we consider the motion of a 2D crystal. Depending on the size of the sliding system, its stiffness and commensurability with the substrate, significant variations are expected. An interesting problem is also the motion of an adsorbate film driven by a constant force. In this case the film can solidify in different phases depending on the coverage and the adsorbate–substrate interaction strength.

16.1 The Prandtl–Tomlinson model in two dimensions

When a nanoasperity ('tip') is elastically driven across a 2D crystal lattice, both the spring force at the slip onset, i.e. the static friction F_s, and the spring force averaged over long sliding distances, i.e. the kinetic friction F_k, depend on the pulling direction φ (Fig. 16.1(a)). To estimate the dependence of $F_s(\varphi)$, the first step is to determine the equilibrium position $\mathbf{r} \equiv (x, y)$ of the tip as a function of the position $\mathbf{R} \equiv (X, Y)$ of the spring support. If $U(\mathbf{r}; \mathbf{R})$ is the sum of the tip–surface interaction potential $U_{\text{int}}(x, y)$ and the elastic potential $U_{\text{el}} = (1/2)k(\mathbf{r} - \mathbf{R})^2$, where k is the spring constant, the equilibrium condition is simply

$$\nabla U = 0, \tag{16.1}$$

where the gradient is calculated with respect to the coordinates x, y for a fixed support position. Assuming that the support moves with time as $\mathbf{R} = \mathbf{v}t$, where $\mathbf{v}$ is the (constant) driving velocity, the value F_s is reached when the equilibrium of the system becomes unstable, i.e. when at least one of the eigenvalues $\lambda_{1,2}$ of the Hessian matrix

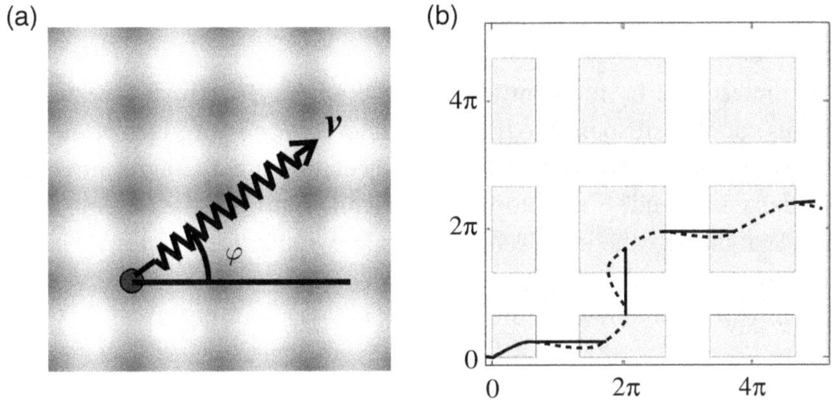

Figure 16.1 (a) Surface potential with square symmetry (16.3). (b) Stability regions (gray areas), trajectories (continuous curve) and equilibrium positions of a particle elastically driven on it (dashed curve).

$$\mathcal{H} = \begin{pmatrix} \dfrac{\partial^2 U_{\text{int}}}{\partial x^2} + k & \dfrac{\partial^2 U_{\text{int}}}{\partial x \partial y} \\ \dfrac{\partial^2 U_{\text{int}}}{\partial y \partial x} & \dfrac{\partial^2 U_{\text{int}}}{\partial y^2} + k \end{pmatrix} \tag{16.2}$$

becomes negative.

As an example Fig. 16.1(b) shows the stability regions for a simple potential U_{int} obtained by overlapping two plane waves (with periodicity a) rotated by 90° with respect to each other:

$$U_{\text{int}} = -U_0(\cos x + \cos y) \tag{16.3}$$

(for the sake of simplicity, we have replaced $2\pi x/a$ and $2\pi y/a$ with x and y in the argument of the trigonometric functions). It is straightforward to see that the stability regions are defined by the conditions $1 + \eta \cos x > 0$ and $1 + \eta \cos y > 0$, where η is defined as in Section 15.1. Analytic expressions for $F_s(\varphi)$ can be in principle found for any shape of the interaction potential $U_{\text{int}}(x, y)$.

As opposed to the static friction, the kinetic friction F_k can in principle be determined only numerically by solving the equation of motion (15.1) generalized to 2D:

$$m\ddot{\mathbf{r}} + m\gamma\dot{\mathbf{r}} + k(\mathbf{r} - \mathbf{v}t) + \nabla U_{\text{int}}(\mathbf{r}) = 0.$$

The interaction potential (16.5) is a noticeable exception, as outlined below.

Square lattices

If U_{int} is defined by Eq. (16.3) the condition $\nabla U = 0$ at different times t leads to the implicit relation

$$\frac{y + \eta \sin y}{x + \eta \sin x} = \tan \varphi, \tag{16.4}$$

which is represented by the continuous curve in Fig. 16.1(b). Inside the stability regions this curve corresponds to the tip trajectory. When the edges of these regions are reached, it can be shown that the tip is ejected along a straight line parallel to the x or the y axis and, if the motion is critically damped, the landing position is simply determined by the intersection of the line with the curve (16.4) [318].

On a surface potential that has large corrugations ($\eta \gg 1$), simple analytical expressions can be obtained. In this case Eq. (16.4) simplifies to

$$\frac{\sin y}{\sin x} = \tan \varphi \tag{16.5}$$

and the spring force $\mathbf{F} \equiv -\nabla U = \nabla U_{\text{int}}$ or, in components,

$$F_x = \frac{ka}{2\pi}\eta \sin x, \qquad F_y = \frac{ka}{2\pi}\eta \sin y.$$

The static friction can be expressed as a function of the pulling direction using Eq. (16.5):

$$F_s(\varphi) = \left.\sqrt{F_x^2 + F_y^2}\right|_{x=x_c} = \frac{ka}{2\pi}\frac{\eta}{\cos \varphi} \tag{16.6}$$

if $0° < \varphi < 45°$, or

$$F_s(\varphi) = \frac{ka}{2\pi}\frac{\eta}{\sin \varphi}$$

if $45° < \varphi < 90°$ (see Fig. 16.2). The force F_s has a minimum when $\varphi = 0°$ and a maximum when $\varphi = 45°$, which shows up as a cusp.

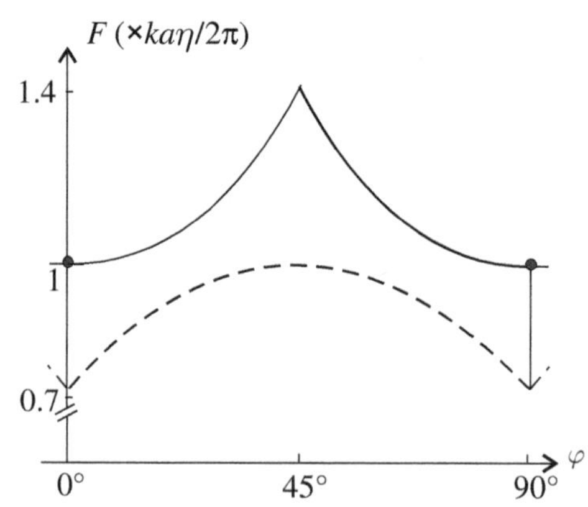

Figure 16.2 Angular dependence of the static (continuous curve) and kinetic friction (dashed curve) corresponding to the potential in Fig. 16.1(a) and evaluated according to the equations (16.6) and (16.7).

Due to the absence of cross terms in the potential (16.3), the average values of the x and y components of the spring force are the same: $\overline{F_x} = \overline{F_y}$ (if $\varphi \neq 0°$, 90°). When $\eta \gg 1$, these values are equal to $ka/2\pi$. Since the kinetic friction is given by the projection of the average spring force $\overline{\mathbf{F}}$ along the pulling direction, it is easy to see that

$$F_k = \frac{ka}{2\pi}\eta\cos(45° - \varphi). \tag{16.7}$$

For both static and kinetic friction forces, the ratio between the extremal values is $F_{\max}/F_{\min} = \sqrt{2}$, and a considerable anisotropy is expected.

Hexagonal lattices

The simplest surface lattice with hexagonal periodicity is defined by overlapping three plane waves, rotated by 60° (Fig. 16.3):

$$U_{\text{int}} = -\frac{U_0}{2}\left[\cos\left(x - \frac{y}{\sqrt{3}}\right) + \cos\left(x + \frac{y}{\sqrt{3}}\right) + \cos\left(\frac{2y}{\sqrt{3}}\right)\right]. \tag{16.8}$$

The same procedure adopted for the square lattice leads to precise analytical expressions for the angular dependence of the static friction also in this case [106]. In particular, when $\eta \gg 1$:

$$F_s = \frac{ka}{2\pi}\frac{\eta}{\sqrt{3}}\left(\cos^2 x + 4\cos x\cos\frac{y}{\sqrt{3}} + 7\cos^2\frac{y}{\sqrt{3}} - 4\cos^2 x\cos^2\frac{y}{\sqrt{3}}\right.$$
$$\left. - 4\cos x\cos^3\frac{y}{\sqrt{3}} - 4\cos^4\frac{y}{\sqrt{3}}\right), \tag{16.9}$$

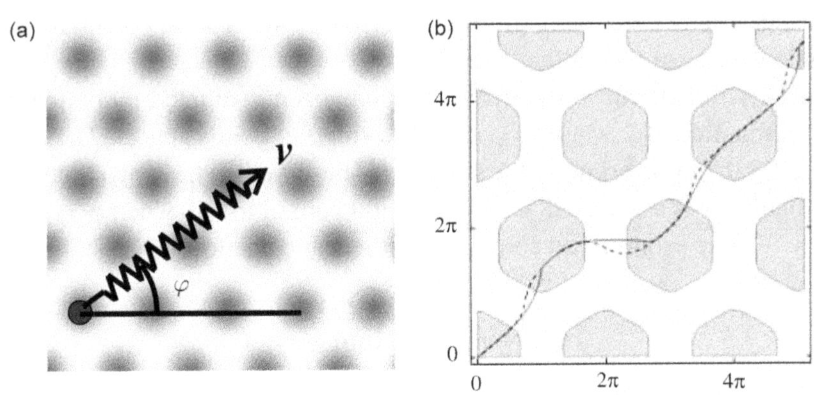

Figure 16.3 Same as Fig. 16.1 for the surface potential with hexagonal symmetry (16.8).

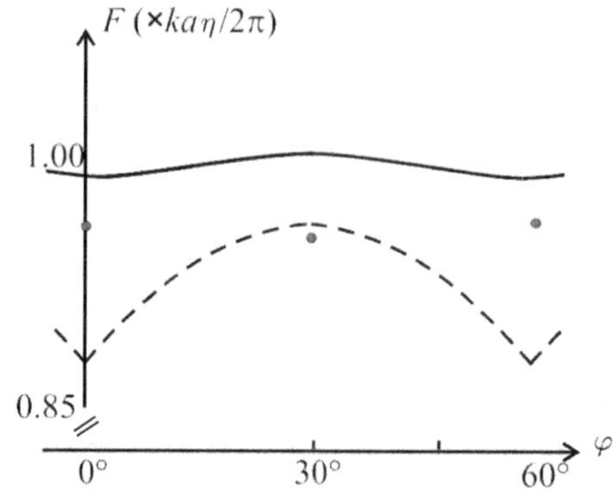

Figure 16.4 Angular dependence of the static (continuous curve) and kinetic friction (dashed curve) corresponding to the surface potential in Fig. 16.3(a).

where, at the critical position,

$$\cos x = \cos\frac{y}{\sqrt{3}} - 2\cos^3\frac{y}{\sqrt{3}} \pm \sqrt{1 - 4\cos^4\frac{y}{\sqrt{3}} + 4\cos^6\frac{y}{\sqrt{3}}}. \tag{16.10}$$

The tip trajectory (when $\eta \gg 1$) is given by

$$\tan\varphi = \frac{\sin\dfrac{y}{\sqrt{3}}}{\sin x\cos\dfrac{y}{\sqrt{3}}}\left(\cos x + 2\cos\frac{y}{\sqrt{3}}\right), \tag{16.11}$$

and an explicit but cumbersome representation of F_s vs. φ is again possible. From the expressions (16.9)–(16.11), calculated for $\varphi = 0°$ and $\varphi = 30°$, the ratio between the extremal values of F_s turns out to be $F_{max}/F_{min} \approx 1.04$. Small variations of $F_s(\varphi)$ are also estimated for lower values of η. Thus, as opposed to the case of a square lattice, the static friction force on the hexagonal potential (16.8) is almost independent of the pulling direction.

In Fig. 16.4 the kinetic friction force F_k has been evaluated numerically and plotted as a function of φ when $\eta = 100$. Two singular points appear. The first one, at $\varphi = 0°$, has the same origin as the singularity observed on the square lattice. The second singularity is associated with the straight path of the tip when $\varphi = 30°$. The anisotropy is more pronounced compared to the static friction.

16.2 Structural lubricity

As shown by Müser *et al.* in a simple model based on geometrically interlocking asperities, the contact between two commensurate surfaces results in a finite static friction force F_s [228]. The value of F_s decreases exponentially with the length of

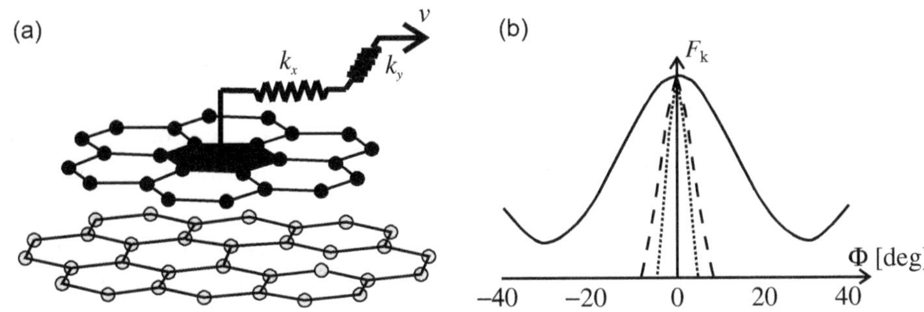

Figure 16.5 (a) A rigid flake consisting of $N = 24$ atoms is connected by two springs to a support moving in the x direction. The substrate is modeled as an infinite layer of rigid graphene. (b) Friction as a function of the orientation angle for symmetric flakes with size $N = 6$ (continuous curve), 64 (dashed curve), and 150 (dotted curve). Adapted from [335] with permission from the American Physical Society.

the common period and is further reduced if the substrates are not rigid. If the contacting surfaces are incommensurate, and the substrates are rigid, the static friction $F_s = 0$. If the substrates are not rigid, a transition from stick–slip to a state of negligible friction is expected when the load decreases or the stiffness increases [226]. In this case the lateral forces on the single atoms cancel out, leading to a dramatic reduction of friction at finite speed v, and to complete vanishing when $v \to 0$. Although this situation is often called 'superlubricity', following Müser [227] we prefer to use the term *structural lubricity* to describe this state of motion and avoid questionable analogies to superconductivity and superfluidity.

Based on the previous reasoning, it is not surprising that the friction varies significantly with the relative orientation of the contacting surfaces and the pulling direction. As an example, consider a rigid hexagonal flake sliding over a hexagonal lattice according to the geometry in Fig. 16.5(a) [335]. The friction force peaks at the values of the orientation angle Φ corresponding to a commensurate contact between flake and surface.[1] Between two consecutive peaks, ultralow friction is observed (Fig. 16.5b). Note that the angular width $\Delta\Phi$ of the friction peaks depends on the size of the flake as $\tan(\Delta\Phi) = a/d$, where d is the diameter of a flake, and a is the lattice constant.

If two amorphous but smooth rigid surfaces slide past each other, numerical simulations by Müser *et al.* showed that $F_s \propto \sqrt{N}$, where N is the number of atoms at the interface [228]. In this way the friction force per atom tends to zero if the area of contact becomes infinite.

[1] Note that Φ is different from the pulling angle, which is fixed at 0°.

Colloidal systems

The kinks and antikinks predicted by the FK model in 1D (Section 15.7) are also expected in 2D. In an original experiment, Bohlein *et al.* succeeded in visualizing these effects in real time on a 2D crystal of charged polystyrene spheres suspended in water and driven across commensurate and incommensurate static potentials [27]. The potentials were generated by interfering laser beams, which allowed them to change the symmetry of the potential at will. The external (lateral) force $F_{\rm ext}$ was simply accounted for by translating the sample cell at constant velocity. The frictional response of the colloidal particles was found to depend only on the number and density of the kinks, which propagated through the monolayer along the direction of the applied force. Remarkably, these excitations were also observed on quasi-periodic potentials.

16.3 Sliding of adsorbate layers

At a finite coverage θ, the sliding of an adsorbate layer can only be investigated (theoretically) by numerical simulations. The goal is to solve the equation of motion

$$m\ddot{\mathbf{r}}_i + m\gamma\dot{\mathbf{r}}_i = -\frac{\partial U_{\rm int}}{\partial \mathbf{r}_i} - \frac{\partial U_{\rm ad}}{\partial \mathbf{r}_i} + \mathbf{F}_{\rm ext}, \tag{16.12}$$

where m is the mass of an adsorbate particle, $U_{\rm int}$ is a periodic potential describing the interaction between adsorbate and substrate, $U_{\rm ad}$ is a potential describing the interaction of the adsorbate particles (a sum of LJ pair potentials) and $\mathbf{F}_{\rm ext}$ is an external force. At a finite temperature T a stochastically fluctuating force is added to the right hand side of Eq. (16.12). Depending on $U_{\rm int}$, θ and T different 'phases' of the adsorbate layer can be distinguished. Two examples of incommensurate and commensurate solid structures on a square lattice are shown in Fig. 16.6 [243]. Incommensurability is observed if the corrugation of $U_{\rm int}$ is low. While an incommensurate structure can slide with negligible activation barriers, a commensurate structure is strongly pinned.

The dependence of the drift velocity $v_{\rm d}$ on the average stress $\sigma = F_{\rm ext}/\theta$ can have two qualitatively different forms [243]. Figure 16.7(a) shows the relation $v_{\rm d}(\sigma)$ for an adsorbate layer in the fluid state. This is the same response as that expected in the dilute limit (Fig. 15.10(a)). Note that this behavior is always observed in some parts of the (θ, T) phase diagram. The drift velocity is non-zero for arbitrary small values of σ, which is what is expected for a fluid. Furthermore, no hysteresis is observed. If the adsorbate layer is in a solid state which is commensurate with the substrate, or at least pinned by it (see below), the behavior in Fig. 16.7(b) is observed. In this case, a minimum stress $\sigma_{\rm a}$ is required to initiate

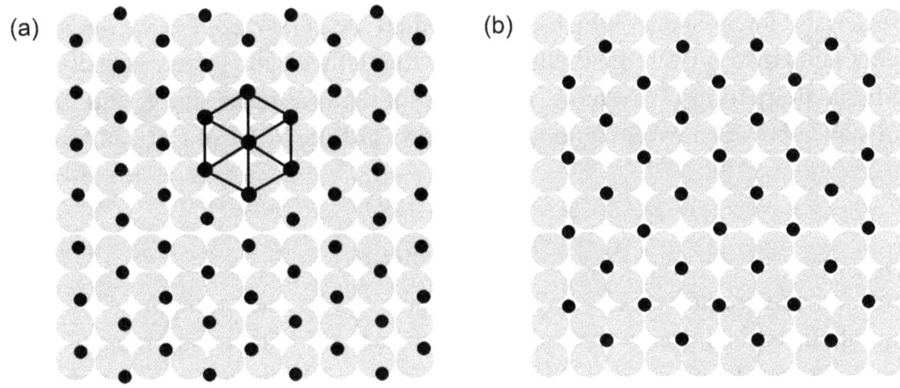

Figure 16.6 (a) Incommensurate and (b) commensurate solid structures formed by an adsorbate monolayer on a square lattice. Adapted from [243] with permission from the American Physical Society.

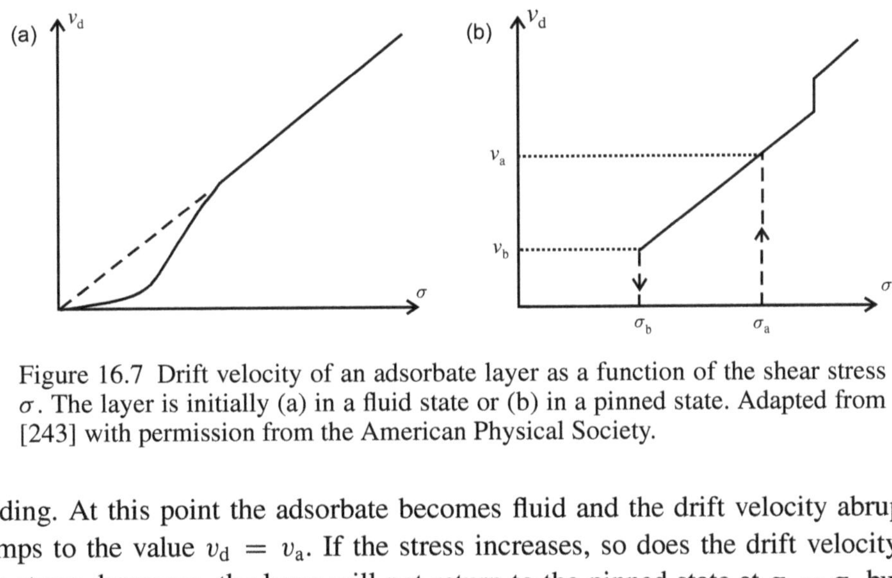

Figure 16.7 Drift velocity of an adsorbate layer as a function of the shear stress σ. The layer is initially (a) in a fluid state or (b) in a pinned state. Adapted from [243] with permission from the American Physical Society.

sliding. At this point the adsorbate becomes fluid and the drift velocity abruptly jumps to the value $v_d = v_a$. If the stress increases, so does the drift velocity. If the stress decreases, the layer will not return to the pinned state at $\sigma = \sigma_a$ but to that at a lower value σ_b. Two different effects contribute to this hysteretic behavior. A resolidification at $\sigma = \sigma_a$ is prevented (i) by the increased temperature of the adsorbate while sliding and (ii) by the drag force exerted by the rest of the fluid on the nucleation islands. Considering only the latter effect, it can be proven that $\sigma_b \approx \sigma_a/2$ [243].

Role of defects

An incommensurate solid structure can also be pinned by a defect, e.g. by a step or a foreign chemisorbed atom. It turns out that the pinned structures have a characteristic linear size ξ, which may be called the *elastic coherence length*. This concept is mediated from the theory of flux line lattices and charge density waves and, by analogy to those systems, if the 2D solid has a linear size L and the average distance between two neighboring defects is of the order of l, it can be proven that [258]

$$\xi \approx \frac{mc^2 l}{U_{\text{def}}\sqrt{4\pi \ln(L/\xi)}},$$

where c is the sound velocity in the film and U_{def} is the strength of the defect potential. In typical quartz crystal microbalance (QCM) measurements on inert gas monolayers (Section 17.3), ξ is expected to be of the order of a few tens of mm [295, section 8.5].

17

Instrumental and computational methods in nanotribology

In this chapter we introduce the methods conventionally used to explore friction on the nanoscale. The leading position among the instrumental setups is held by the atomic force microscope. Here we will briefly illustrate the type of forces sensed by this instrument and its basic modes of application. Other experimental techniques in nanotribology are the surface force apparatus, the quartz crystal microbalance and also, to some extent, scanning tunneling microscopy and transmission electron microscopy. Virtual experiments rely on molecular dynamics simulations. A short introduction to this method will be followed by a series of numerical results reproducing friction and wear measurements at the atomic level.

17.1 Atomic force microscopy

In a typical *atomic force microscope* (AFM) [24] a sharp micro-fabricated tip is scanned over a surface. Standard AFM tips are made of silicon or silicon nitride, but tips can be also coated to allow a large variety of material combinations. The probing tip is attached to a cantilever force sensor, the sensitivity of which can be well below 1 nN. Images of the surface topography are recorded by controlling the tip–sample distance in order to maintain a constant (normal) force. This is made possible by using piezoresistive cantilevers, or, most commonly, by a light beam reflected from the back side of the cantilever into a photodetector, which allows one to monitor the cantilever bending (Fig. 17.1). The lateral force between tip and surface is responsible for the cantilever torsion and can be measured if the photodetector is equipped with four quadrants. If this is the case the AFM can be used as a *friction force microscope* (FFM), see Appendix A. The design of a home-built AFM, optimized for friction measurements in ultra-high vacuum (UHV), is shown in Fig. 17.2.

The tip–sample force can be related not only to the static bending or torsion of the cantilever. In dynamic AFM techniques [217] the cantilever is excited in

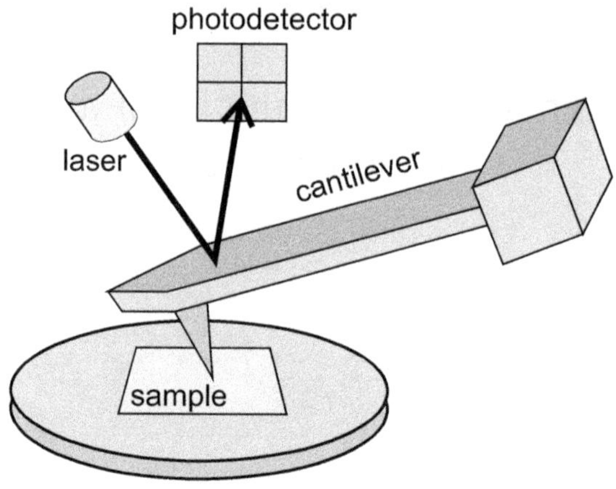

Figure 17.1 Schematic diagram of a beam-deflection atomic force microscope.

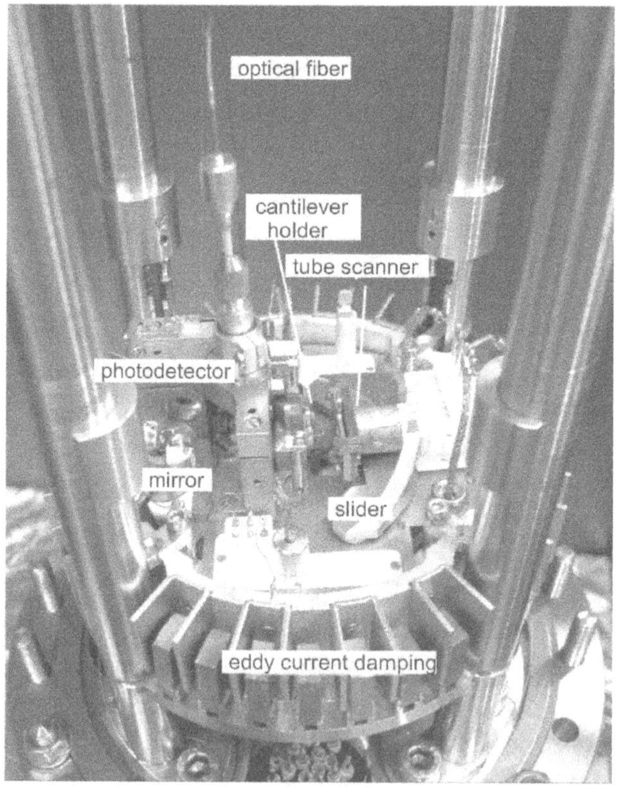

Figure 17.2 UHV-AFM designed and installed at the University of Basel (photo by Dr. Oliver Pfeiffer).

the vicinity of a mechanical resonance, and the tip–sample interaction is estimated from the oscillation amplitude or the shift of the resonance frequency. In this way sliding contact between tip and sample is avoided, and damage to tip and surface is considerably reduced.[1]

The imaging process in AFM takes place continuously above the surface. The tip is usually scanned at a constant velocity forwards and backwards in the so-called 'fast' scan direction, then the motion is stopped, the tip is displaced by a short distance along the 'slow' scan direction, which is perpendicular to the fast scan direction, and the process is repeated several times to produce a two-dimensional topography or force map.

Relevant forces in AFM

The most important interactions in AFM are due to short-range chemical forces. Chemical forces are sensitive to single atoms and are responsible for atomic resolution. They also define the atomic structure of tip and surface, and cause atomic displacements when the tip is brought in close proximity to the surface.

The van der Waals (vdW) forces are due to the electromagnetic interaction of fluctuating dipoles in the atoms forming tip and surface. These forces are extremely weak on the atomic level. However, they are generally attractive, which can result in forces of several nanonewtons when the small interactions between individual atoms of tip and sample are summed up. In this way the vdW forces can exceed the chemical forces and dominate the tip–surface interaction. Van der Waals forces are always present independently of the tip and surface conditions or the environmental conditions of the experiment. In the case of a sphere close to a flat surface, the vdW force is given by [148]

$$F_{\mathrm{vdW}}(z) = \frac{HR}{6z^2},$$

where H is the *Hamaker constant* (dependent on the materials, and usually of the order of 10^{-19} J), R is the tip radius, and z is the distance between tip and surface. In the case of a conical tip terminated by a spherical cap and a flat surface [124]:

$$F_{\mathrm{vdW}}(z) = -\frac{H}{6}\left(\frac{R^2}{z^2} + \frac{\tan\alpha^2}{z+R_\alpha} - \frac{R_\alpha}{z(z+R_\alpha)}\right),$$

where α is the half-angle of the cone and $R_\alpha = R(1-\sin\alpha)$.

The cleavage process used to prepare atomically flat surfaces often results in charges trapped at the sample surface. Other surface preparation techniques such as ion sputtering in UHV may also lead to charging effects. If localized charges are present at the tip apex, electrostatic forces are generated, the strength and distance dependence of which is given by the Coulomb's law. Electrostatic forces are also

[1] This may not be the case if a contact resonance is excited while the sliding is occurring; see Section 18.2.

acting between charged surfaces and conductive tips. Considering the tip–surface system as a capacitor with a distance-dependent capacitance C, these forces are given by

$$F_{\rm el} = \frac{\partial C}{\partial z}(V_{\rm bias} - V_0)^2,$$

where $V_{\rm bias}$ is the voltage applied between tip and surface and V_0 is the contact potential difference produced by the different work functions of tip and surface.

Other contributions to the tip–surface interaction may come from magnetic forces and capillary forces.

17.2 Other scanning probe modes

Dynamic-mode AFM

In *non-contact* (NC) *mode* the probing tip of an AFM oscillates with an amplitude of a few nm at the resonance frequency f_0 of the cantilever. The oscillation is usually applied by a piezoactuator mounted at the cantilever base. Silicon cantilevers with normal spring constants $k_{\rm N}$ of few tens of N/m and resonance frequencies $f_0 \sim 10^5$ Hz are typically used, with corresponding Q factors $\sim 10^4$ (in UHV).

The system formed by cantilever and tip can be represented as a damped harmonic oscillator:

$$m\ddot{z} = -k_{\rm N}\left[z - A_{\rm exc}\cos(\omega t + \varphi)\right] - m\gamma\dot{z} + F(z), \tag{17.1}$$

where m is the effective mass of the cantilever, z is the vertical position of the tip, the damping coefficient γ is related to the internal friction of the material, and $F(z)$ is the force between tip and sample surface. For rectangular cantilevers m is approximately one fourth of the cantilever mass [125]. The excitation has an amplitude $A_{\rm exc}$ and the response has a phase lag φ. The (normal) friction force $-m\gamma\dot{z}$ is compensated by the driving force $F_{\rm exc} = kA_{\rm exc}\cos(\omega t + \varphi)$, so that Eq. (17.1) simplifies to

$$m\ddot{z} = -k_{\rm N}z + F(z).$$

The excitation needed to keep the oscillation amplitude can be seen as the *damping signal* of the NC-AFM.

If the tip oscillations are small compared to the characteristic decay length of $F(z)$, a linear expansion of $F(z)$ is possible, leading to a proportionality relation between the shift Δf of the resonance frequency and the force gradient in the z direction [3]:

$$\frac{\Delta f}{f_0} = -\frac{1}{2k_{\rm N}}\frac{{\rm d}F}{{\rm d}z}.$$

If this is not the case the interaction between tip and sample modifies the harmonic motion only close to the lower turning point of the tip. Assuming that the tip oscillations have an amplitude A, and the frequency shift Δf is small, Giessibl derived the following relation between measurable parameters and the force F averaged over the oscillation cycle [102]:

$$\Delta f = \frac{f_0}{\pi k_N A} \int_0^{2\pi/\omega} \overline{F}(z_0 + A \sin \omega t) \sin \omega t \, dt. \tag{17.2}$$

The integral on the right hand side of Eq. (17.2) can be calculated for large oscillation amplitudes, assuming different force–distance relations. If the distance between tip and surface at closest approach is smaller than the tip radius, the long-range interactions are dominated by the spherical cap of the tip. In this case the frequency-shift Δf_{el} due to the electrostatic interaction and the frequency shift Δf_{vdW} due to the vdW interaction are given by [124]

$$\begin{aligned} \Delta f_{el} &= -\frac{f_0}{k_N A^{3/2}} \frac{\pi \varepsilon_0 R (V_{bias} - V_{cpd})^2}{\sqrt{2 z_{min}}}, \\ \Delta f_{vdW} &= -\frac{f_0}{k_N A^{3/2}} \frac{HR}{12\sqrt{2}(z_{min} A)^{3/2}}, \end{aligned} \tag{17.3}$$

where z_{min} is the closest distance between the surface and the mesoscopic part of the tip, R is the tip radius and H is the Hamaker constant. Note that the force vs. distance curves $F(z)$ can be reconstructed from the measured $\Delta f(z)$ curves without any assumption about the force law. An iterative method introduced by Dürig is described in [81].

While the NC mode is commonly adopted in UHV, the preferred technique in ambient conditions is the *tapping mode*. In this case the oscillation amplitude A of the probing tip is used as feedback parameter while the cantilever is vibrated close to its resonance frequency. The energy dissipated during one oscillation cycle can be estimated from the driving amplitude A_d and the phase shift φ as

$$\Delta E \approx \pi k_N \left(A A_d \sin \varphi - \frac{A^2}{Q} \right), \tag{17.4}$$

where k_N is the normal stiffness and Q is the quality factor of the cantilever. Note that the oscillation amplitude is larger than in the NC mode (of the order of 100 nm) and intermittent contact occurs.

Scanning tunneling microscopy

In a *scanning tunneling microscope* (STM) piezoelectric transducers bring a sharp metallic tip down to a distance of a few tenths of a nm from a conducting surface, where the wave functions of the electrons of tip and surface overlap [23]. A bias voltage, V_{bias}, applied between tip and sample causes electrons to tunnel from the tip to the surface or vice versa, depending on its sign. The resulting tunneling current, I_{t}, can range from a fraction of a pA to a few nA depending on the materials, distance and bias voltage.

As a first approximation, the tunneling current decays exponentially with the tip–sample distance z:

$$I(z) \propto V_{\mathrm{bias}} \rho(\varepsilon_{\mathrm{F}}) \mathrm{e}^{-2\kappa z}.$$

In this formula $\rho(\varepsilon_{\mathrm{F}})$ is the local density of states at the Fermi level and $\kappa = \sqrt{2m\Phi/\hbar^2}$, where Φ is the height of the tunneling barrier. If the current I_{t} is kept constant by a feedback loop while scanning, a constant charge density surface can be mapped. Although STM is not adequate to measure mechanical forces, its resolution is usually higher than AFM. For this reason, the two techniques can be alternated, for instance in nanomanipulation experiments on metals.

17.3 Other experimental techniques in nanotribology

Surface force apparatus

The *surface force apparatus* (SFA) is based on two curved, molecularly smooth surfaces immersed in a liquid or in a controlled atmosphere [149]. The material of choice is usually mica, since it can be easily cleaved into atomically flat surfaces over macroscopic areas. When the surfaces are brought into contact, the gap caused by a thin lubricant film can be measured by optical interferometry. In particular, when the wavelength of the white light passing through the system matches the local separation, fringes of color are created and the contact geometry can be imaged.

The SFA can resolve distances of the order of 0.1 nm and forces of about 10 nN, and the contact diameter is usually of the order of few tens of μm. If a tangential force is applied, sliding friction can also be studied. The SFA is ideal for studying molecular-level structural and rheological properties of liquids and tribological properties of lubricants under compression. The main drawback is the trapping of impurities in the gap. If the size of the contaminants is below the light wavelength, the interference fringes are not modified, and the distance measurements become unreliable.

Quartz crystal microbalance

Consider a quartz disk, the faces of which are covered by two thin metal films. Since quartz is piezoelectric, transverse vibrations with a characteristic frequency

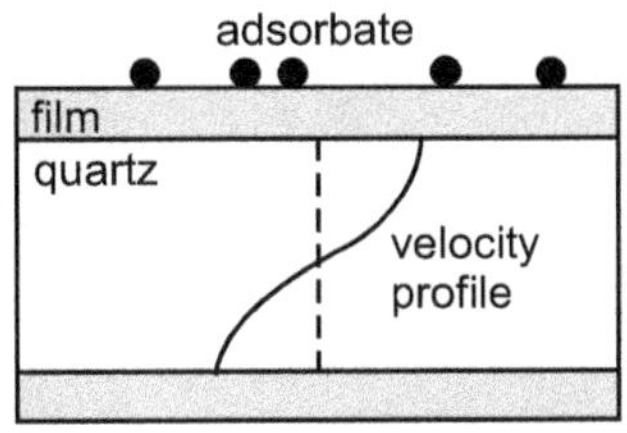

Figure 17.3 Schematic of a quartz crystal microbalance for measuring the sliding friction of adsorbate layers.

ω_0 can be excited by applying an ac voltage. If the voltage is suddenly switched off, the amplitude of the oscillations will decay exponentially with a rate Γ. The presence of molecules on the surfaces of the metal films (Fig. 17.3) causes small shifts, respectively $\Delta\omega$ and $\Delta\Gamma$, in the frequency response and decay rate even for a small fraction of monolayer. If m is the mass of the vibrating crystal, m_{ML} is the mass of an adsorbate monolayer and γ is the damping coefficient for the sliding adsorbate (considered as a rigid body), it can be proven that [245, section 8.6]

$$\frac{\Delta\omega}{\omega} = -\frac{m_{ML}}{m}\frac{\gamma^2 P_1}{\omega^2 + (\gamma P_1)^2}, \qquad \Delta\Gamma = -\frac{2\omega}{\gamma P_1}\Delta\omega,$$

where P_1 is the fraction of adsorbates in the first layer in direct contact with the substrate. The previous relations allow us to estimate the quantities γ and P_1 and, consequently, the friction force on thin physisorbed layers of simple inert atoms or molecules, such as Kr, Xe or N_2, when they slide on metals such as Au or Ag. Although the *quartz crystal microbalance* (QCM) was already a well-established technique for film thickness measurements, its application to nanotribology was pioneered by Krim and coworkers [170] in the 1980s.

Electron microscopy

In spite of their versatility, AFM, SFA and QCM do not allow us to visualize the structure of nanocontacts as they form and slide. This problem can be partially overcome by the live view of a nanoasperity inside a scanning electron microscope (SEM) or a transmission electron microscope (TEM). The TEM was first coupled to nanoindentation measurements by Minor *et al.* [222] and subsequently used to characterize abrasive wear with Au, W and Si tips sliding on Si, graphite and diamond respectively [287, 216, 150]. In the last of those studies, lattice resolved images of the worn tip (Fig. 17.4) did not give any evidence of dislocations or defects, strongly suggesting that the Si atoms are removed one by one (see also Section 20.1).

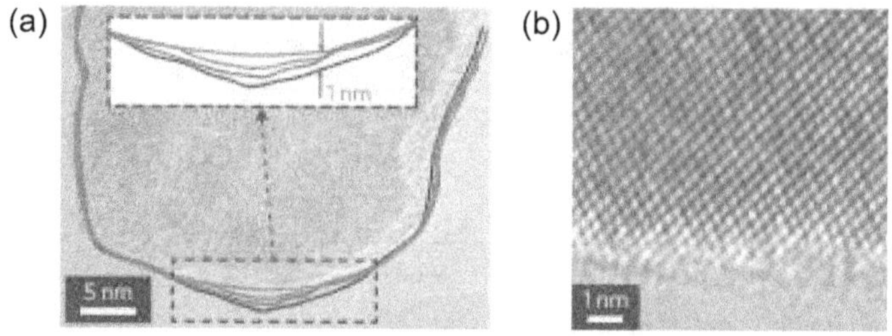

Figure 17.4 (a) Successive TEM profiles of an Si tip scraped against a diamond surface. (b) Lattice resolved image of the worn tip. Reproduced from [150] with permission from Macmillan Publishers Ltd.

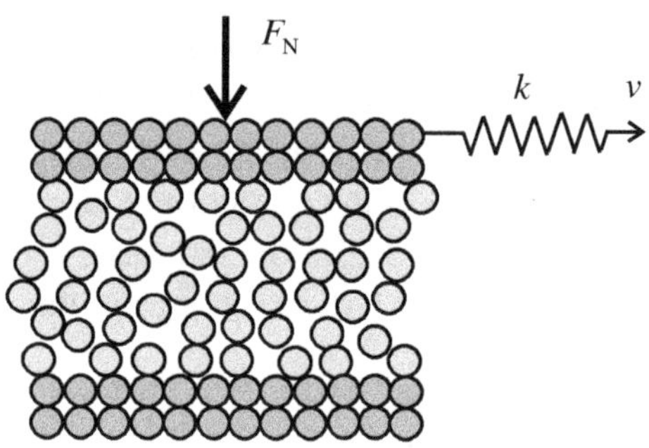

Figure 17.5 Sketch of a molecular dynamics simulation of a boundary lubricated interface under shear.

17.4 Molecular dynamics and nanotribology: methods

Molecular dynamics (MD) *simulations* can be considered as 'computational experiments', where the dynamics of the atoms in a sliding system is studied numerically by solving ad-hoc Newton (or Langevin) equations of motion. In such a way friction, adhesion and wear processes can be explored by a suitable choice of geometry, boundary conditions and, above all, interaction potentials. A typical 'setup' is shown in Fig. 17.5. Additionally, a thermostat can be introduced to eliminate the Joule heat and obtain a steady state of motion. As a result, important physical quantities such as the instantaneous and the average friction forces, the mean velocity of the slider, the heat flow and different correlation functions can be calculated.

A major problem in MD simulations is the fact that the total energy U of the system depends on the states of the electrons. The Car–Parrinello method [43] is not applicable, since it only handles few hundreds of atoms on time scales well below 1 ns. For this reason the interactions are described by empirical 'force-fields' and, as a result, the description remains rather qualitative. This is especially the case for wear processes, where atoms may suddenly change their coordination, chemistry and even charge. At the time of writing, it is estimated that a simulation involving 10^5 atoms can reproduce a sliding process lasting 1 μs in approximately one day [334].

If the simulations aim to predict the response of generic atoms, one can adopt the Lennard–Jones (LJ) potential:

$$V_{\mathrm{LJ}} = 4\varepsilon\left[\left(\frac{\sigma}{r}\right)^{12} - \left(\frac{\sigma}{r}\right)^{6}\right], \tag{17.5}$$

where the first term on the right reproduces the Pauli repulsion at very short distances and the second one describes the vdW attraction. In Eq. (17.5) ε is the depth of the potential well and σ is the finite distance at which the potential is zero. Metals are better described by taking into account the interaction between each atom and its surrounding free electrons, as done in the 'embedded atom method' [63]. This method is computationally efficient and has been successfully applied to study various phenomena related to atomic friction. Covalent bonds can be accounted for by the Stillinger–Weber model [320], which was originally conceived for diamond structured silicon. In this case, the potential is determined by the stretching and bending of the bonds and sometimes also by their torsion. The disadvantage is that only one equilibrium configuration is captured. This is not the case for the so-called 'bond order potentials', which introduce a bond order parameter and can simultaneously describe different stable states. Even if computationally very expensive, several MD simulations of friction have been performed in this way. Ionic bonds are reproduced with long-range Coulomb forces, which significantly increases the computational time. Among the few attempts to simulate friction on ionic materials, it is worth mentioning the work by Wyder *et al.* [343], who used a combination of short range Buckingham and Coulomb potentials with each ion modeled as a spring-coupled system of positively charged core and negatively charged shell (the so-called 'shell model').

If tip and surface are made of the same material, which is the case if some substrate atoms are picked up during initial contact, MD simulations are considerably simplified since all the interactions in the sliding system are modeled by the same potential. If this is not the case, cross potentials describing the interaction between dissimilar species need to be introduced. If the frictional response is dominated by vdW forces and no wear occurs, the LJ model can be an adequate choice.

17.5 Molecular dynamics and nanotribology: results

Friction on the atomic scale

Atomistic simulations of friction were pioneered by Landman *et al.* [178]. The authors were able to reproduce the stick–slip motion of realistic Si tips on Si samples, and also the necking of sample material when the tip was retracted. Molecular

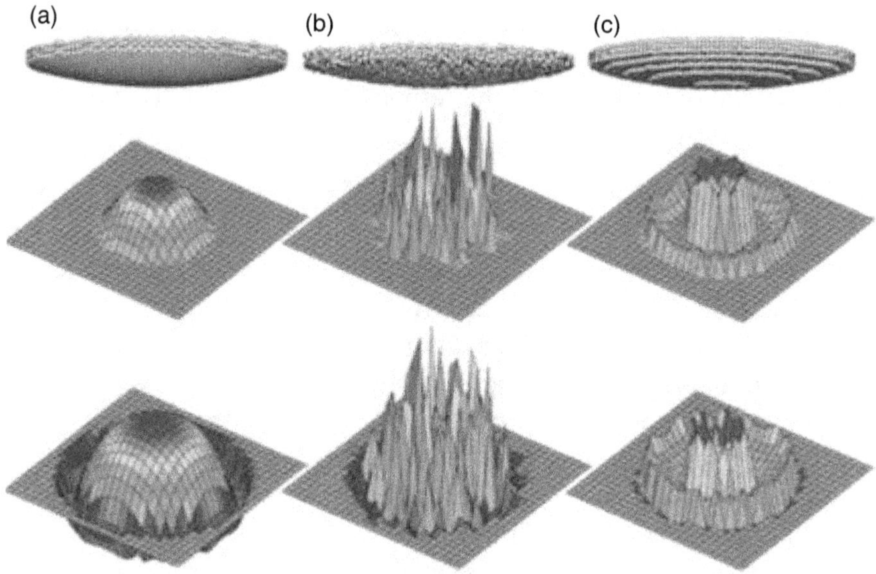

Figure 17.6 Geometry (top row) and stress distribution without adhesion (middle row) and with adhesion (bottom row) for (a) a crystalline tip, (b) an amorphous tip and (c) a stepped tip pushed against a rigid flat surface. Reproduced from [196] with permission from Macmillan Publishers Ltd.

dynamics simulations have also shown that continuum mechanics may fail to capture the properties of nanoscale contacts. This can be seen in Fig. 17.6, where the same global geometry and loading conditions, but different atomic arrangements, result in contact areas and stress variations spanning more than two orders of magnitude. While a crystalline tip is well described by the Hertz theory, without adhesion, and by the Maugis–Dugdale theory, with adhesion, the compressive stress in the contact of an amorphous tip presents strong fluctuations, and a stepped tip shows compressive stress peaks at step edges. The definition of the contact area A is thus very critical in these cases. For instance, we may define a distance within which a couple of atoms are assumed to be in contact, and take A as the area of the region circumscribing those atoms. In this way, the contact distance obviously depends on the type of bonding at the interface.

To simulate the normal force applied while scanning, the uppermost layers of the model tip are usually treated as a rigid body onto which a constant load is applied. As an example, Fig. 17.7(a) shows the simulated load dependence of the friction force when a truncated cone-shaped Pt tip forms an incommensurate contact with a Au(111) substrate. The friction increases slowly with load till wear occurs and atoms are exchanged between tip and substrate. This results in a larger contact area and in a rearrangement of the atoms in the contact to more stable configurations.

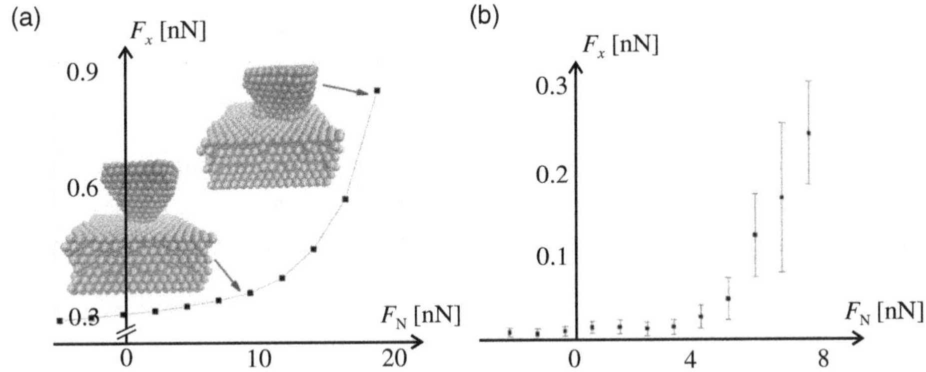

Figure 17.7 (a) MD simulations and (b) AFM measurements of the load dependence of friction for a Pt/Au(111) system. Reproduced from [76] with permission from AIP Publishing.

The considerable increase of friction accompanying these processes has indeed been observed in AFM measurements (Fig. 17.7(b)) but cannot be described by continuum mechanics.

The main problem when studying the velocity dependence of friction by MD simulations is the fact that speeds below few m/s are not accessible. These values are well above typical speeds at which atomic stick–slip is resolved by AFM (nm/s to μm/s). The reason for that is the typical time scales in MD, which are of the order of 1 fs. Furthermore, phenomena such as thermally activated hopping are not effective at high speeds, which makes any attempts to extrapolate the model predictions on the velocity dependence of friction quite doubtful. A possible solution consists of accelerating the simulations during the long stick phases separating rapid slip events.

On the other side, an advantage of MD simulations over the experiments is the fact that the temperature can be controlled very easily. On the variety of substrates examined so far, including metals (Cu and Au), diamond and alkylsilane monolayers, a decrease of friction with temperature has always been observed. Due to the finite size of the systems addressed by MD simulations, boundary conditions need to be properly chosen to avoid unphysical effects. For instance, the thermal energy generated in the slip cannot be simply reflected at a fixed boundary or re-enter the system via periodic boundary conditions. The problem is solved by introducing a numerical thermostat. In this way energy is extracted such that the temperature of the system fluctuates around a given value. In MD simulations of atomic friction, the thermostat is applied only to the atoms far from the contact region. In this way the dynamics of the contacting atoms is not disturbed, although the heat generated during sliding is effectively dissipated.

As described in the previous chapter, the friction force between two crystalline surfaces is a function of the misfit angle between the substrates and the pulling direction. However, MD simulations have shown that this dependence is made smoother by surface irregularities and thermal vibrations. As an example, Qi *et al.* observed that, at room temperature, a rotation of 45° can change the friction by two orders of magnitude on an atomically flat Ni(100)/Ni(100) contact, but only by a factor of four if one of the surfaces has a roughness of 0.08 nm [278].

Nanowear

The long time scales which characterize wear processes and the large amounts of material involved make any attempt to simulate these mechanisms on a computer extremely challenging. In spite of that, MD can provide useful insight into the mechanisms of removal and deposition of single atoms by a nano-indenter, which cannot be directly visualized by AFM. Complex processes like abrasive wear and nanolithography can only be investigated using approximations in classical mechanics.

The first MD simulations showing material transfer during tip retraction were performed by Landman *et al.* [178] and by Nieminen *et al.* [234] on Si/Si and Cu/Cu contacts respectively. Livshits and Shluger observed that an AFM tip undergoes a process of self-organization when scanning alkali halide surfaces [187]. The tip contamination caused by the adhesion of surface atoms may improve the resolution of crystal surfaces, if the adsorbed material forms stable structures on the tip. In a series of MD simulations on Cu surfaces, Sørensen *et al.* used (111)- and (100)-terminated Cu tips and also amorphous structures obtained by annealing the tip at high temperature [314]. The lateral motion of the neck formed while scanning the surface revealed stick–slip due to combined sliding and stretching, and ruptures caused by deposition of debris onto the substrate (Fig. 17.8).

Only a few examples of plowing wear simulations on the atomic scale have been reported. The first of these investigations was performed by Belak and Stowers on a Cu surface using a rigid C tip [17]. The interactions within metal atoms were modeled with an embedded atom potential while LJ potentials were used between C and Cu atoms. The tip was first indented and then pulled over the surface. In 2D simulations Hertzian behavior was observed up to a load $F_\mathrm{N} \approx 2.7$ nN and an indentation of about 3.5 Cu layers. At this point a series of single dislocation edges was created along the easy slip planes. After indenting six Cu layers, the tip was slid parallel to the original surface. The work per unit volume of removed material was found to scale as $\delta^{-0.6}$, where δ is the depth of cut. Interestingly, the same power law was observed in machining of Cu with a nanotip [223], although macroscopic experiments typically give an exponent of -0.2. In 3D simulations,

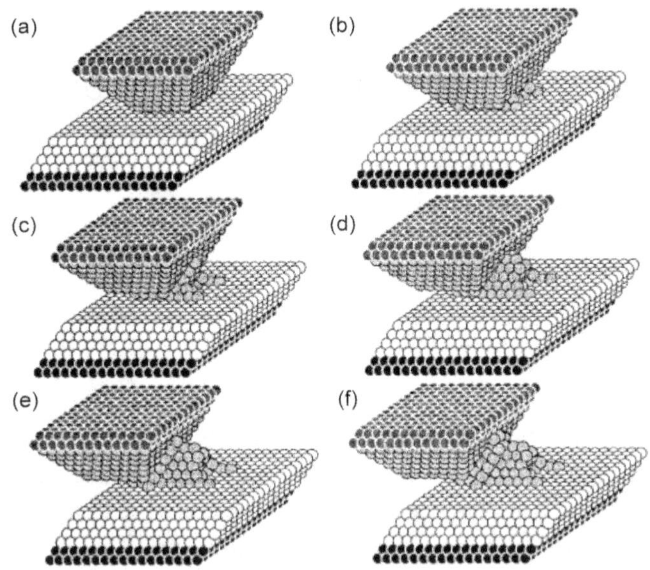

Figure 17.8 Snapshot of the neck formed during the scanning of a Cu(100)-terminated tip on a Cu(100) surface. Reproduced from [314] with permission from the American Physical Society.

long-range elastic deformations were also included. After an indentation of only 1.5 layers, a small dislocation loop was formed, resulting in plastic deformation. While dislocations spread several hundred lattice constants in 2D, this was not the case in 3D, where they remained confined to distances of a few lattice constants. The tip was also moved laterally at $v = 100$ m/s during indentation, which resulted in negligible friction till the onset of plastic yield. At this point, the friction force suddenly jumped to values comparable to the normal force, corresponding to a coefficient of friction $\mu \approx 1$.

The nanoindentation and sliding of sharp and blunt Ni tips on atomically flat Cu surfaces were studied by Buldum *et al.* using MD simulations [38]. Quasi-periodic variations of the lateral force were observed with the sharp tip. In this case one layer of the asperity was deformed to match the substrate during the first slip and then two asperity layers merged into one through structural transition during the second slip. In the case of the blunt tip, the stick–slip appeared much less regular.

18

Experimental results in nanotribology

In this chapter we will discuss a selection of experimental observations of friction on the nanoscale, obtained by atomic force microscopy and related techniques. After presenting high resolution friction maps on different materials, we will compare the load, velocity and temperature dependence of friction detected in the experiments to the predictions of the Prandtl–Tomlinson model. The comparison will be extended to simple experiments showing the effect of contact vibrations and friction anisotropy on crystalline samples.

18.1 Friction measurements on the atomic scale

The first lattice resolved maps of stick–slip were acquired by Mate *et al.* [206] just one year after the atomic force microscope was invented [24]. In their experiment, Mate *et al.* used a tungsten wire as a probing tip and detected lateral forces on a graphite surface using non-fiber interferometry. Since graphite is stable, chemically inert and easy to cleave along atomic planes, it is an ideal material for this kind of measurement. The pioneer work by Mate *et al.* was followed by experiments on ionic crystals (NaCl, KBr etc.), metals (Cu, Au, Al, W, Pt, Pd and Ag) and covalent materials: semiconductors, carbon-based materials (e.g. graphite, diamond, and diamondlike carbon), organic materials and many oxides.

The ultra-high vacuum (UHV) environment reduces the influence of contaminants on the sample surfaces and results in more precise and reproducible results. An atomic-scale friction map, acquired with a silicon tip sliding on a NaCl(001) cleavage surface in UHV (lattice constant $a = 0.564$ nm), is shown in Fig. 18.1(a). The spring force F grows up to a maximum value, corresponding to the static friction $F_s \approx 0.4$ nN at which the tip suddenly slips. After that, the tip quickly rebinds to a neighboring unit cell on the crystal surface. The process is repeated several times along each scan line, reproducing the structure of the surface lattice.[1] However, the value of F_s while crossing the centers of the unit cells in Fig. 18.1(a)

[1] A noticeable exception is given by the measurements on NaF(001) by Ishikawa *et al.*, who were able to resolve both ionic species depending on the applied load [147].

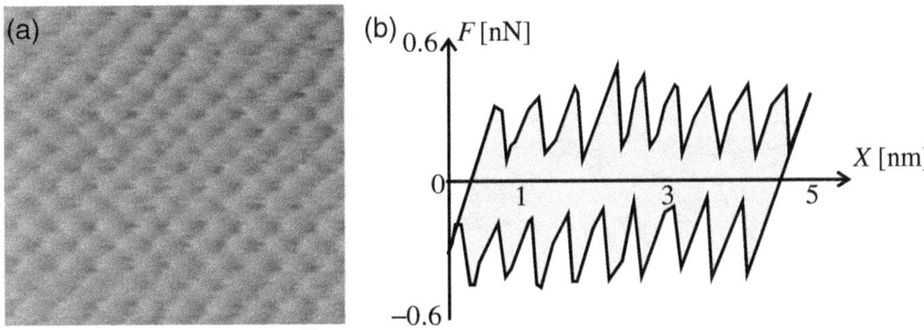

Figure 18.1 (a) Lateral force map on a NaCl(001) surface in UHV and (b) cross-section (forward and backwards) through the centers of the unit cells. Scan size: 5 nm; normal force value: $F_N = 0.65$ nN. Adapted from [107] with permission from the American Physical Society.

is not always the same. This is due to thermally activated hopping, as discussed in Section 15.3. As shown in Section 15.1, the maximum value of the static friction and the slope of the turning points of the $F(x)$ curves can be used to determine the corrugation U_0 of the tip–surface interaction potential and the effective lateral stiffness k of the system. From Fig. 18.1(b) we estimate $U_0 \approx 0.22$ eV and $k \approx 1$ N/m.

Howald *et al.* [142] were able to measure atomic-scale friction on the reconstructed Si(111)7 × 7 surface after coating the tip with a polymer (PTFE), which has lubricant properties and does not react with the dangling bonds of the substrate. On the other hand, uncoated Si tips and tips coated with Pt, Au, Ag, Cr, Pt/C damaged the sample irreversibly. Different reconstruction domains on semiconducting surfaces of InSb(0001) and Ge(001) [117], and variations of the friction force while crossing the step edges of NaCl(001) and Ge(001) [319] have been recently observed using FFM (Fig. 18.2). Atomic-scale friction was also resolved on metal surfaces in UHV [18].

If ultrasharp probing tips are used (with radii of curvature of the order of 1 nm), additional features in the atomic-scale maps of the friction force can be captured. This is proven by the images in Fig. 18.3, which refer to a KBr film heteroepitaxially grown on an NaCl(001) surface [202]. A so-called Moiré pattern is seen, which is caused by the mismatch between the ultrathin films imaged by AFM and the substrate lattice. In the case of KBr/NaCl(001) the ratio between the lattice constants of substrate and adsorbate is approximately 6:7. Note that the Moiré pattern was not visible when the loading force was reduced close to the pull-off value and the tip interacted only with the top layer of KBr. Another impressive example (with hexagonal Moiré pattern) was reported by Filleter *et al.* by scanning a graphene film on SiC(0001) [92].

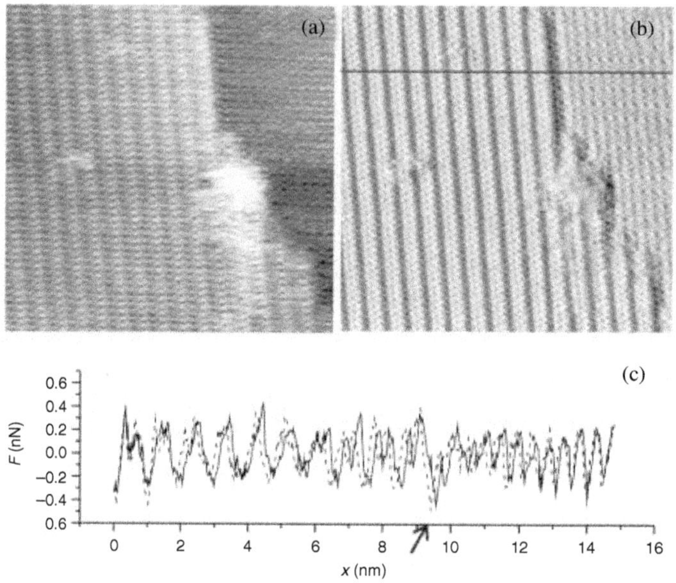

Figure 18.2 (a) AFM topography and (b) lateral force map acquired on two terraces of a (2 × 1) reconstructed Ge(001) surface separated by a monatomic step edge. (c) Section along the black line in (b). Adapted from [117] with permission from the American Physical Society.

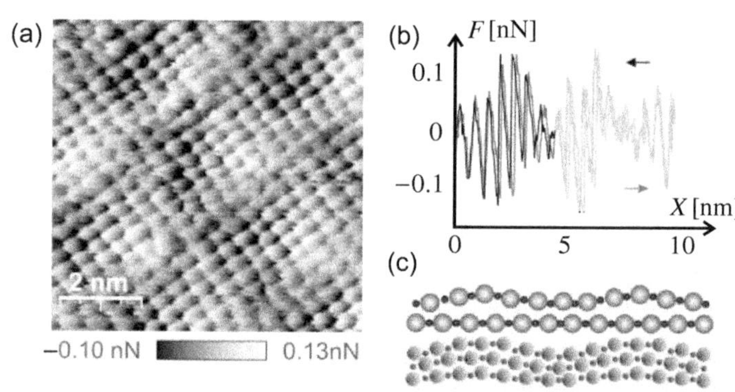

Figure 18.3 (a) Lateral force map of the Moiré superstructure formed by KBr on a NaCl(001) surface and (b) cross-sections (forward and backward) through the Moiré pattern. (c) Interface structure as calculated by Monte-Carlo simulations. Adapted from [202] with permission from the American Physical Society.

Another excellent environment for high-resolution friction force microscopy is liquid solutions, where capillary forces are eliminated and attractive forces become very small. Figure 18.4 shows a dolomite (104) surface and a thin film of Cu phthalocyanine (Pc) molecules deposited on it [236]. The structures of the dolomite

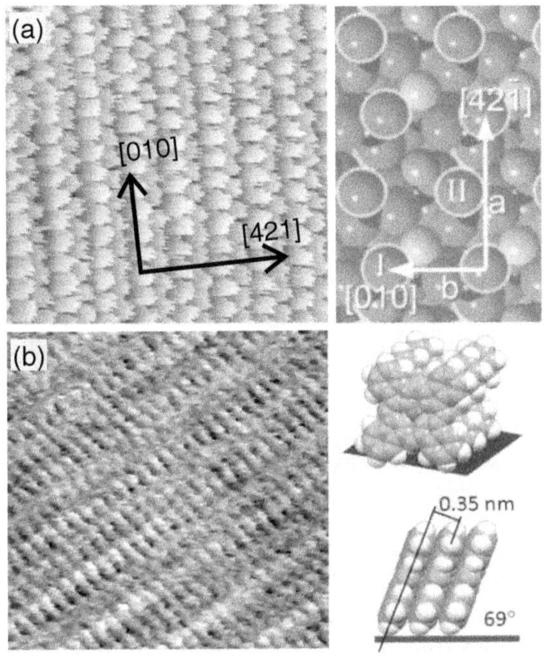

Figure 18.4 Lateral force maps of (a) a dolomite (104) cleavage surface and (b) CuPc molecules grown on it. Both images were acquired in deionized water at room temperature. The structure of the mineral surface, with the protruding oxygen atoms enhanced by open circles, and the molecular stack geometry are shown on the right side. Adapted from [236] with permission from The Royal Society of Chemistry.

face and the stack geometry of the organic molecules, as determined from single crystal X-ray diffraction, are also sketched. Note that two pinning sites per unit cell are resolved in Fig. 18.4(a), corresponding to oxygen atoms protruding out of the dolomite surface. The atoms are resolved with different contrast, depending on the scan direction, which can also be explained using the PT model [272]. In Fig. 18.4(b) two benzene rings per molecule are distinguished while sliding over the CuPc film. Such a resolution could not be achieved in ambient conditions using the same setup and probing tips.

18.2 Lateral and normal stiffness

In a first approximation, the effective lateral stiffness[2] $k_{\mathrm{L,eff}}$ of an FFM originates from two contributions:

$$\frac{1}{k_{\mathrm{L,eff}}} = \frac{1}{k_{\mathrm{L}}} + \frac{1}{k_{\mathrm{L,con}}}, \tag{18.1}$$

[2] This quantity is simply denoted k in Chapter 15 and Section 18.1.

where k_L is the lateral spring constant of the micro-fabricated cantilever (Appendix A) and $k_{L,con}$ is the lateral force per unit displacement causing an elastic deformation of the contact region. Assuming that contact mechanics holds at a single asperity level, the contact stiffness[3] $k_{L,con} = 8G^*a$, where a is the radius of the contact area and the effective shear modulus G^* of the materials forming tip and substrate is defined by Eq. (5.8). The contact stiffness $k_{L,con}$ can be further separated into the stiffness of the tip body, k_{tip}, the stiffness of the tip apex, k_{apex}, and the stiffness caused by the deformation of the substrate, k_{sub}.

As mentioned in Appendix A, standard cantilevers have a lateral stiffness k_L of the order of 10–100 N/m. Using the finite element method and realistic geometric parameters measured by TEM, Lantz *et al.* estimated that, for silicon and silicon nitride tips, the values of k_{tip} are comparable to those of k_L [181]. Due to its nanoscale sharpness, the tip apex is expected to be much more compliant. It is responsible for the fine structures and elongated slip duration observed in atomic friction maps [201]. Both k_{apex} and k_{sub} play a key role in determining $k_{L,eff}$, but no convincing ways to separate their contributions have been proposed so far.

Compared to the lateral stiffness, the normal stiffness of a nanocontact, $k_{N,eff}$ is more difficult to determine. A possible method is to track the normal resonance frequency f_N while scanning. Atomically resolved measurements of this kind were reported by Steiner *et al.* [317]. Figure 18.5 shows the variation of f_N along the [100] direction of a NaCl(001) surface in UHV together with the corresponding variation of the lateral force F. Note that an atomic defect (indicated by an arrow) is resolved in the frequency signal, but not in the lateral force signal. The normal stiffness k_N can be estimated from the formulas in Section 7.1, modified to account for the inclination of the cantilever with respect to the sample surface (approximately 15°). As a result, $k_{N,eff} \sim 10$ N/m, i.e. two orders of magnitude more than the spring constant k_N of the free lever.

18.3 Load dependence of nanoscale friction

The general trend observed in AFM experiments is that the friction force increases with the normal load F_N. However, different behaviors have been recognized. A linear load dependence was reported on various substrates including gold and alkylthiol molecules [98, 270]. The power law predicted by the DMT model (Section 10.2) was observed on different carbon forms using well-defined spherical tips with radii of curvature of tens of nanometers [305]. In this case the tip terminations were produced by contaminating standard silicon probes with amorphous carbon inside a TEM. Other AFM measurements performed in UHV on alkali

[3] It is also assumed that the tip radius is much larger than a and the tip motion has a negligible influence.

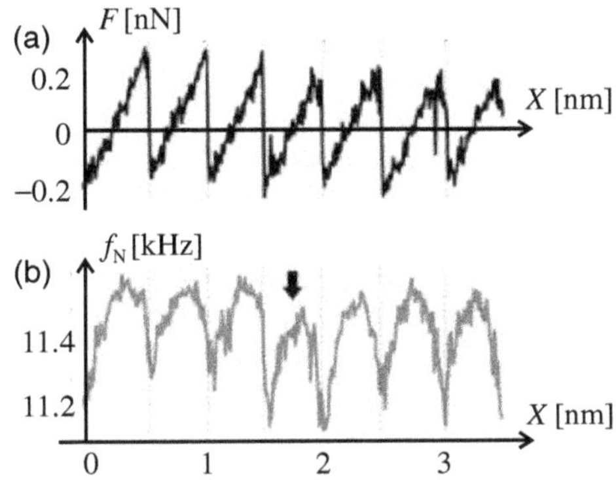

Figure 18.5 (a) Lateral force profile and (b) corresponding variation of the contact resonance frequency recorded while scanning a NaCl(001) surface in contact mode. Adapted from [317] with permission from IOP Publishing.

halides, mica and various metals could be better reproduced with the JKR model [218, 46, 274]. A reason for the variety of the behaviors reported experimentally is the delicate balance between chemical and mechanical properties of the contacting surfaces. While the elasticity and roughness determine the response of chemically inert surfaces, the formation and rupture of bonds while sliding may prevail on chemically active surfaces. On metal surfaces, nanojunctions can also be formed and wear may occur.

The situation is better defined in UHV. In this environment Socoliuc *et al.* were able to recognize the transition from stick–slip to smooth sliding predicted by the PT model when the characteristic parameter $\eta < 1$ (section 15.1) on a NaCl(001) surface (Figures 15.2(b) and 15.3(b)) [312]. A similar transition was observed in manipulation experiments of single Pb atoms by STM [12]. In this case, the stick–slip was found to disappear after bringing the probing tip close to the atoms, which weakened the Pb – substrate interaction.

18.4 Velocity dependence of nanoscale friction

The logarithmic velocity dependence of friction caused by thermally activated hopping of the nanotip (Section 15.3) has been measured, starting from lattice resolved images, on Cu(111) and NaCl(100) surfaces in UHV [18, 107]. From a comparison of Fig. 18.6 with Eq. (15.17), a characteristic velocity $v_0 \approx 2.5$ µm/s and a corresponding attempt frequency $f_0 \approx 66$ kHz can be estimated when sliding on NaCl. A similar logarithmic dependence (without lattice resolution) was also reported in earlier measurements on a polymer layer grafted on silica [29] and interpreted within the Eyring model introduced in Section 24.1.

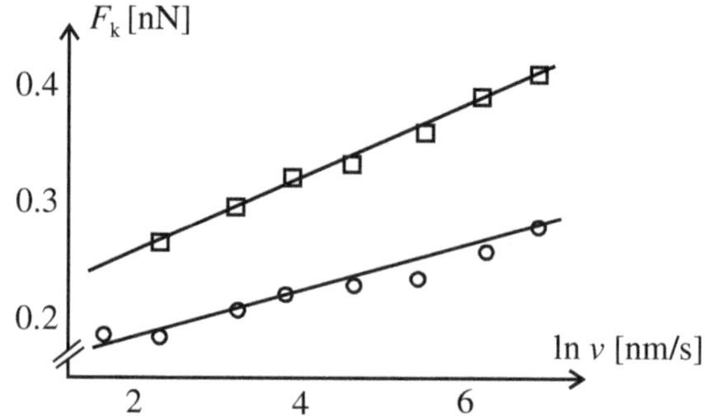

Figure 18.6 Kinetic friction force F_k as a function of the sliding velocity for a silicon tip in contact with an NaCl(100) surface. The measurements were performed for two different values of the normal force $F_N = 0.65$ nN (squares) and 0.44 nN (circles). Adapted from [107] with permission from the American Physical Society.

The friction force can present a different behavior, and the slope of the $F(\ln v)$ curve can even change sign, in a humid environment. This can be attributed to the formation of water menisci by thermally activated capillary condensation, as discussed in Section 24.2. However, a logarithmic decrease of friction can be also associated with chemical modifications. This happens in systems forming cross-linked structures that can be broken by the applied load, such as surfaces terminated by -OH, -COOH and -NH_2 groups [52]. At slow velocities there is more time to form bonds between tip and surface, which results in larger friction.

If the scan velocity increases, thermally activated processes are less important, and, beyond a critical value, the friction force becomes independent of the velocity, as seen in a series of measurements between Si tips and diamond, graphite and amorphous carbon surfaces with scan velocities above 1 μm/s [353]. The transition from a logarithmic increase to a friction plateau was recognized on a mica surface and reproduced using Eq. (15.18) [291].

18.5 Temperature dependence of nanoscale friction

Apart from a slight logarithmic dependence on the scan velocity, thermally activated stick–slip results in a strong decrease of the friction with the temperature T, as seen from Eqs. (15.17) and (15.18). A significant decrease of friction as $1/T$ was indeed observed by Zhao *et al.* in a series of AFM measurements on graphite in UHV conditions and in a wide temperature range (140–750 K) [351], but no comparison with the PT model at finite temperature was attempted. In another series of UHV experiments, Jansen *et al.* measured the temperature dependence of atomic-scale friction from cryogenic conditions to a few hundred Kelvin on silicon,

SiC, ionic crystals and graphite [151]. When the samples were cooled down from room temperature, a substantial agreement with the thermally activated PT model was found down to a peak or a plateau, which, depending on the material, appeared around 50–200 K. Below these values, the friction was found to decrease with temperature, which was attributed to the competition between thermally activated formation and rupture of chemical bonds [11].

18.6 Effect of contact vibrations

Developing strategies for reducing sliding friction is important for proper functioning of micro- and nanoelectromechanical systems (MEMS and NEMS). In this context, traditional lubricants cannot be used, since the viscosity of mineral oils dramatically increases when the lubricant molecules are confined into nanometer-sized interstices, as discussed in Section 23.1. Different strategies, such as mechanical excitations, need to be explored. Ultrasonic vibrations have been used for years to modify the frictional behavior of materials at a macroscopic scale and their application at the nanoscale looks promising.

AFM experiments on alkali halide surfaces in UHV have shown that the friction can be efficiently reduced if mechanical resonance modes of the nano-junctions formed while sliding are excited [313]. Lantz *et al.* [182] showed how the abrasive wear of a silicon tip sliding over several hundred meters can be prevented by this strategy. The state of *dynamic superlubricity* so-achieved has been also exploited to acquire lattice-resolved FFM maps of crystal surfaces without damaging the samples [109]. An example, referring to graphite, is shown in Fig. 18.7.

In the previous cases, the actuation was applied perpendicularly to the sliding plane (out-of-plane). Nevertheless, a reduction of friction is also observed if

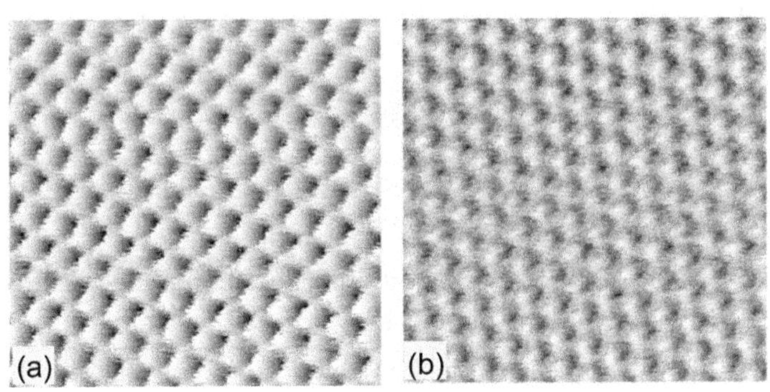

Figure 18.7 Lateral force maps acquired on a graphite surface in UHV: (a) in contact mode, (b) by exciting the tip at the contact resonance frequency while scanning. Frame sizes: 3 nm. Reproduced from [109] with permission from IOP Publishing.

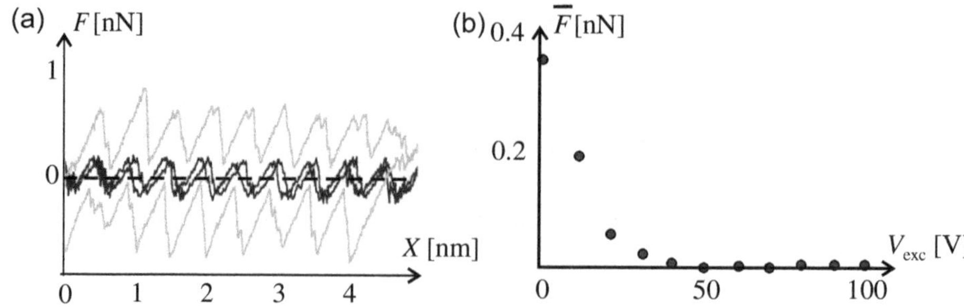

Figure 18.8 Effect of lateral vibrations on atomic-scale friction as measured on an NaCl(001) surface by a room temperature UHV-AFM: (a) friction force loops with (dark curves) and without (light curves) actuation; (b) average friction force as a function of the excitation amplitude. Reproduced from [294] with permission from AIP Publishing.

in-plane vibrations are excited. The transition from stick–slip to ultra-low friction is indeed attested by the two friction loops on an NaCl(001) surface in Fig. 18.8 [294]. In the absence of vibrations, the usual sawtooth pattern is observed (Section 18.1). When the lateral vibrations are excited the forward and backward curves become closer and completely overlap if the excitation amplitude is large enough (dark curves in Fig. 18.8). In this case the average friction force F_k becomes negligible. Note that the slip towards the new equilibrium position is accompanied by a series of back-and-forth jumps induced by the thermal vibrations at the temperature $T = 300$ K of the measurements (as seen with higher temporal resolution). The dependence of the average friction force on the excitation amplitude is shown in Fig. 18.8(b). The force $\overline{F}$ decreases gradually, as observed when normal oscillations are applied. A comparison between Fig. 15.8 and Fig. 18.8 shows that an excitation of 50 mV applied to the piezoactuator corresponds to lateral oscillations of approximately half the lattice constant, suggesting a simple method for calibration of the driving amplitude.

18.7 Friction anisotropy

The importance of the misfit angle when two flat surfaces slide past each other was first demonstrated experimentally with the surface force apparatus [212, 140]. The friction force between two mica sheets was indeed found to increase when the surfaces formed a commensurate contact. In STM measurements with a monocrystalline tungsten tip on Si(001), Hirano *et al.* observed what they called 'superlubricity' in the case of incommensurate contact [141]. Structural lubricity has also been reported on other materials including MoS_2, Si/W, Ti_3SiC_2 and graphite.

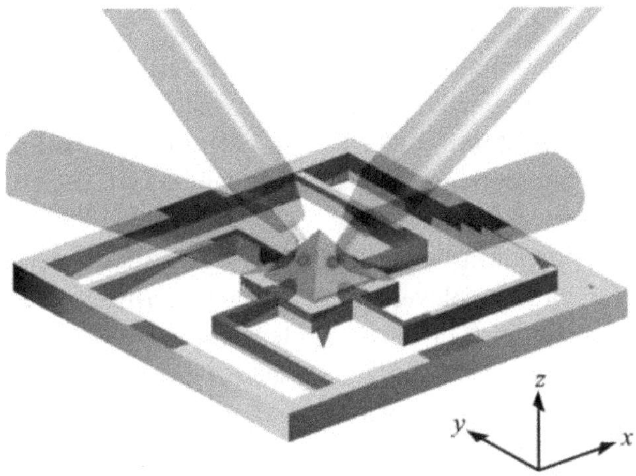

Figure 18.9 Sketch of the tribometer used by Dienwiebel *et al.* to measure friction anisotropy on an atomic scale. Reproduced from [72] with permission from the American Physical Society.

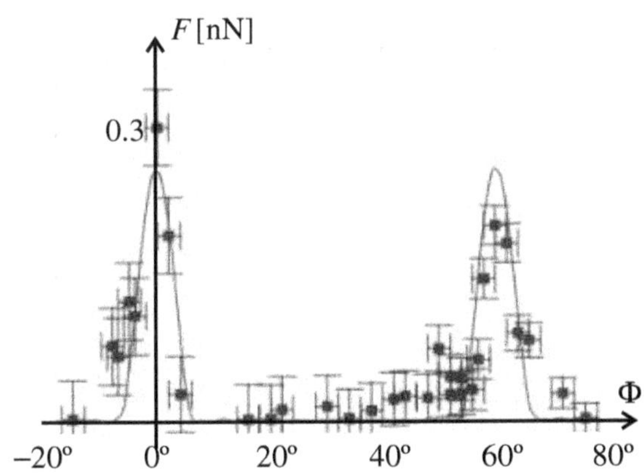

Figure 18.10 Angular dependence of friction on graphite, as measured with the setup in Fig. 18.9. Adapted from [72] with permission from the American Physical Society.

One of the most convincing evidences of structural lubricity was given by Dienwiebel *et al.* by means of an original setup consisting of four optical interferometers directed towards a pyramid holding a tungsten wire (Fig. 18.9) [72]. Thanks to the symmetric design of the instrument the normal and lateral forces could be detected with minimal crosstalk. In this way lateral forces below 50 pN were measured between a graphite flake attached to the wire and an atomically flat graphite surface, but only if the two surface lattices were misaligned. In a narrow range of angles centered at 0° and 60° the force peaked at values of about 0.25 nN, as shown in Fig. 18.10. A comparison with the simulations described in Section 16.2 allowed them to estimate that the flake attached to the tip was formed by approximately 80–90 C atoms. However, structural lubricity was also found to disappear after repeated scanning. A reasonable explanation is a sudden rotation of the graphite flake leading to alignment with the substrate [90]. Other experiments with nano-islands and carbon nanotubes are discussed in Section 19.1 and 19.3 in the context of nanomanipulation.

To conclude this chapter it is worth mentioning another experiment on flower-shaped islands of a lipid monolayer on mica, which consisted of domains with different molecular orientations. In this case Liley *et al.* observed a significant angular dependence of the friction force, reflecting the tilt direction of the alkyl chains of the monolayer. Friction anisotropy was also reported on graphene/Si and attributed to a rippling process caused by the probing tip (see also Section 20.2) [55].

19

Nanomanipulation

One of the greatest advantages of the AFM, compared to other imaging techniques, is the possibility of modifying the morphology of a surface while scanning it. This possibility is well exemplified by the number of nanolithography and nanomanipulation experiments reported in the literature. However, to use these techniques for practical applications, the motion of the probing tip must be controlled in order to pattern the surfaces in a desired way or to rearrange the manipulated objects in a desired configuration. In the case of nanomanipulation, assembling nanoparticles in a well-defined arrangement is usually a difficult and time-consuming task. Friction and adhesion forces between particles and substrates indeed play a major role in the motion on the nanoscale and, even if one works in a controlled environment, the size of the nanoparticles is usually comparable to that of the tip apex, which makes any attempt of controlling the manipulation process quite challenging.

19.1 Contact mode manipulation

One of the first examples of AFM manipulation was reported by Lüthi *et al.* [199], who succeded in moving compact C_{60} islands on a NaCl(001) surface by pushing them with the probing tip (Fig. 19.1). From the area of the islands determined by the AFM topographies and the values of the kinetic friction force recorded while sliding, a shear stress between C_{60} and NaCl of the order of 0.1 MPa was estimated. A larger shear stress, caused by the static friction and more difficult to quantify, accompanied the onset of motion. In another experiment, Sheehan and Lieber recognized the importance of the misfit angle in the manipulation of MoO_3 islands on an MoS_2 surface [307]. In this case, the islands could only move along low index directions of the substrate. Another experiment on Sb islands manipulated on the same substrate is presented in Section 19.3. Note that, instead of pushing a nanoisland, the tip can also be positioned on top of it. In this case, the friction force

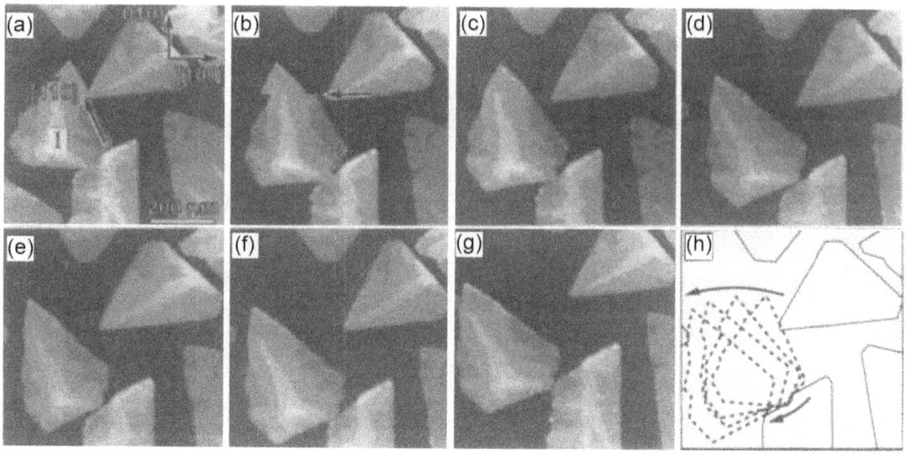

Figure 19.1 Sequence of AFM topography images of a C_{60} island manipulated on a NaCl(100) substrate in contact mode. Frame size: $530\times530\text{mm}^2$. Reproduced from [199] with permission from AAAS.

between tip and island must be larger than the friction between island and substrate in order to move tip and island together [75].

Manipulation of nanocrystals in water has been reported by Pimentel *et al.* [271]. In this environment the shear strength accompanying the detachment of calcite islands from dolomite and kutnahorite (104) surfaces was estimated to be 7 and 130 MPa respectively, in line with the different lattice mismatch between adsorbate and substrates.

Carbon nanotubes have also been manipulated using AFM. In their experiments on multiwalled nanotubes on graphite, Falvo *et al.* were able to distinguish between sliding and rolling motion [89]. A dramatic increase of the lateral force was observed in the directions in which the hexagonal surfaces of the two materials formed a commensurate contact. Indications of superlubricity have been recognized in the telescopic extension and retraction of multiwalled nanotubes, which showed no signs of wear after many repetitions [61].

19.2 Dynamic mode manipulation

Nanomanipulation can also be performed using the AFM in tapping mode. In this case, as shown by Aruliah *et al.*, it is possible to relate the energy dissipation (17.4) to the friction forces between particles and substrate [8]. As first demonstrated by Ritter *et al.* with latex spheres on graphite [293], one can position the tip at the side of a nanoparticle and increase the oscillation amplitude A till the particle starts moving. By changing the driving amplitude A_d (and hence the power input into the sample) it is possible to switch between an imaging mode and a manipulation mode. Since the lateral force applied to the particle depends on A^2, the accessible dynamic range is quite broad.

Mougin *et al.* used tapping mode manipulation to study the temperature dependence of Au nanospheres on SiO_2 [224]. The amplitude at which the detachment occurred was found to decrease at higher temperatures, showing that the particle detachment is a thermally activated process. The detachment threshold could be further lowered by coating the nanospheres with hydrophobic functional groups. In a similar way, Tripathi *et al.* investigated the static friction between Au clusters and graphite [333].

Note that the tapping mode cannot be used in UHV, and, in these conditions, the preferred dynamic AFM technique is the frequency modulation mode. However, this mode is very sensitive to external perturbations and the manipulation of nanoparticles with linear size beyond 10 nm usually results in the breakdown of the cantilever oscillations. This is not the case for single atoms [331] or organic molecules [180].

19.3 Nanoparticle trajectories during AFM manipulation

When nanoparticles are manipulated by AFM, the probing tip is usually hit towards the center of mass of the particle to be displaced. However, it is not easy to estimate the position of this point while imaging, especially if the particle has an irregular shape or its size is comparable with the tip apex. Furthermore, the alignment between tip and particle can be affected by thermal drift. These problems can be overcome if the particle is repeatedly hit from the side using a zigzag scan pattern. To prove that, we will discuss, more generally, how the scan pattern defines the direction of motion and the angle of rotation of the nanoparticles. If the particle (e.g. an atomically flat island) forms an extended contact with the substrate, the friction can in principle be estimated from the trajectories recorded during nanomanipulation.

Nanospheres

Consider a rigid sphere of radius R_p manipulated in tapping mode [281]. The tip has a conical shape with half-angle γ and is ended by a spherical cap with radius R_t. Along the scan path, the tip hits the particle as shown in Fig. 19.2(a). The sections of tip and particle parallel to the xy plane at the point of contact are circles with radii R_1 and R_2 respectively (Fig. 19.2(b)). The centers of the two circles will be denoted by O and P.

In a raster scan pattern (Fig. 19.3(a)) the tip moves forth and back along the x axis, it is displaced by a given distance b along the y axis, and moves again forth

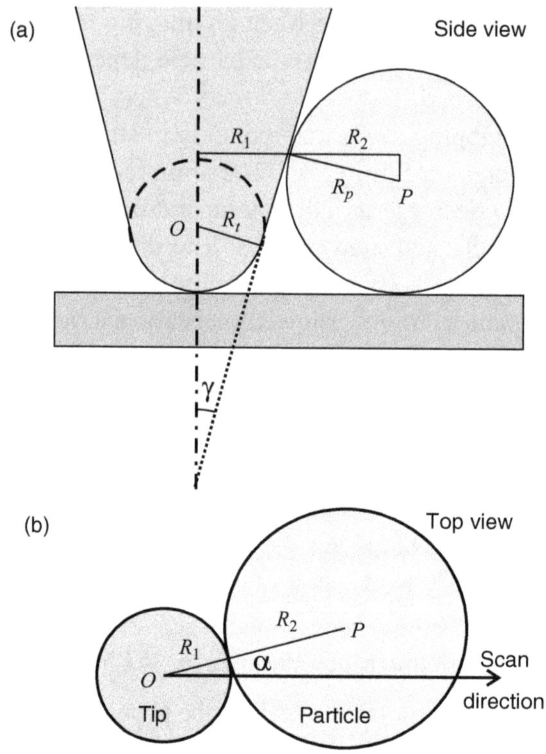

Figure 19.2 (a) Side view and (b) top view of a conical AFM tip colliding with a nanosphere.

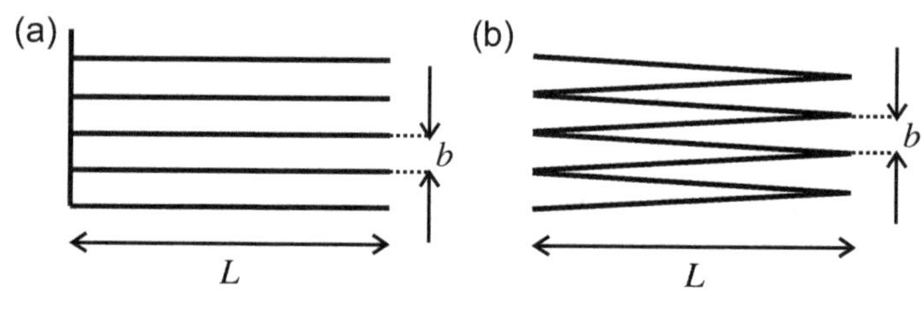

Figure 19.3 Most typical scan patterns adopted in AFM: (a) raster scan path and (b) zigzag scan path.

and back along x and so on. Suppose that the friction between sphere and substrate can be neglected when the tip hits the sphere, but it is high enough to stop the sphere immediately after the contact with the tip is lost. In this case the sphere displacement after each hit is determined by the equations

$$\frac{dy}{dx} = \tan\alpha, \qquad \frac{dy}{d\alpha} = R\cos\alpha, \tag{19.1}$$

where $R = R_1 + R_2$ and α is the angle between OP and the x direction.

Dividing the equations (19.1) by each other, we can relate the displacement of the sphere along x to the variation of α. At the end of the scan line, the total displacement Δx of the sphere is obtained by integrating the result between the initial and the final values of α, which we will denote α_0 and α_f respectively. The value of α_0 depends on the initial position of the sphere, and is not defined on the very first scan line. In the next scan lines, it is easy to see that α_0 is always the same, and is equal to

$$\alpha_0 = \arcsin(1 - b/R)\,. \tag{19.2}$$

When the sphere is completely displaced by the tip, OP forms an angle $\alpha_f = \pi/2$ with respect to the x direction. The total displacement of the sphere in the fast scan direction is thus

$$\Delta x = R\int_{\alpha_0}^{\pi/2} \frac{\cos^2\alpha}{\sin\alpha}\,d\alpha = -R\left(\cos\alpha_0 + \log\tan\frac{\alpha_0}{2}\right),$$

i.e. it is a function of the spacing b between consecutive scan lines (for a given tip shape). With the exception of the first scan line, the corresponding displacement along the slow scan direction is $\Delta y = b$. Thus, the angle of motion θ of the sphere (with respect to the x direction) is given by

$$\tan\theta \equiv \frac{\Delta y}{\Delta x} = -\frac{b}{R\left(\cos\alpha_0 + \log\tan\frac{\alpha_0}{2}\right)}. \tag{19.3}$$

To conclude the derivation, we have to consider the 3D shapes of tip and particle. With the geometry in Fig. 19.2(a), we distinguish two cases. If $R_p < R_t(1 - \sin\gamma)$ the particle is pushed by the spherical part of the tip and, from geometric considerations, R is twice the geometric average of the tip and particle radii: $R = 2\sqrt{R_t R_p}$. If $R_p > R_t(1 - \sin\gamma)$, the particle is pushed by the conical part of the tip, and

$$R = R_p(1 + \sin\gamma)\tan\gamma + R_t\frac{1 - \sin\gamma}{\cos\gamma} + R_p\cos\gamma.$$

The expression for R so obtained can be substituted into Eqs. (19.2) and (19.3) to get an analytic relation between the angle of motion θ and the 'spacing' b. In Fig. 19.4 this relation is used to fit experimental data on Au nanospheres with radius $R_p = 25$ nm manipulated by a tip with radius $R_t = 10$ nm and half-angle $\gamma = 5°$ on a silicon wafer [281].

A different relation is obtained if the tip moves along a zigzag scan path (Fig. 19.3(b)). In this case θ depends on the x coordinate of the nanosphere, and hence changes in each scan line. However, it can be proven that $\theta \to 90°$ if the spacing $b \to 0$ [281]. This suggests that, in order to move a sphere in a desired

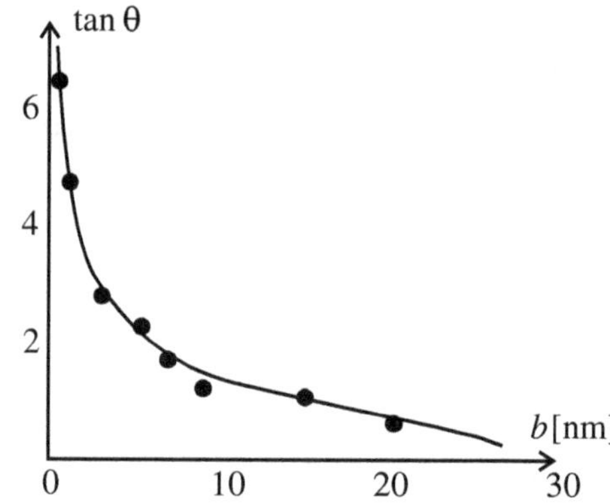

Figure 19.4 Direction of motion (filled circles) of Au nanospheres manipulated on an SiO2 surface in tapping mode AFM. The continous line represents the best fit of the experimental data with Eqs. (19.2) and (19.3). Adapted from [281] with permission from IOP Publishing.

direction, it is sufficient to scan the tip along a 'dense' zigzag pattern perpendicular to this direction.

Note that the model goes beyond the restrictive hypothesis that the particles stop immediately after losing contact with the tip. This has been shown by numerical simulations, where the distance d covered by the nanoparticles after being hit by the tip was introduced. If the friction between particles and substrate decreases, and hence d increases, the trajectory becomes more irregular, but the relation $\theta(b)$ remains essentially unchanged [282].

Other shapes

The previous analysis can be extended to particles with different shapes. For instance, the center of mass of a rigid nanowire of length L moves in the direction defined by the angle

$$\theta = \arccos(b/L)$$

if the raster scan pattern is adopted [111]. On each scan line the wire flickers around a configuration perpendicular to this direction.

The rotation accompanying the manipulation of non-spherical particles has been studied on star-shaped symmetric islands [105]. In Fig. 19.5 the 'angular velocity' $d\varphi/dN$, defined as the average angle of rotation of the islands per scan line, is compared to the direction θ of the center of mass. Depending on the values of b the rotation is 'quantized' at the angles defining the symmetry of the island, as

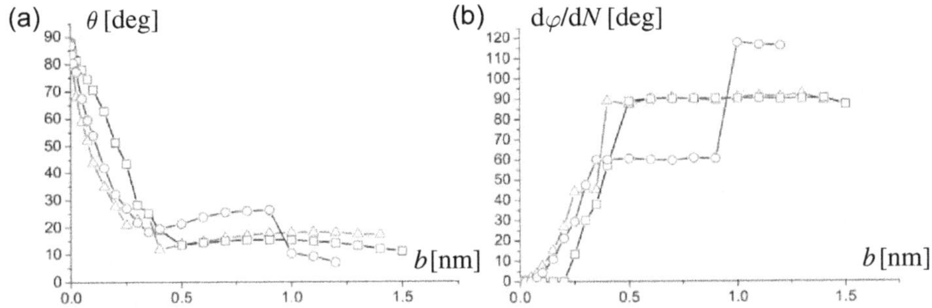

Figure 19.5 (a) Angle of motion and (b) angular velocity of star-shaped islands with 4, 6 and 8 branches (squares, circles and triangles respectively) as a function of distance between consecutive scan lines. Adapted from [105] with permission from Beilstein Institut.

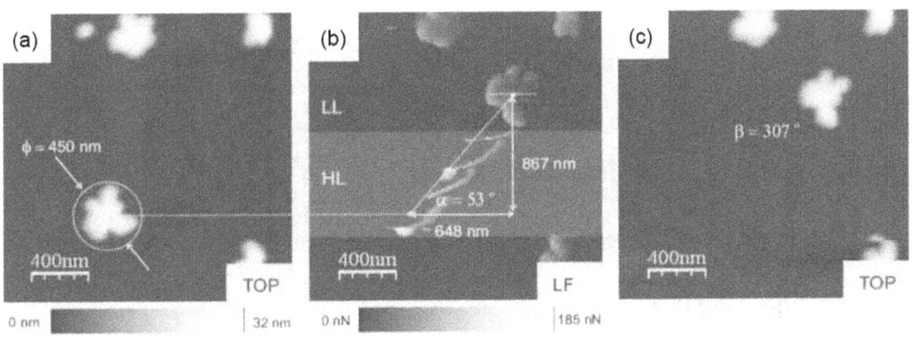

Figure 19.6 Manipulation of an Sb island in contact mode on an MoS_2 surface by contact mode AFM: (a) topography image before manipulation, (b) lateral force map during manipulation and (c) topography image after manipulation. In the 'HL' region the normal load was increased by 65 nN and the nanoisland was translated and rotated. In the 'LL' region the load was returned to the standard imaging value. Frame size: $2 \times 2\ \mu m^2$. Reproduced from [235] with permission from IOP Publishing.

shown by the horizontal plateaus corresponding to 45, 60, 90 and 120 degrees in the curves in Fig. 19.5(b).

The simultaneous translation and rotation of irregularly shaped islands can be recognized in the AFM images accompanying the manipulation of Sb islands on a MoS_2 surface in Fig. 19.6 [235]. The island profiles have been digitized and fed as an input to a collisional algorithm based on the previous assumptions. However, since the manipulation was performed in contact mode, the friction F_{fric} between island and substrate could not be neglected when the tip was pushing the islands. A shear strength between Sb and MoS_2 of the order of 0.2 MPa could be estimated by varying the value of F_{fric} in the simulations till the translation and rotation observed experimentally were simultaneously reproduced.

19.4 Lifting up molecular chains

The AFM can also be used to pick up a nano-object lying on a substrate and measure the normal and lateral forces accompanying the detachment. An impressive example has been provided by Kawai *et al.* [165], who managed to lift up isolated polyfluorene chains from a herringbone-reconstructed Au(111) surface in NC mode (Section 17.2). The experiment was performed in UHV at very low temperature (4K). Under these conditions, it was possible to measure the force gradients accompanying the detachment of the chains with extremely high accuracy (Fig. 19.7). The primary observation was a remarkable modulation of the normal force during detachment of single fluorene groups. This modulation could be precisely related to the adhesion energy of the groups using the extended FK model introduced in the end of Section 16.2. A small modulation of the force

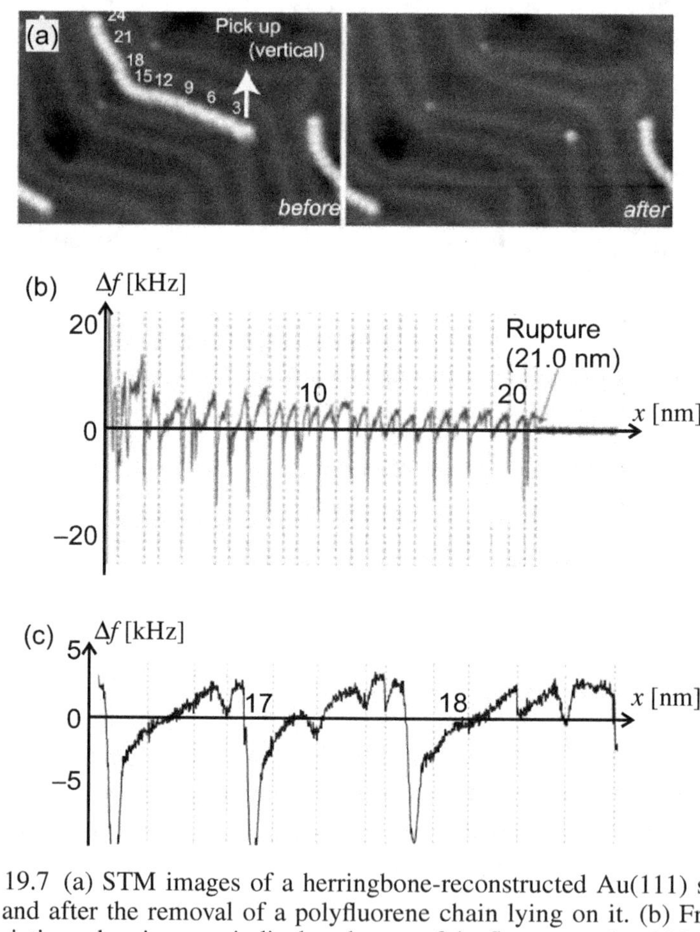

Figure 19.7 (a) STM images of a herringbone-reconstructed Au(111) substrate before and after the removal of a polyfluorene chain lying on it. (b) Frequency shift variations showing a periodic detachment of the fluorene units. (c) Magnified section showing minor variations caused by the sliding of the chain. Reproduced from [165] with permission from the National Academy of Sciences, USA.

gradient caused by the sliding on the gold surface was also observed. This secondary modulation indicates that the sliding is superlubric, as expected from the incommensurability between the lattice constant of the substrate and the equilibrium distance between consecutive fluorene groups. The AFM measurements also showed that the stiffness of the chemical bonds between groups is relatively large (around 200 N/m), which makes elastic deformation to accommodate the molecule on the substrate not favorable.

20

Wear on the nanoscale

In this chapter we present a series of illustrative AFM experiments on ionic crystals and layered materials demonstrating the onset of abrasive wear on the nanoscale. The results support the idea that the bond breaking in the worn surfaces is thermally activated. The energy dissipation can be estimated from the lateral force variations recorded while scraping. If a micrometric region is repeatedly worn off, wavy surface patterns can be formed on a variety of materials including polymers, metals and ionic crystals. An extension of the Prandtl–Tomlinson model with a variable interaction potential corresponding to the surface profile can partially reproduce this result. Still, geometric features such as the orientation of the wave profiles and distortions at the edges of the worn regions are not well understood.

20.1 Wear on the nanoscale

Figure 20.1 shows high-resolution FFM images of the damage produced on a KBr(001) surface repeatedly worn off by the probing tip in UHV [108]. The left side of the image corresponds to a groove obtained after 256 'scratches' with a normal force $F_N = 21$ nN. The small hill of material piled up at the end of the groove is imaged under different magnifications (and much lower values of F_N). The hill is formed by a few monolayers, into which the atoms recrystallized with the same structure as the undamaged surface. The topmost layers of the hill are nevertheless quite unstable and the hill is progressively reduced with time. While scanning the groove back and forth for about 30 minutes, the average lateral force $\langle F \rangle$ was found to saturate asympotically with the number of scans N:

$$\langle F \rangle (t) = F_\infty + (F_0 - F_\infty) e^{-N/N_0},$$

where $F_0 = 1.0$ nN, $F_\infty = 14.5$ nN and $N_0 = 5.45 \times 10^3$.

Although it is not easy to understand how the wear process is initiated and how the material is displaced by the tip, interesting hints come from the lateral force

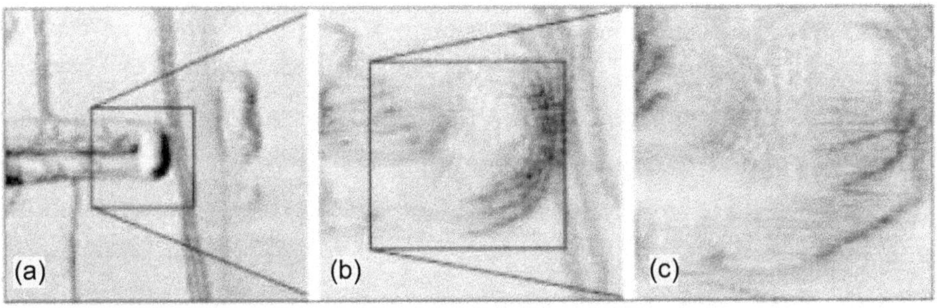

Figure 20.1 Series of lateral force images at the end of a groove previously scraped on a KBr(001) surface by contact mode AFM. Normal force value (for imaging): $F_N = 1$ nN. Frame size: (a) 115 nm, (b) 39 nm, (c) 25 nm. Reproduced from [108] with permission from the American Physical Society.

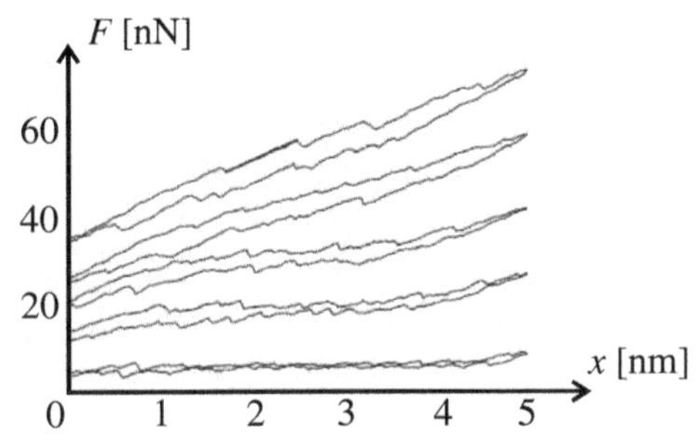

Figure 20.2 Lateral force loops acquired while scanning for the 100th time 5×5 nm^2 large areas on KBr(100) with increasing values of F_N: 5.7, 10.0, 14.3, 18.6 and 22.8 nN. Reproduced from [108] with permission from the American Physical Society.

profiles recorded during scraping. Figure 20.2 shows some friction loops acquired while scanning 5×5 nm^2 square areas with increasing values of F_N. The total energy dissipation is obtained as the mean lateral force multiplied by the travelled distance. The result are the pits in Fig. 20.3. The number of atoms removed by the tip could be estimated after imaging the damaged area with lower values of F_N. In the present case, up to 70% of the energy loss went into wearless friction. Figures 20.2 and 20.3 show how the lateral force and the damage increased with increasing load. On the other hand, changing the scan velocity between 25 and 100 nm/s did not result in visible variations of the pit shapes.

Assuming that the bond breaking accompanying the scraping process is a thermally activated process governed by the Arrhenius equation, the overall wear rate can be written as

$$\Gamma = \Gamma_0 \exp\left(-\frac{\Delta U - \sigma \Delta V}{k_B T}\right), \tag{20.1}$$

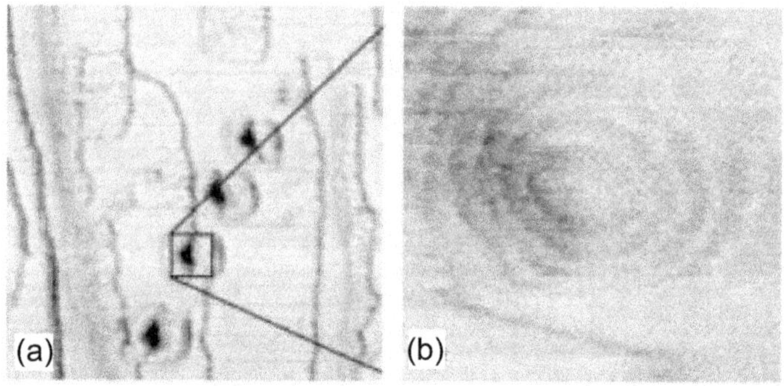

Figure 20.3 Lateral force images of the pits and surrounding mounds corresponding to the areas in Fig. 20.2. Frame size: (a) 150 nm, (b) 17 nm. Reproduced from [108] with permission from the American Physical Society.

where Γ_0 is an effective attempt frequency, ΔU is the (stress-free) energy of activation, σ is the stress component lowering the energy barrier, and ΔV is the so-called activation volume. Note that σ depends on the geometry, the way the forces are applied and the materials of both tip and surface. Equation (20.1) is consistent with original measurements by Jacobs and Carpick on the progressive wear of silicon nanotips [150]. In this case the authors visualized the damage and quantitatively estimated it using an in-situ TEM, which allowed then to resolve worn volumes of few tens of nm^3. Remarkably, they also observed that the wear of silicon against diamond is not accompanied by fracture or plastic deformation.

Another interesting experiment was performed by Kopta and Salmeron using AFM on mica [168]. In this case the authors assumed that the 0.2 nm deep scars in Fig. 20.4 resulted from the growth of defects accumulated beyond a critical concentration. Based on Eq. (20.1) one can assume that, at a given value of F_N, the number of defects created in the contact area $A(F_\mathrm{N})$ is

$$N_\mathrm{def}(F_\mathrm{N}) = t_\mathrm{res} n_0 A(F_\mathrm{N}) f_0 \exp\left(-\frac{\Delta E}{k_\mathrm{B}T}\right),$$

where t_res is the time of residence of the tip, n_0 is the surface density of atoms, and f_0 is the attempt frequency to overcome the load-dependent energy barrier, ΔE, required to break an Si-O bond. When the density of defects reaches a critical value, a hole is nucleated. The lateral force accompanying the creation of a hole could also be estimated as

$$F = A(F_\mathrm{N} - F_0)^{2/3} + C F_\mathrm{N}^{2/3} \mathrm{e}^{B F_\mathrm{N}^{2/3}}, \qquad (20.2)$$

where A, B, C and F_0 are constants. The first term on the right hand side of Eq. (20.2) gives the wearless dependence of the friction predicted by the DMT model (Section 10.2). The second term is the contribution of the defect production. A good agreement between Eq. (20.2) and the experiments can be seen in Fig. 20.5.

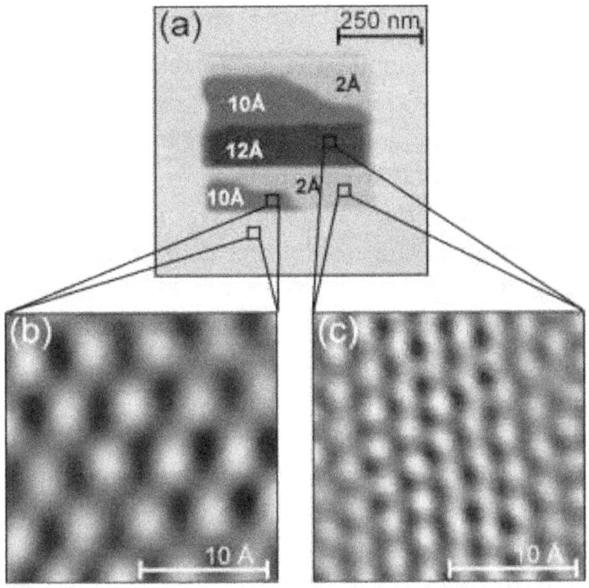

Figure 20.4 (a) AFM topography of a region on muscovite mica previously scanned with a normal force $F_N = 230$ nN. (b), (c) Fourier-filtered images of areas abraded at different depths, corresponding to crystal planes with different atomic arrangements. Reproduced from [168] with permission from AIP Publishing.

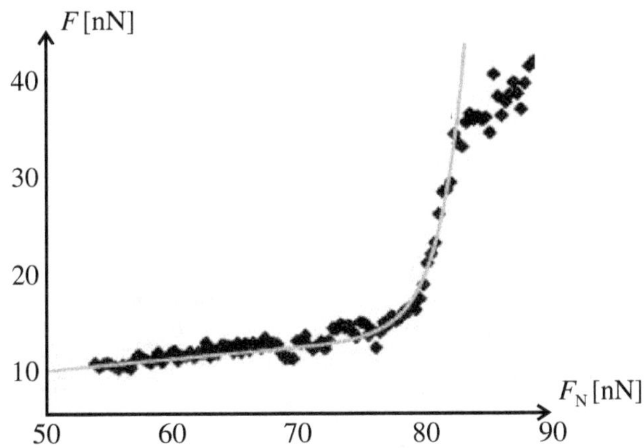

Figure 20.5 Friction vs. load curve during the generation of a hole in the mica surface. The rapid increase corresponds to the removal of a monolayer. Adapted from [168] with permission from AIP Publishing.

20.2 Surface rippling

When the shear stress on a compliant surface exceeds the yield strength of the material, a periodic wrinkle pattern is often observed. Similar phenomena have also been recognized at the nanometer scale on polymers [184], metals, ionic crystals [311] and semiconductors [326]. In these cases, the mechanical stress can be efficiently provided by an AFM tip elastically driven at constant velocity along the surface. An example of ripples formed by repeatedly scanning a straight line on a KBr(001) surface in UHV is given in Fig. 20.6 [311]. The scan line was oriented along the (100) direction of the surface and scanned hundreds of times forth and back with a normal force $F_N = 25$ nN (the image in Fig. 20.1 is actually the end part of one such groove). 'Travelling ripples' have also been reported using circular scan patterns [110]. In general, the ripple periodicity λ is slightly larger than the linear size of the tip apex, and changes with the material properties. This can be seen, for instance, in the rapid increase of λ when a polymer surface is heated above the glass transition temperature and the material enters the rubbery state [118, 304, 110].

To explain the occurrence of this kind of pattern various models have been proposed. Elkaakour *et al.* assumed that the rippling of polymer surfaces is caused by a peeling process in which the material is pushed ahead of the contact by crack propagation [85]. Filippov *et al.* reproduced the profile in Fig. 20.6(b) assuming that the material is removed atom-by-atom and randomly displaced aside at an angle depending on the tip shape [91]. In this case, the elastic spring restoring the tip apex in real AFM experiments was also taken into account. As seen in Section 15.1, the stiffness k of this spring can discriminate between stick–slip motion and continuous sliding on a crystal lattice. Since stick–slip occurs also when the ripples

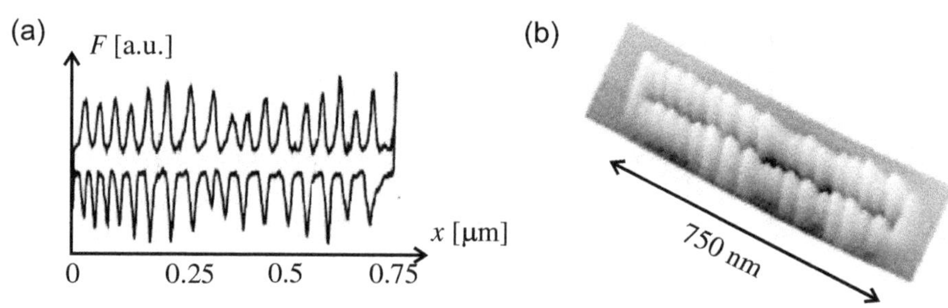

Figure 20.6 (a) Cross-section of the lateral force produced during the repeated scraping a 750 nm long groove on a KBr(001) surface. (b) Topography image of the groove after 512 scratches. The ripples have an average periodicity λ of about 40 nm and a corrugation of about 2.4 nm. Reproduced from [311] with permission from the American Physical Society.

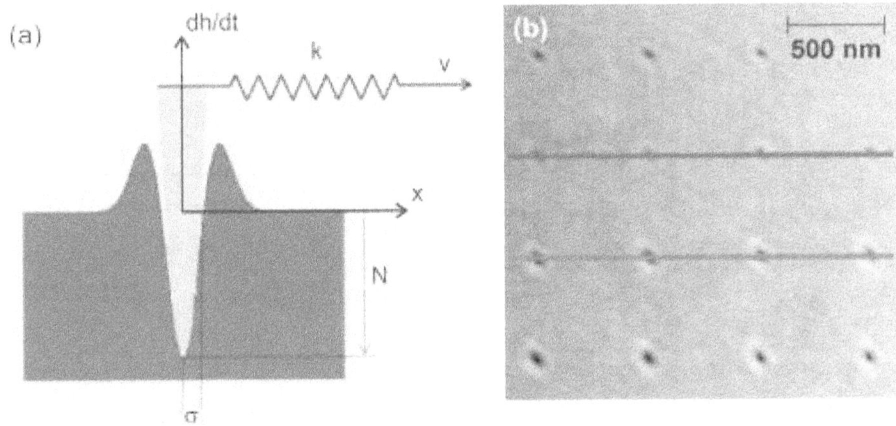

Figure 20.7 (a) The PT model as applied to the formation of surface ripples. A rigid tip is pulled by a spring of stiffness k (driven at a velocity v) and, at the same time, is indenting a compliant surface, evolving with time according to the profile in the figure. (b) A solvent-enriched polystyrene surface after indentation with normal forces of 400 nN (upper two rows) and 1200 nN (lower two rows) (courtesy of Dr. Franco Dinelli).

are formed (Fig. 20.6(a)), it seems to be possible to interprete the ripple evolution within the PT model.

Suppose that, in 1D, the tip–surface interaction potential U_{int} introduced in Section 15.1 changes with time according to the equation

$$\frac{\mathrm{d}U_{\text{int}}}{\mathrm{d}t} = N\left(-G_{x_0}(x) + \frac{1}{2}G_{x_0+2\sigma}(x) + \frac{1}{2}G_{x_0-2\sigma}(x)\right), \tag{20.3}$$

where $G_{x_0}(x)$ is a Gaussian profile with half-width σ centered at x_0 (Fig. 20.7(a)). This shape is suggested by the typical footprints left by the tip when indenting a polymer surface without scanning (Fig. 20.7(b)). The factor 1/2 corresponds to the assumption that the mass density remains constant in the indentation, and no material is displaced far away from the indentation site. Assuming that $U_{\text{int}}(x,t)$ resembles the surface profile $h(x,t)$, the parameter N can be interpreted as the indentation rate of the surface.

Equation (20.3) can be introduced into the PT model, with the tip position $x_0(t)$ defined as the minimum of the total potential $U(x,t)$ and $U_{\text{int}}(x,0) = 0$. In the present case the model parameters are the indentation rate N, the driving velocity v, the lateral stiffness k and the 'tip radius' σ. As a result, ripple profiles are obtained in a certain range of parameter values consistent with the AFM measurements (Fig. 20.8(a)). The tip, while pushed against the surface with a normal force F_N, builds up two hills ahead and behind the indentation pit. At the same time

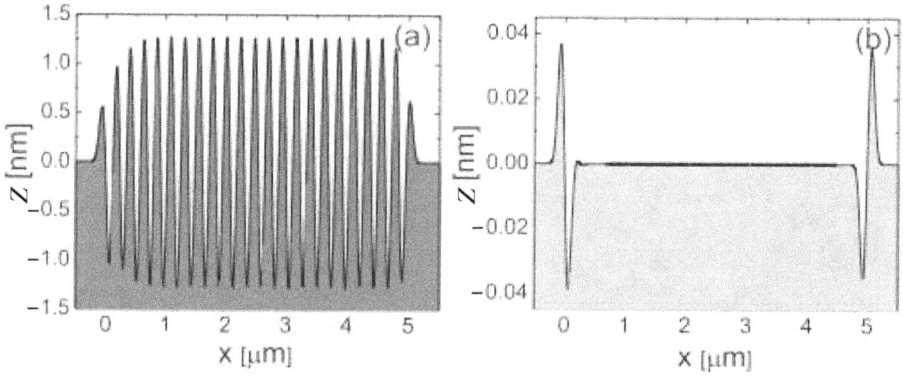

Figure 20.8 Surface profiles obtained with the modified PT model described in the text. Parameter values: lateral stiffness $k = 2$ N/m, tip half-width $\sigma = 50$ nm, scan velocity $v = 10$ μm/s, and indentation rate N(a) 100 nm/s, (b) 10 nm/s.

the lateral force increases, since the spring is continuously elongated. If the spring force reaches a certain threshold (depending on the corrugation of the ripples), the equilibrium becomes unstable and the tip suddenly hops beyond the hill ahead and sticks again into a new equilibrium position on the surface. The process is repeated several times.

If the indentation rate N is decreased the ripple pattern may disappear, as shown in Fig. 20.8(b). In this case the depth of the indentation pit and the height of the side hills are too low to prevent sliding, and the tip follows the spring support without stopping. Consequently, no ripples are formed and only two pairs of pits with corresponding hills at the beginning and the end of the scanned line are left as a result. This mechanism has a certain analogy with the transition from stick–slip to continuous sliding observed in atomic-scale friction experiments when the normal force is reduced below a critical value (Section 15.1).

Still, the previous 1D model cannot describe the patterns observed when a 2D region of the substrate is worn off [229]. As an example, Fig. 20.9 shows the ripple pattern formed after scanning an array of parallel lines at close distance from each other, and increasing F_N in well-defined geometric areas. The angle between the ripples and the fast scan direction increases with F_N and boundary effects are clearly visible. Here, we can notice a remarkable analogy with the direction of motion of rigid nanoparticles manipulated by AFM (Section 19.3). In the first scan line the material builds up in front of the tip until the mechanical equilibrium becomes unstable and the tip jumps over the mound so formed. In the next forward scan (at a distance b from the first one) the upper part of the bump is pushed again at an angle θ, which depends on b, as seen in Fig. 20.10. This results in the formation of a wavy pattern perpendicular to the flux direction defined by θ. The precise

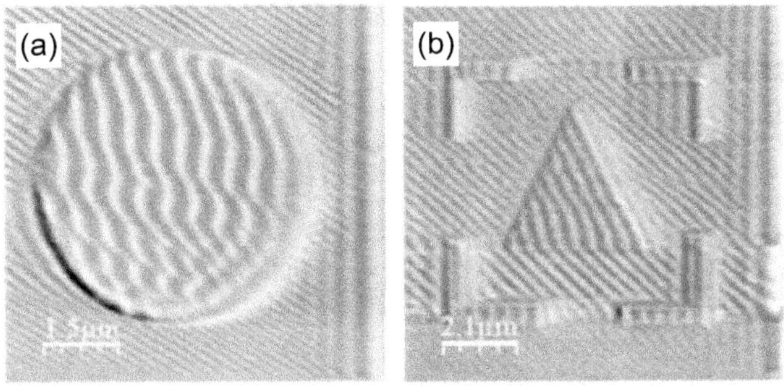

Figure 20.9 Ripple patterns on (a) a circular area and (b) a triangle surrounded by four L-shaped regions. The gray scale covers about 30 nm on both images. Reproduced from [229] with permission from IOP Publishing.

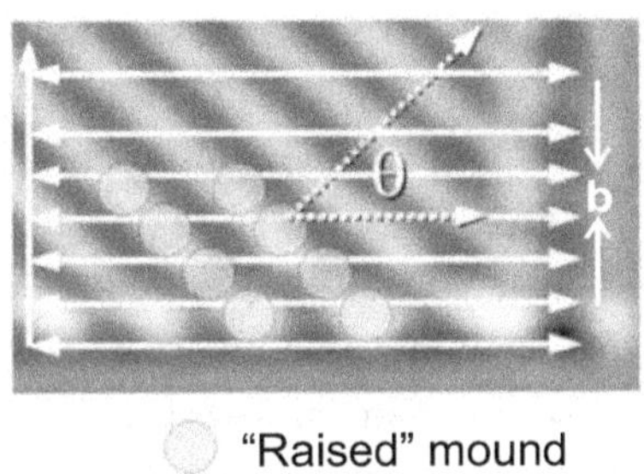

Figure 20.10 Analogy between surface rippling and nanoparticle manipulation by AFM. Similarly to an ensemble of nanospheres (Fig. 19.3(a)), a compliant polymer surface is reshaped perpendicularly to a direction θ defined by the distance b between consecutive scan lines. Reproduced from [229] with permission from IOP Publishing.

relation between θ and b depends on the material properties of the polymer and it is, at the present, unknown. Different ripple orientations were also observed by Schmidt *et al.* [304], who reported a branching of the ripple pattern corresponding to a zigzag scan path on a thin polystyrene film above T_c. Since the polymer surfaces are continuously reshaped when the tip slides over them, a detailed analysis of these processes is more difficult than for rigid particles manipulated on a solid surface.

21

Non-contact friction

The friction force observed between two moving bodies separated by a vacuum gap is called *non-contact friction* and was investigated for the first time by non-contact force microscopy setups [68, 78, 119, 322]. Non-contact friction forces are often in the range of 10–100 aN and corresponding damping coefficients, Γ, of the order of 10^{-13} kg/s. In some cases, e.g. on charge density wave systems, giant non-contact friction is observed, where friction coefficients of the order of 10^{-5} kg/s are measured [300, 179]. Non-contact friction forces are several orders of magnitude smaller than contact-friction forces, therefore ultrasensitive force detection is required to investigate them. To sense the smallest possible forces between two bodies the internal friction force of the force sensor has to be minimized. At separations below 1 nm, a rapid increase of frictional forces is observed. This regime is also called *near-contact friction*, where typical forces are of the order of some fractions of nano newtons. In this regime, chemical forces and tunneling currents are observed as well.

When the excitation of a damped oscillator is stopped, the amplitude of the oscillator will decay in time, which is accompanied by the conversion of kinetic energy into heat. The energy transfer lasts until the cantilever system reaches its thermodynamic equilibrium. In this state, stationary fluctuations from the mean value $\langle x \rangle$ are observed. Both decay time τ and equilibrium fluctuations $x(t)$ contain information about the dissipative process. The equation of motion of a linear damped harmonic oscillator can be written as

$$m_{\text{eff}} \frac{\mathrm{d}^2 x}{\mathrm{d}t^2} + \Gamma \frac{\mathrm{d}x}{\mathrm{d}t} + \omega_0^2 x = F_{\text{ext}}(t), \qquad (21.1)$$

where k is the spring constant, ω_0 the angular resonance frequency, Γ the friction coefficient, m_{eff} the effective mass and F_{ext} the external force.

The external force F_{ext} can be regarded as a superposition of a non-stochastic force and a stochastic force. Since Eq. (21.1) is a linear differential equation, both

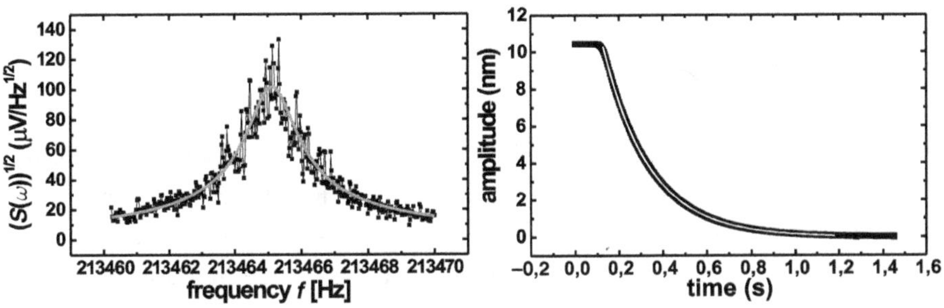

Figure 21.1 Left: example of a thermal spectrum, which has been fitted by (21.7), leading to the value $Q = 142\,914$. Right: oscillation of the cantilever as a function of time. This is called the ring-down method. The curve was fitted by (21.8) and the value $Q = 142\,280$ was found, in good agreement with the value from the thermal spectrum.

contributions can be treated separately. An experimentally accessible quantity to describe the dissipation process of a vibrating cantilever at its resonance frequency is its decay time τ. The simplest way to measure the decay time is a ring-down measurement (see Fig. 21.1). An exponential fit of the data yields the decay time. The decay time can be used to determine the quality factor Q:

$$Q = \tau\omega_0/2. \tag{21.2}$$

Alternatively, the friction coefficient Γ can be determined from the spring constant k and decay time τ:

$$\Gamma = \frac{k}{\omega_0 Q}. \tag{21.3}$$

Then, the non-conservative friction force F can be derived as

$$F = -\Gamma v, \tag{21.4}$$

where v is the velocity of the oscillator.

To give an idea about the orders of magnitude, parameters for typical force sensors are as follows. A soft free cantilever with a resonance frequency of 2.7 kHz, a vibration amplitude of x_0 = 20 nm, a Q-factor of 5×10^5 and $\Gamma = 10^{-14}$ kg/s experiences friction forces of $F = \Gamma\omega_0 x_0 = 3.4 \times 10^{-18}$ N. The dissipated power $P = \Gamma v^2 = 1.15 \times 10^{-21}$ W (7.17×10^{-3} eV/s). For conventional non-contact force sensors with a resonance frequency of 300 kHz, $\Gamma = 10^{-11}$ kg/s and a typical amplitude of 1 nm, the friction force is $F = 1.8 \times 10^{-14}$ N, which corresponds to a dissipated power of 3.5×10^{-17} W (221 eV/s).

The experimentally determined friction coefficient Γ can be modeled as a superposition of internal damping of the cantilever Γ_0 and damping due to dissipative interactions between the probing tip and sample Γ_s:

$$\Gamma = \Gamma_0 + \Gamma_s. \tag{21.5}$$

Approaching the free cantilever close to a flat surface opens new dissipative channels due to the long-ranging electromagnetic field between probing tip and surface, which give rise to the friction coefficient Γ_s. An important limitation is that soft cantilevers jump into contact when the attractive force gradient is larger than the spring constant, which happens on almost all surfaces because of the attractive van der Waals forces. An alternative way has been recently introduced by the use of the so-called *pendulum geometry* [296]. Spring constants of the order of mN/m are used, which greatly improves the force sensitivity to the level of attonewtons close to the surface. The cantilever is oriented perpendicular to the surface, which avoids a jump-into contact.

The first experimental demonstration was reported by Denk and Pohl, who analyzed the resonance of an oscillating cantilever with a metallic tip in electrostatic interaction with a heterostructured semiconducting sample [68]. In this experiment, the cantilever oscillation was damped by Joule dissipation of charge carriers which were moved by the oscillating electric field produced by the tip vibration. The authors pointed out that the damping deduced from the resonance analysis can also be obtained from the excitation amplitude A_{exc} needed to maintain a constant oscillation amplitude.

21.1 Experimental methods to measure non-contact friction

In non-contact force microscopy measurements the power dissipation P_0, caused by internal friction in the freely oscillating cantilever, is given by

$$P_0 = 2\pi f_0 \frac{kA^2/2}{Q}, \tag{21.6}$$

with f_0 the eigenfrequency of the freely oscillating cantilever. This dissipation is independent of the sample and cannot be avoided. It produces a background signal in which variations in the dissipation have to be detected. The extra dissipation P_s caused by tip–sample interaction can be calibrated by comparison with the intrinsic dissipation, once the Q-factor of the free oscillation is known.

There are different ways to determine the Q-factor of the cantilever far from the surface. One possibility is to record the amplitude spectrum of the thermal noise of the cantilever. The form of this spectrum (see Fig. 21.1) is given by

$$S(\omega) = \frac{2k_B T \omega_0^3}{Qk\left[(\omega^2 - \omega_0^2)^2 + \omega_0^2\omega^2/Q^2\right]}, \tag{21.7}$$

where $S(\omega)$ is the spectral amplitude density, $\omega = 2\pi f$ the angular frequency, and k the spring constant.

An accurate method for cantilevers with high Q-factor is the ring-down method. The amplitude A is measured as a function of time t after stopping the excitation. An example is shown in Fig. 21.1. The curve is analyzed with the equation

$$A(t) = A(0)\mathrm{e}^{-\pi(f_0/Q)t} \, . \tag{21.8}$$

Finally, the Q-factor can also be determined in a phase variation experiment. The frequency f and the excitation amplitude A_{exc} are recorded as a function of the phase φ while the oscillation amplitude is kept constant. The advantage of this procedure is that it can be applied with the tip in close proximity to the surface [189]. The relation between frequency and phase is given by

$$f = f_0\left(1 - \frac{1}{2Q}\tan\varphi\right) \, . \tag{21.9}$$

Phase variation experiments, as shown in Fig. 21.2, are in good agreement with thermal noise spectra and ring-down experiments. At close separations, non-linear effects may lead to distortions or broadening of the spectrum, which may lead to wrong Q-values for the thermal spectrum method (see [322]).

A convenient way to interpret the local measurements of the excitation signal A_{exc} is to calculate the power dissipation. As suggested by Cleveland *et al.* [58] and Gotsmann *et al.* [120], the power P_{s} dissipated by the interaction between tip and sample is given by the difference between the power which is delivered by the piezoactuator to the cantilever base P_{in} and the power which is used by the intrinsic damping of the cantilever (background dissipation) P_0:

$$P_{\mathrm{s}} = P_{\mathrm{in}} - P_0 \, . \tag{21.10}$$

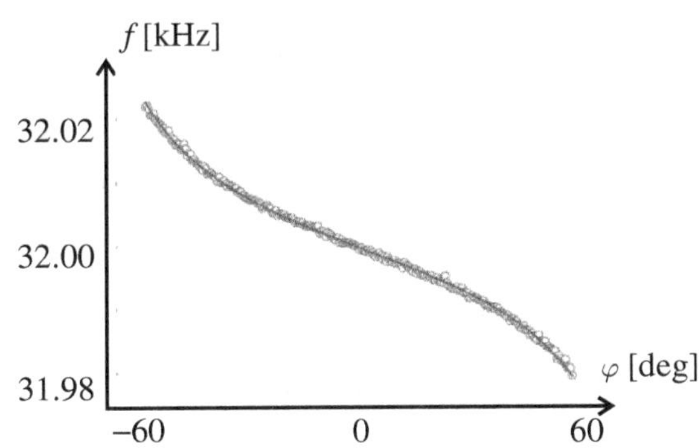

Figure 21.2 Phase variation experiment, in which (21.9) is used to determine the Q-factor.

This power dissipation can be determined from the measurement of the excitation signal A_{exc}:

$$\overline{P_s} = \frac{1}{2}\frac{kA^2\omega}{Q_0}\left(\frac{Q_0 A_{\text{exc}}\sin\varphi}{A} - \frac{\omega}{\omega_0}\right). \tag{21.11}$$

Since frequency shifts of the cantilever are relatively small, the term ω/ω_0 can be approximated by 1. Furthermore, the phase shift is given by $\varphi = 90°$ for dynamic force microscopy, which leads us to the equation

$$\overline{P_s} = \frac{1}{2}\frac{kA^2\omega}{Q_0}\left(\frac{Q_0 A_{\text{exc}}}{A} - 1\right) = \frac{1}{2}\frac{kA^2\omega}{Q_0}\left(\frac{A_{\text{exc}}}{A_{\text{exc},0}} - 1\right), \tag{21.12}$$

where $A_{\text{exc},0}$ denotes the excitation amplitude required to drive the oscillation with amplitude A far from the sample.

21.2 Internal friction of cantilevers

In order to measure non-contact friction, it is important to use cantilevers with small internal friction, which improves the force sensitivity. Therefore, a short summary of mechanisms of internal friction of cantilevers is discussed in this section. The temperature dependence of resonance frequency of cantilevers is rather well understood [127]. Geometry changes due to thermal expansion can be neglected. However, the temperature dependence of the Young's modulus, $E(T)$, is given by the *Wachtman formula*:

$$E(T) = E_0 - BT\exp\left(-\frac{T_0}{T}\right), \tag{21.13}$$

where T_0 is related to the Debye temperature of the sensor material. With Eq. (21.13) the temperature-dependent resonance frequency can be calculated:

$$\omega_n = \alpha_n^2 \frac{t}{L^2}\sqrt{\frac{E}{12\rho}}, \tag{21.14}$$

where $\alpha_1 = 1.875$ for the first eigenmode, t is the thickness, L the length and ρ the mass density. In case of silicon cantilevers, the experimental frequency vs. temperature data are well fitted with $T_0 = 317$ K [127]. According to $D = T_0/2$ a Debye temperature of $D = 634$ K is determined, which is in good agreement with literature values of $D = 645$ K for silicon.

In contrast, the damping of cantilevers is still rather poorly understood. Several contributions have to be distinguished:

1. damping due to thermoelastic damping;
2. damping due to bulk losses;
3. damping due to surface losses;

4. damping due to acoustic emission into the bulk;
5. losses due to the clamping;
6. viscous damping due to the presence of gases or liquids.

As far as ultrasensitive measurements under ultrahigh vacuum conditions are concerned, the influence of viscous damping at pressures below 10^{-6} mbar can be neglected. The influence of clamping can be optimized by rigid holders and the exclusion of glues with high damping rates. Damping due to acoustic emission is also found to be negligible in most practical cases. Therefore, the first three mechanisms are the most important ones.

Thermo-elastic damping

The conduction of heat is an important energy loss mechanism. Periodical compression and expansion of oscillating micromechanical elements is associated with heat flow between compressed and expanded areas. The *Zener model* is a continuum model of this thermo-elastic damping mechanism [350, 185]. The internal friction is given by

$$Q^{-1} = \frac{\alpha^2 T E}{\rho c_\mathrm{p}} \frac{\omega\tau}{1 + (\omega\tau)^2}, \tag{21.15}$$

where α is the thermal expansion coefficient, c_p the specific heat capacity and ρ is the mass density. The relaxation time τ is given by

$$\tau = \frac{t^2}{\pi^2} \frac{\rho c_\mathrm{p}}{\kappa}, \tag{21.16}$$

where κ is the thermal conductivity. Typical parameters for silicon at room temperature are E=1.68 GPa, $\alpha = 2.54 \times 10^{-6}\ \mathrm{K}^{-1}$, $\rho = 2.33 \times 10^3\ \mathrm{kgm}^{-3}$, $c_\mathrm{p} = 711\ \mathrm{J\,kg}^{-1}\mathrm{K}^{-1}$ and $\kappa = 150\ \mathrm{Wm}^{-1}\mathrm{K}^{-1}$. The temperature dependence of the Young's modulus is small compared to the strong variations of thermal expansion (zero crossings at 20 K and 125 K). The thermal conductivity in the bulk varies between 100 and 5000 $\mathrm{Wm}^{-1}\mathrm{K}^{-1}$. One should also take into account that the thermal conductivity is reduced due to phonon-boundary scattering for thicknesses of the order of microns below 100 $\mathrm{Wm}^{-1}\mathrm{K}^{-1}$ at temperatures below 30 K [9]. At present, many experimental data indicate that thermo-elastic damping is the dominant loss mechanism at room temperature [198]. At temperatures below 200 K other channels start to dominate, which may be related to bulk or surface losses.

Bulk and surface losses

The scattering of elastic waves with defects on the surface or in the bulk is an important loss mechanism. The oscillation of the cantilever leads to a time-dependent local stress field. The energy landscape of the defects is changed by this stress field. Instabilities of these defects may occur, where atoms jump from one equilibrium position to another position. The energy difference between equilibrium positions is the activation energy. Therefore, damping vs. temperature curves show activation peaks, also called Debye peaks. So far, most of the experimental work is limited to silicon cantilevers which exhibit the highest Q-factor of available cantilevers. Typical Q-factors are between 10^4 and 5×10^5. Comparable cantilevers made of Si_3N_4 or SiO_2 show much smaller Q-factors of 100 to 1000. Therefore, we conclude that bulk losses are dominant for these amorphous structures. In the case of silicon, bulk or surface losses may become dominant at temperatures below 200 K. At 160 K a peak is observed, which may be related to such an activation peak with an activation energy of 0.25 eV. Unfortunately, the nature of these defects in silicon is still poorly understood. Simple defects, such as vacancies or interstitials, are ruled out because of their high activation barriers [127]. Recently, it has been observed that the 160 K peak can be reduced strongly by annealing under vacuum conditions [133]. It is also observed that the peak does not shift with the resonance frequency, which is not in agreement with the simple activation energy model. The authors suggest that the 160 K peak is related to an adsorbate layer. Another peak at 30 K shifts with the resonance frequency and seems to be in better agreement with a Debye peak [133].

Coating of cantilevers leads to a strong increase of dissipation. The polycrystalline nature of these metallic films implies grain boundaries, where increased phonon scattering leads to an increase of damping losses. Other surface coatings, such as silicon oxide, or adsorbates, such as H_2O or hydrocarbons, lead to rather large damping losses. Yang *et al.* annealed extremely small cantilevers (length $<$ 80 μm) [345]. They used rather high annealing temperatures (1000 °C), which was sufficient to remove the oxide layers. Recently, it has been shown that annealing at temperatures below 600 °C of rather large silicon cantilevers (length of 400 or 500 μm and thickness of 0.5 to 1.5 μm) under ultrahigh vacuum (UHV) conditions can also lead to a reduction of dissipation [283]. The quality factor of 6.2×10^4 could be improved by an order of magnitude after 6 hours annealing at temperatures below 600°C. Further annealing improved the quality factor to 1.24×10^6. The annealing temperature was too low to remove the oxide layer. Thus, the removal of weakly bound molecules, such as H_2O, OH ions or hydrocarbons, improves the quality factor. It could also be demonstrated that it is possible to reconstruct the cantilever surface with induced defects with annealing temperatures below 600 °C. Therefore a silicon bar cantilever with a length of 300 μm, a width of 35 μm and a thickness of 1 μm with a quality factor of 3×10^6 was sputtered with argon ions. After the ion bombardment a quality factor of 3×10^3 was measured. A layer thickness of the cantilever was reduced by approximately 7–9 nm. One can assume that by the ion bombardment the surface structure was modified. The cantilever was annealed for

15 hours and the initial quality factor could be reached again. This simple experiment demonstrates that the surface can already be modified at temperatures below 600 °C.

Alternatively, defects on the surface or in the bulk of the cantilever may be reduced by the annealing procedure. It is evident that the internal friction coefficient reduces in a rather continuous way down to 9 K, which is in qualitative agreement with previous studies of internal friction of silicon [350, 127]. Below 9 K a small plateau of constant dissipation is observed. Some of the annealed cantilevers are found to have attonewton-force sensitivity even at room temperature due to Q-value enhancement, which opens new possibilities for experiments, such as magnetic resonance force microscopy at room temperature or cantilever mass spectroscopy. The force sensitivity of commercial silicon sensors with a spring constant of 0.176 N/m changes by an order of magnitude after annealing. Quality factors higher than 6×10^6 can be reached after heat treatment. Cantilevers with mN/m spring constant were fabricated by the use of a dry etching method [183]. For these cantilevers, which have a thickness of 200 nm, the Q-factor can be improved by a factor of 100.

21.3 Origins of non-contact friction

The dissipation between two moving bodies separated by a distance d is due to electromagnetic interactions (see Fig. 21.3). In the case when no external field is applied, *van der Waals friction* occurs. The non-contact friction is related to an asymmetry of the reflection coefficient along the direction of motion. If one body

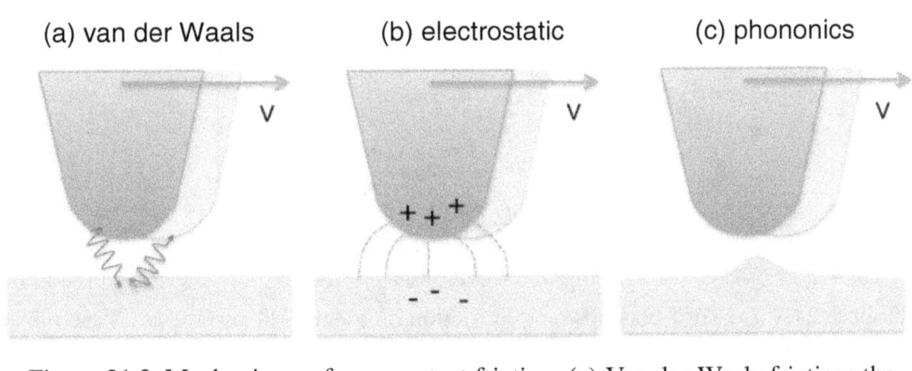

Figure 21.3 Mechanisms of non-contact friction. (a) Van der Waals friction: the returning back-action photon experiences Doppler shift. (b) Electrostatic friction: the induced charge follows the tip motion and experiences Joule losses. (c) Phononic friction: the elastic deformation follows the tip motion and induces phonons.

emits radiation, the waves are Doppler-shifted in the rest-frame of the other body, which will result in a different reflection coefficient. The same is true for the second body. The exchange of Doppler-shifted photons is the origin of van der Waals friction. *Electrostatic friction* occurs when an electrical field is applied between the two surfaces. Mirror charges in combination with the relative movements lead to Joule losses. Adsorbate layers or 2D-systems can lead to high effective conductivity, which enhances this type of electrostatic friction. Static electric fields between two different surfaces can exist without any externally applied voltage due to different work functions of different orientations of the crystallites of a polycrystalline surface. These so-called *patch forces* can lead to electrostatic friction in the case of compensated average contact potential. *Phononic friction* is related to the local deformation of the surfaces, which leads to the creation of phonons.

Stipe *et al.* [322] observed electrostatic dissipation at separations of 1–200 nm by using ultrasensitve force sensors (see Fig. 21.4). A gold tip was attached to

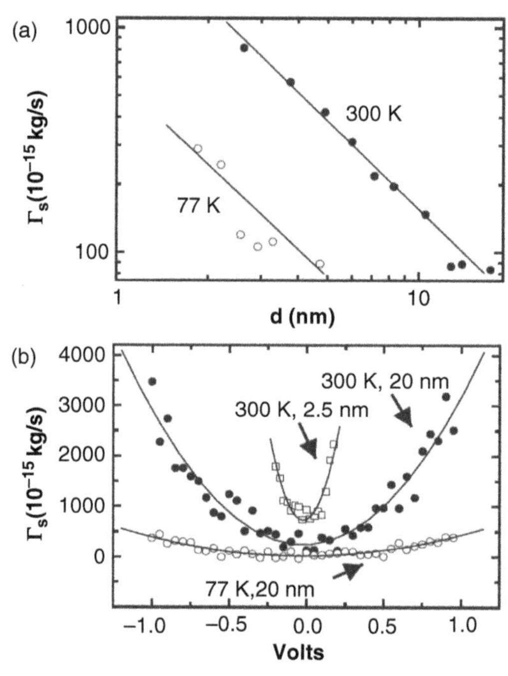

Figure 21.4 (a) Zero-bias tip–sample friction on a Au(111)-surface as a function of distance for temperatures of 300 and 77 K. Data were taken by the ring-down method with an ultrasoft sensitive cantilever, of 0.3 mN/m, and a Au-coated probing tip with an initial amplitude of 10 nm. (b) Friction as a function of bias voltage. Note that friction at 300 K for $d = 20$ nm is approximately six times larger than at 77 K, regardless of voltage. Reproduced from [322] with permission from the American Physical Society.

an ultra-sensitive cantilever in the pendulum geometry. Friction coefficients of the order of 10^{-13} kg/s between tip and metal substrate were observed. An increase of dissipation with increasing temperature was observed. The distance dependence of the friction coefficient was fitted by a power law $\Gamma \propto d^{-n}$ with an exponent $n = 1.3 \pm 0.2$. The measurements of Stipe *et al.* were performed under high vacuum conditions. The friction coefficient fits to a quadratic power law $\Gamma \propto (V_{bias} - V_{cpd})^2$. The friction coefficient of the force sensor changes by approximately 3×10^{-12} kg/s by applying 1 V at room temperature.

Chumak *et al.* have calculated the non-contact friction due to Joule losses [56]. For a clean metallic surface, they derived the formula

$$\Gamma = 3^{1/2} \frac{\sqrt{R} V^2}{2^7 d^{3/2} \pi \sigma}. \tag{21.17}$$

The distance dependence is in agreement with the experimental results from Stipe *et al.* [322]. By assuming a tip radius $R = 1$ μm, the conductivity of gold at 300 K, $\sigma = 4 \times 10^{17}$ s^{-1} $= 4.4 \times 10^7$ Ω^{-1} m^{-1}at a tip sample distance of 10 nm we obtain $\Gamma = 1.4 \times 10^{-23}$ kg/s. This value is ten orders of magnitude smaller than the experimental value. For a cylindrical tip the friction increases by two orders of magnitude, which is still too small to explain the experimental results. Volokitin, Persson and Ueba [336] have then considered the presence of an adsorbed layer. In this case, the electrostatic friction is increased by orders of magnitude. A formula in analogy to Eq. (21.17) is derived, where an effective conductivity is introduced. They find agreement for an effective conductivity $\sigma_{eff} \approx 4 \times 10^9$ s^{-1}. This value is reasonable under the assumption of surface coverage with adsorbates of 10%.

At compensated contact potential the dissipation might be related to van der Waals friction. However, even at compensated contact potential there are some uncompensated charges, which are due to the inhomogenieties of the probing tip and sample. The work functions of different facets, such as (111) or (100), are different. The spherical tips have different crystal orientations, which leads to incomplete compensation and some remaining charges on the facets. Therefore, the minimum of non-contact friction is most probably not related to van der Waals friction, but to fluctuations of charges and associated electrical fields, which give raise to the remaining non-contact friction at compensated contact potential. Volokitin *et al.* [336] have also shown that charge fluctuations in the bulk of dielectrics can give dissipation values comparable to the experimental values of the order of 10^{-12} kg/s. Nonetheless, Volokitin *et al.* [336] have predicted that van der Waals friction can be large enough to be measured by state-of-art non-contact force microscopy.

The quadratic behavior of the friction coefficient as a function of the bias voltage confirms the assumption that the friction is of electromagnetic origin. Several contributions can enhance the friction between tip and sample. Stipe *et al.* [322]

observed that artificial electric fields generated by defects (*E′ centers*) in gamma-irradiated quartz have an influence on the friction coefficient. The higher the defect concentration the higher the value of distance-dependent friction coefficient. Coverage with a 20 nm thick gold film prevents a penetrating of the electrical field into the non-conducting substrate, so that damping is reduced again. No extra damping is observed due to the presence of layers below the gold film.

The temperature dependence shows clearly an enhancement of non-contact friction with increasing temperature. In accordance with the fluctuation dissipation theorem, the force fluctuations are given by

$$S_{\mathrm{F}} = 4\Gamma_{\mathrm{S}} k_{\mathrm{B}} T. \tag{21.18}$$

If we assume that the field fluctuations are small compared to the overall applied field, the force fluctuations are given by:

$$S_{\mathrm{F}} = q^2 S_{\mathrm{E}} = C^2 V^2 S_{\mathrm{E}}, \tag{21.19}$$

which is in agreement with the observed V^2-dependence, if S_{E} is a constant independent of voltage [322]. For $\Gamma = 3 \times 10^{-12}$ kg/s, $C = 10^{-16}$ F one finds $S_{\mathrm{E}}^{0.5} = 2$ $\mathrm{Vm^{-1}Hz^{-0.5}}$, which corresponds to charge fluctuations of the order of 2×10^{-5}e $\mathrm{Hz^{-0.5}}$, comparable to values observed in single-electron transistor experiments.

Electronic versus phononic friction

From the above discussion, we expect that friction on a metal substrate should drop when the metal is cooled below the superconducting critical temperature T_{c}. This effect was first measured by Krim *et al.* using the quartz crystal microbalance (QCM) technique [64]. The temperature variation of the friction coefficient Γ across the superconducting transition was measured by Kisiel *et al.* oscillating a cantilever in close proximity to a Nb film surface like a pendulum [166]. As shown in Fig. 21.5, the damping coefficient is reduced by a factor of 3 when $T < T_{\mathrm{c}}$ at a separation of about 5 nm. The voltage dependence in the metallic state is proportional to $(V - V_{\mathrm{CPD}})^2$, in agreement with the theoretical expectations for metallic surfaces and the previous experimental results. In the superconductive state the non-contact friction is proportional to $(V - V_{\mathrm{CPD}})^4$, which indicates that electronic friction vanishes and phononic friction takes over (see Fig. 21.6). The $(V - V_{\mathrm{CPD}})^4$-dependence was predicted theoretically by Volokitin *et al.* for the phononic case [336]. Furthermore, the distance dependence in the superconductive state shows a more rapid decay of non-contact friction, which is proportional to $d^{-3.8}$. Again, Volokitin *et al.* predicted a comparable distance dependence of the friction coefficient on d^{-4}.

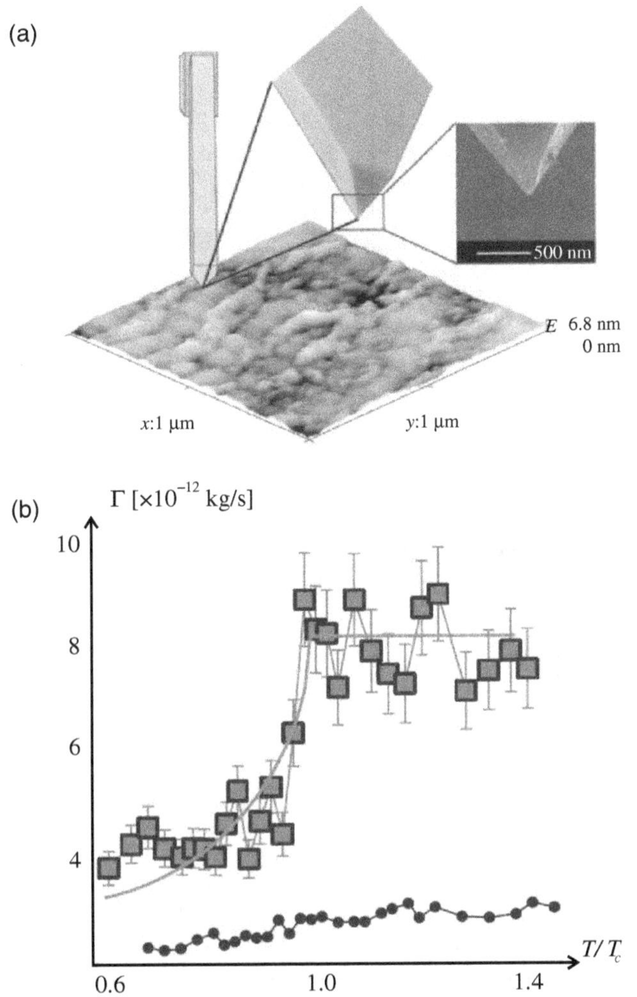

Figure 21.5 (a) AFM topography image of the Nb film studied with a pendulum AFM setup (frame size: 1×1 μm^2). The z-controller was feed-backed to the cantilever excitation signal. A scanning electron micrograph of the ultra-soft Nanosensors Arrow-TL1 cantilever probe is shown in the top right corner. (b) Temperature variation of the damping coefficient across the critical point $T_c = 9.2$ K of Nb. The line is the fit based on the BCS-theory, which implies that this friction coefficient is proportional to the number of unpaired electrons. The black points show the temperature dependence of the internal friction of the cantilever far away from the surface. Reproduced from [166] with permission from the Nature Publishing Group.

Two related experiments may be mentioned. Electronic friction was also demonstrated in doped semiconductors, where the local carrier concentration could be controlled through the application of forward or reverse bias voltages between an

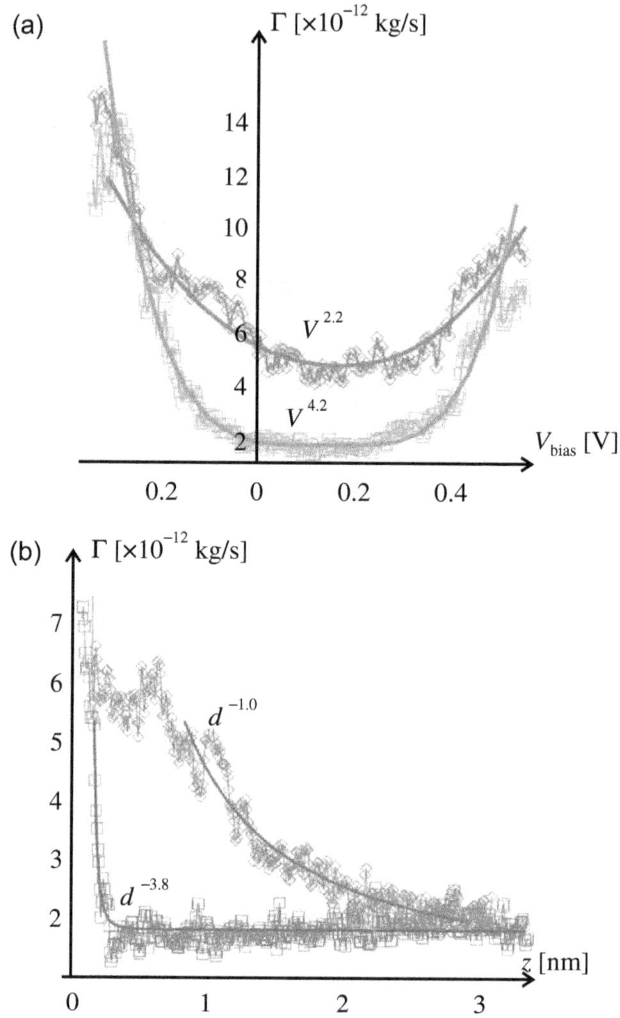

Figure 21.6 (a) Voltage and (b) distance dependence of the friction coefficient Γ in the metallic and superconductive state. The squares and diamonds refer to the superconductor and the metal state respectively. Reproduced from [166] with permission from the Nature Publishing Group.

AFM tip and the sample in the p and n regions [239]. In this case, repulsive contact forces were applied and friction in the contact regime was measured. A related experiment was performed by Cannara *et al.* [42], who investigated hydrogen- and deuterium-terminated diamond and silicon surfaces. In all cases the hydrogenated surfaces exhibited higher friction. In this case, where repulsive contact force were applied, phononic friction seems to dominate and the lower natural frequency of the deuterium atoms reduces the rate of dissipation.

21.4 Giant non-contact friction

Saitoh *et al.* [300] reported the occurrence of giant non-contact friction between the surfaces of $NbSe_2$ and $SrTiO_3$ and a sharp Pt-covered probing tip. At temperatures of 4.2 K, the friction coefficient showed a giant maximum of the order of 10^{-5} kg/s at tip–surface distances of several nanometers. The large distance excludes repulsive contact formation and mechanical instabilities. The maximum is at distances where chemical forces and tunneling currents are still rather small. They observed that the maximum of non-contact friction is drastically reduced at room temperature. The conductivity was found to be of minor importance, because the peaks were found for metals and insulators. In the case of $NbSe_2$, which is superconductive below 7.2 K, the authors found a maximum below and above T_c. However, they found no maximum at room temperature. Saitoh *et al.* suggest that the giant non-contact friction is related to a Debye-like relaxation mechanism with multiple time scales. In the case of of $SrTiO_3$ a broader peak is observed compared to the sharper maximum of $NbSe_2$ at about 2 nm. Langer *et al.* [179] have studied the case of $NbSe_2$ in more detail. They found a mutiplet of dissipation peaks at distances of several nanometers (see Fig. 21.7). They were able to relate the giant dissipation to the existence of charge density waves (CDW). If the temperature is increased above 70 K, where CDW short-range order is known to disappear, the peaks disappeared as well. The authors suggest that the probing tip couples to the extended charge density wave. The probing tip leads to 2π slips of the phase of the CDWs, which leads to hysteresis and dissipation.

21.5 Near-contact friction

At distances of less than 1 nm, where chemical forces start to emerge and tunneling currents flow, a strong increase of the friction coefficient is observed. This type of non-contact friction is orders of magnitudes larger than the non-contact friction due to charge fluctuations or Joule losses. Typically, forces of the order of fractions of nano-Newtons and friction coefficients of the order of 10^{-5} to 10^{-8} kg/s are observed. The rapid increase of the friction coefficient is commonly fitted by an exponential function with decay lengths comparable to the decay lengths of chemical forces or tunneling currents. The origins of this type of friction, often called *near-contact friction*, are discussed below.

An early example was presented by Lüthi *et al.*, where atomic-scale variation of the damping coefficient in a non-contact or near-contact force setup on a Si(111) 7×7 surface were found. The strongest friction coefficient was found at the sites of the corner holes [200].

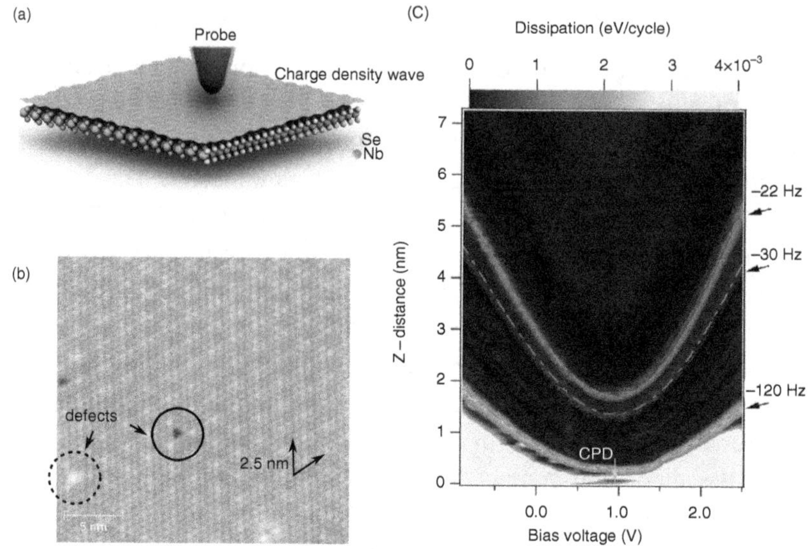

Figure 21.7 Observation of charge density wave on an $NbSe_2$ surface and accompanied non-contact friction. (a) An oscillating AFM tip in proximity to the charge density wave on the $NbSe_2$ surface. (b) Constant current STM ($I = 10$ pA, $V = 5$ mV) image of the $NbSe_2$ surface, showing a hexagonal CDW induced Moiré pattern as well as two types of surface defect – adsorbed CO molecules (dashed circle) and Se atom vacancies (circle). (c) Energy dissipation between the $NbSe_2$ surface and the pendulum AFM tip versus tip–sample distance Z and bias voltage V. The bright regions on the image stand for large non-contact friction. The friction increase is at the same cantilever frequency $f = f_0 - 22$ Hz, $f = f_0 - 30$ Hz, $f = f_0 - 120$ Hz, and the constant frequency contours are shown with chained curves. The measurements were acquired at $T = 6$ K. Reproduced from [179] with permission from the Nature Publishing Group.

The sharp impact of the short-range forces acting at the lower turning point of the tip oscillation may create phonons in the sample. This pathway to dissipation has been studied within the framework of continuum mechanics [80], the fluctuation–dissipation theorem [99], and molecular dynamics simulations [1]. The values found in these studies are orders of magnitude smaller than the experimentally measured dissipation. Nevertheless, phonon excitation by the tip may play an important role for soft materials and force microscopy modes including repulsive contact formation.

The experimental results of dissipation force microscopy might also be influenced by non-linear effects [100]. A non-linear force law can cause a frequency spectrum with manifolds, where only one branch is accessible for the experiment. These asymmetric frequency spectra might mimic an increase in the Q-factor. The examination of frequency spectra could disentangle this effect from real dissipation effects. Experiments by Erlandsson [87] indicate the non-linear character of the resonance, while phase variation experiments by Loppacher *et al.* [189] are in accordance with the harmonic approximation.

Another possibility to explain the changes in the driving force at the resonance frequency is the excitation of higher oscillation modes by the short force pulse of the short-range forces. However, Pfeiffer *et al.* have shown that the higher bending modes are negligible with good accuracy during constant amplitude operation [269]. On the other hand, higher harmonics nf_0 may play a more important role. These harmonics are difficult to quantify due to their artificial appearance in the frequency spectrum caused by control electronics.

Sasaki *et al.* proposed that atomic-scale instabilities on the tip or on the sample may be important for understanding dissipation force microscopy [303]. In close analogy to the Tomlinson model in FFM, a second minimum of the potential is formed during the approach, where a tip or sample atom jumps out of its equilibrium position. During retraction the atom jumps back to the original position. This type of mechanism may explain the rather large energy losses without involving permanent changes to the tip–sample geometry. This explanation was further confirmed by a study of Alireza Ghasemi *et al.* [4], where the potential energy landscape of realistic silicon probing tips was investigated by ab initio calculations. Many energetically close local minima were found. When thermal excitation is present, configurations driven into metastability by the tip motion can access

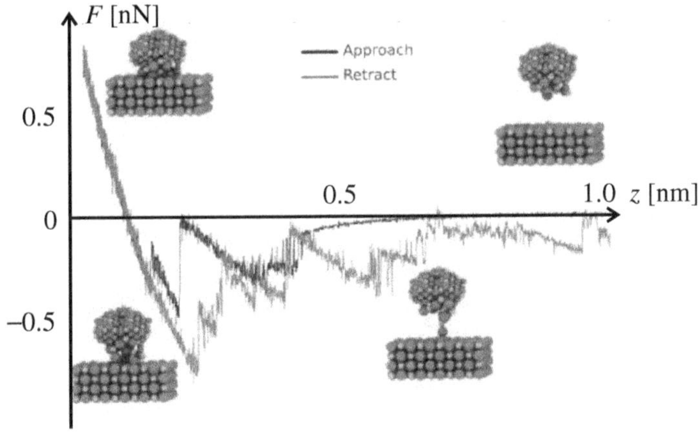

Figure 21.8 Typical force curve calculated with a maximum approach distance of 0.1 nm above a KBr(001) surface. Approach and retract curves are represented by thick and thin lines, respectively. Strong hysteresis is found due to the formation of atomic wires. A few snapshots of the system are shown along the force curve to indicate the indentation process. Reproduced from [164] with permission from the American Physical Society.

other energy structures, which then leads to hysteresis and energy loss. A comparative study of non-contact force spectroscopy and molecular dynamics simulations on KBr(001) surfaces has shown that the quasi-static force vs. distance curves hide stochastic dissipation processes. Due to the hysteresis of approach and retraction curves the measurement of the damping coefficient shows additional dissipation averaged over a few cycles. The formation of atomic chains is found to be an important process for the dissipation processes, where some fraction of eV per cycle are observed [164] (see Fig. 21.8).

Dissipation force microscopy is an operational mode with great relevance to atomic-scale studies of surfaces, particularly with respect to emerging nanomechanical technologies. Furthermore, the monotonic increase in dissipation with decreasing distance makes it a candidate for replacing the frequency shift as control parameter [152]. However, one has to keep in mind that frequency detection was invented for high Q-factor systems because of its fast response compared to amplitude detection schemes.

Part IV

Lubrication

22

Drag in a viscous fluid

The friction force on an object moving in a viscous fluid (so-called 'drag') has a completely different character from the friction on the same object sliding on a solid substrate. The parasitic drag is 'tuned' by the shape of the object ('form drag') and also by the contact between the fluid and the surface of the body ('skin friction'). In a first approximation the parasitic drag is proportional to the square of the velocity. Furthermore, the lift force created on a streamlined body such as a wing can also cause friction ('induced drag'). Here, we will introduce the most important expressions for the drag forces. The corresponding derivations can be found in textbooks on advanced fluid dynamics such as [176]. The wave drag caused by the shock waves formed at transonic and supersonic speed will be not discussed.

22.1 The Navier–Stokes equation

The motion of an incompressible viscous fluid is described by the *Navier–Stokes equation* [230, 323]

$$\rho\left(\frac{\partial \mathbf{v}}{\partial t} + \mathbf{v}\cdot\nabla\mathbf{v}\right) = -\nabla p + \eta\nabla^2\mathbf{v}, \tag{22.1}$$

where ρ and $\mathbf{v}$ are the density and velocity of the fluid, p is the pressure and η is the *dynamic viscosity*.[1] Equation (22.1) is obtained from the Newton equation with the addition of a diffusing viscous term. Typical values for the viscosity of various fluids at room temperature are listed in Table 22.1. A brief discussion on the viscosity of gases, as estimated with the kinetic theory, is presented in Appendix B.

Equation (22.1) is accompanied by the equation of continuity

$$\nabla\cdot\mathbf{v} = 0,$$

[1] The ratio of the dynamic viscosity to the fluid density is the *kinematic viscosity* $\nu = \eta/\rho$.

Table 22.1 *Typical values of dynamic viscosity (in mPa·s)*

Fluid	η
Air	0.017
Water	0.9
Ethanol	1.2
Mercury	1.5
Glycerol	1.2×10^3

which simply states that the mass of the fluid is conserved. Furthermore, appropriate boundary conditions must be satisfied. A common assumption is the *no-slip condition*, according to which the velocity $\mathbf{v}$ is zero on the solid walls limiting the fluid. Although the no-slip condition is an excellent approximation for most engineering applications, it can be violated in micromechanical devices with very smooth surfaces (see Appendix C).

In this chapter we will also assume that the fluid is *Newtonian*, i.e. that η (at a fixed temperature) does not depend on the rate of change of the velocity at which a fluid layer flows over an adjacent one[2] (the so-called *shear rate* $\dot{\gamma}$). This hypothesis holds well for liquids like water, benzene and light oils, whereas more complex fluids present a non-Newtonian behavior.[3]

Using the Navier–Stokes equation it can be proven that the kinetic energy per unit volume of the fluid is dissipated at a rate

$$\frac{\mathrm{d}E_{\mathrm{kin}}}{\mathrm{d}t} = -\frac{\eta}{2}\left(\frac{\partial v_i}{\partial x_j} + \frac{\partial v_j}{\partial x_i}\right)^2 \tag{22.2}$$

and the components of the friction force on the unit area of a bounding wall are

$$\tau_{\mathrm{fric},i} = -\eta\left(\frac{\partial v_i}{\partial x_j} + \frac{\partial v_j}{\partial x_i} - \frac{2}{3}\delta_{ij}\frac{\partial v_k}{\partial x_k}\right) n_i, \tag{22.3}$$

where $\mathbf{n}$ is a unit vector directed normally into the solid surface.

Viscosity of slurry

A slurry is a mixture of liquid and fine solid particles which retains fluid properties. If the ratio c between the volume occupied by the particles and the total volume of the suspension is extremely low, the viscosity differs by the amount $\Delta\eta = 5c\eta/2$ from the viscosity η of the original fluid. This result, which is strictly

[2] We assume that the fluid flows in parallel layers without lateral mixing (*laminar flow*).This is always the case if the velocity is low enough.

[3] For instance, animal joints are lubricated with a slightly viscous fluid, which is made non-Newtonian by the addition of long-chain polymers.

valid for spherical particles, can be derived from the Navier–Stokes equation using an elegant procedure proposed by Einstein [83]. However, things change if the suspension is dense. As demonstrated by experiments on polystyrene spheres in water [53], in the absence of shear stress spherical particles arrange in close-packed hexagonal layers. If a very low shear stress τ is applied the system solidifies in a polycrystalline state. At higher values of τ a sudden transition may occur into a new state where the particles are arranged in sliding layers, the fraction of which increases with the shear stress.

22.2 Flow of a viscous fluid

As a first example of application of the Navier–Stokes equation, consider the flow of a fluid confined between two parallel planes which are separated by a distance h and move transversally with a relative velocity V. In this case it is easy to prove that the velocity of the fluid will increase linearly with the height z as

$$v(z) = Vz/h$$

(Fig. 22.1(a)). The friction force on the unit area of the lower wall is obtained from Eq. (22.3) as

$$\tau_{\text{fric}} = \eta \, (\partial v_x/\partial z)|_{z=0} \tag{22.4}$$

and, in the present case, it is proportional to V:

$$\tau_{\text{fric}} = \eta V/h. \tag{22.5}$$

If the planes are fixed, but the pressure p changes along the direction of flow, the velocity distribution has a parabolic profile with maximum value at half the distance between the two planes (Fig. 22.1(b)):

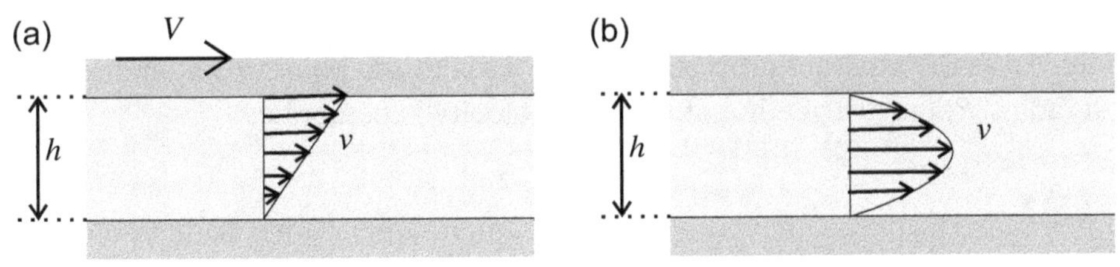

Figure 22.1 Laminar shear of fluid (a) between two plates in relative motion with a constant velocity V and (b) between two fixed plates with a gradient of pressure along the flow direction.

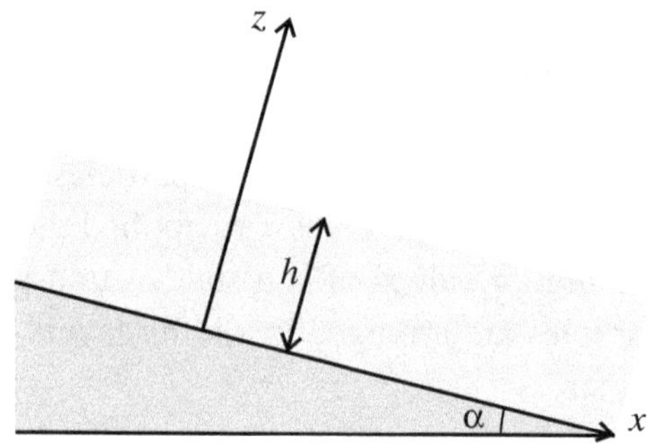

Figure 22.2 Flow along an inclined plane.

$$v(z) = -\frac{1}{2\eta}\frac{\mathrm{d}p}{\mathrm{d}x}z(h-z).$$

In this case the friction on the walls is $\tau_{\text{fric}} = -(1/(2h))(\mathrm{d}p/\mathrm{d}x)$.

If a fluid layer with thickness h flows along a plane which is inclined by an angle α, as in Fig. 22.2, it is not difficult to see that the velocity distribution is

$$v(z) = \frac{\rho g \sin\alpha}{2\eta}z(2h-z),$$

where g is the acceleration of gravity. Thus, the friction force (per unit area) is $\tau_{\text{fric}} = \rho g h \sin\alpha/\eta$ and the amount of fluid (per unit width) flowing along the plane in the time unit (*volumetric flow rate*) is

$$Q \equiv \int_0^h v\,\mathrm{d}z = \frac{\rho g h^3 \sin\alpha}{3\eta}.$$

Flow through a pipe

The viscous flow through a pipe of length L and arbitrary cross-section is determined by the two-dimensional equation

$$\nabla^2 v = -\frac{\Delta p}{\eta L}, \tag{22.6}$$

where Δp is the pressure drop between the ends of the pipe. In a cylindrical pipe with radius R, the distribution of the fluid velocity is parabolic:

$$v(r) = \frac{\Delta p}{4\eta L}(R^2 - r^2). \tag{22.7}$$

Integrating Eq. (22.7) through the cross-section of the pipe, the volumetric flow rate is found to be proportional to the fourth power of the radius:[4]

[4] Equation (22.8) represents the famous *Poiseuille's law* [273].

$$Q = \frac{\pi \Delta p}{8\eta L} R^4. \tag{22.8}$$

The friction on the unit area of the wall is

$$\tau_{\text{fric}} = \frac{R}{2\eta} \frac{\Delta p}{L}.$$

Analytical expressions can be also derived for different cross-sections. For instance, if the pipe has a rectangular cross-section with width a and height b:

$$Q = \frac{\Delta p}{12\eta L} ab^3. \tag{22.9}$$

Rotary flow

The motion of a fluid between two coaxial cylinders rotating with different angular velocities Ω_1 and Ω_2 is known as *Taylor–Couette flow*. In this case the velocity distribution in the fluid depends on the radial distance r from the common axis as

$$v(r) = \frac{\Omega_2 R_2^2 - \Omega_1 R_1^2}{R_2^2 - R_1^2} r + \frac{(\Omega_1 - \Omega_2) R_1^2 R_2^2}{R_2^2 - R_1^2} \frac{1}{r}.$$

The moment of the friction forces acting on the unit length of each cylinder is

$$M_{\text{fric}} = \frac{4\pi\eta(\Omega_1 - \Omega_2) R_1^2 R_2^2}{R_2^2 - R_1^2}. \tag{22.10}$$

Equation (22.10) is important in rotational viscometers.

22.3 Motion in a viscous fluid

When a rigid body moves in a viscous fluid with a velocity $\mathbf{V}$, a resistive *drag* force $\mathbf{F}_{\text{fric}}$ parallel to $\mathbf{V}$ appears. This problem is equivalent to that of the fluid flowing around the fixed body, where $\mathbf{V}$ is the velocity of the main stream. In general, the flow of a viscous fluid past a rigid body is characterized by the *Reynolds number* [285]

$$\mathrm{R} = \frac{\rho V L}{\eta},$$

where L and V are the characteristic length and velocity of the problem. If R is sufficiently small, a steady flow is stable. However, when R increases, it can eventually reach a critical value $\mathrm{R_c}$ beyond which the flow becomes unstable. In this case the fluid velocity varies very irregularly at each point in the fluid, and vortices of various sizes are superimposed on the mean flow (so-called *turbulence*). The value of R_c depends on the geometry of the problem and is usually of the order of 10^5.

If R is small the left hand side in Eq. (22.1) is negligible and the Navier–Stokes equation takes a simple form:

$$\eta \nabla^2 \mathbf{v} - p = 0. \tag{22.11}$$

In this case dimensional arguments imply that, in a first approximation, the drag force depends on $\mathbf{V}$ by a relation of the form $F_i = \eta a_{ij} V_j$, where a_{ij} is a symmetric tensor determined by the shape of the body.

If the fluid flows around a sphere of radius R the velocity distribution obtained from Eq. (22.11) is described in spherical coordinates by the relations

$$v_r = V \cos\theta \left(1 - \frac{3R}{2r} + \frac{R^3}{2r^3}\right),$$
$$v_\theta = -V \sin\theta \left(1 - \frac{3R}{4r} - \frac{R^3}{4r^3}\right)$$

and is plotted in Fig. 22.3. The drag on the sphere is given by the famous *Stokes' law* [324]:

$$F_{\text{fric}} = -6\pi\eta R V. \tag{22.12}$$

Considering the contribution of the velocity field at large distances from the body, a correction $\Delta F_{\text{fric}}/F_{\text{fric}} = (3/8)$R must be introduced in Eq. (22.12) [238]. It is also interesting to observe that the drag force on a spherical bubble is two thirds of the force acting on a rigid sphere with the same radius [128, 298].

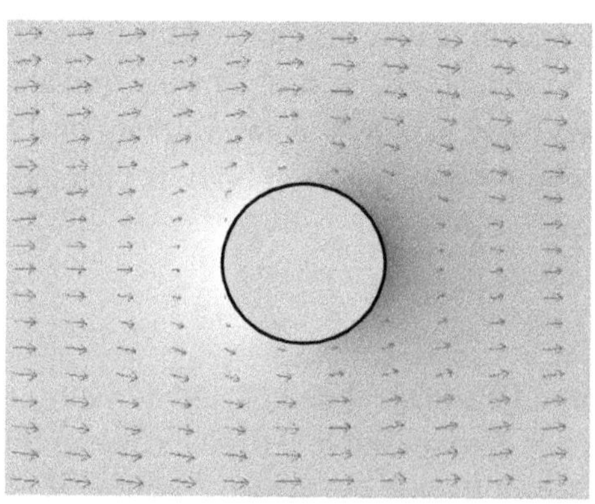

Figure 22.3 Velocity distribution of a viscous liquid flowing around a rigid sphere (cross-section).

If the fluid flows parallel to the plane of a disk of radius R the drag force is [174]

$$F_{\text{fric}} = -\frac{32}{3}\eta R V.$$

However, if the fluid flows perpendicularly to the plane, the force is 3/2 times larger:

$$F_{\text{fric}} = -16\eta R V.$$

If the disk is approaching a flat plane with a velocity V the drag force on the moving disk is

$$F_{\text{fric}} = -\frac{3\pi\eta V R^4}{2h^3}, \tag{22.13}$$

where h is the separation between the two objects. We will come back to Eq. (22.13), and its limitations when $h \to 0$, in Section 24.4.

The two-dimensional problem of the viscous flow around a cylinder was first solved by Lamb [173]. The drag force on the unit length of the cylinder depends on the logarithm of the velocity V of the main stream as

$$F_{\text{fric}} = -\frac{4\pi\eta R V}{1/2 - C - \log(RV/4\nu)},$$

where $C \approx 0.577$ is Euler's constant.

The moment of the friction forces acting on a rigid body rotating in a fluid can be also estimated analytically in a few simple cases. If a sphere of radius R rotates with with an angular speed $\mathbf{\Omega}$, the velocity distribution of the fluid around the sphere is

$$\mathbf{v} = (R/r)^3 \mathbf{\Omega} \times \mathbf{r},$$

and the frictional torque exerted by the fluid is [176, par. 20]

$$M_{\text{fric}} = -8\pi\eta R^3 \Omega. \tag{22.14}$$

Equation (22.14) can be seen as a generalization of Stokes' law for rotational motion. Note also that a fluid confined in a rotating spherical cavity moves rigidly with the cavity.

Finally, the moment of the friction forces on a rotating disk was calculated by von Kármán [337]:

$$M_{\text{fric}} = -1.94 R^4 \sqrt{\rho\eta\Omega^3}.$$

The rotation is accompanied by a constant axial flow from infinity, as shown in Fig. 22.4. Note that the friction torque is not proportional to the angular velocity Ω.

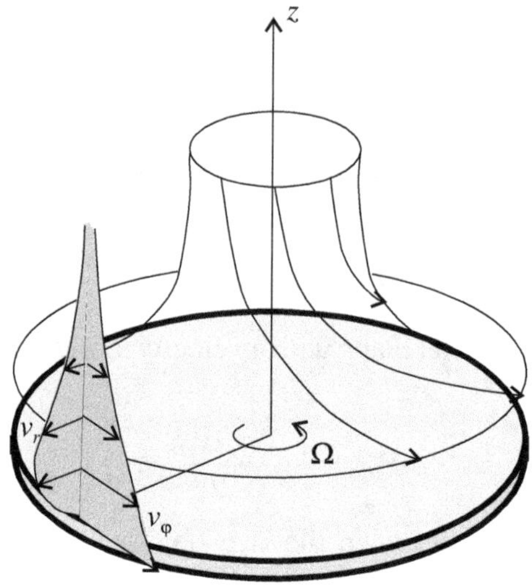

Figure 22.4 Flow in the region close to a disk rotating in a viscous fluid.

However, this is the case for a cylinder rotating about its axis, as seen from Eq. (22.10):

$$M_{\text{fric}} = -4\pi\eta R^2\Omega.$$

22.4 Boundary layers and skin friction

If the Reynolds number R is large, the velocity of a fluid decreases to zero in a thin *boundary layer* adjoining the walls confining it. In a boundary layer the Navier–Stokes equation can be simplified, and approximated expressions for the fluid flow can be derived.

If the fluid flows along a semi-infinite plate, as in Fig. 22.5, the thickness of the laminar boundary layer increases as the square root of the downstream coordinate x:

$$\delta_{\text{lam}} \sim \sqrt{\frac{x\eta}{\rho V}},$$

where V is the velocity of the main stream. The velocity distribution in the layer can be found numerically by solving a non-linear differential equation for the *stream function* $\psi(x, y)$ defined by the relations

$$v_x = \frac{\partial\psi}{\partial z}, \qquad v_z = -\frac{\partial\psi}{\partial x}.$$

The friction force on the unit area of the plate (so-called *skin friction*), as obtained from Eq. (22.4), is [25]

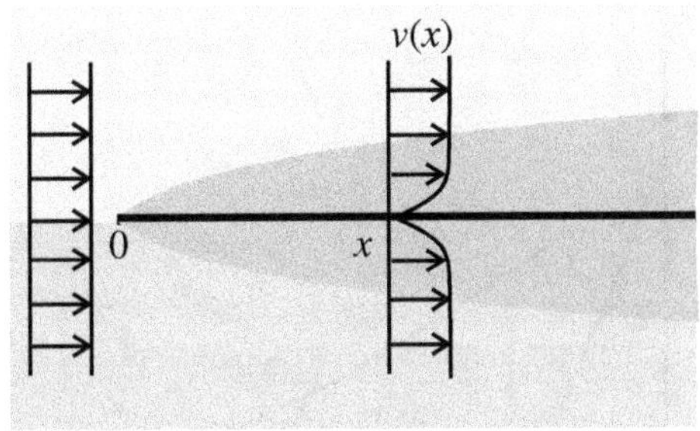

Figure 22.5 Laminar boundary layer (darker area) on a semi-infinite plate.

$$\tau_{\text{fric}}(x) = 0.322\sqrt{\eta\rho V^3/x}.$$

In a turbulent boundary layer the mean velocity v_x parallel to the wall can be estimated by dimensional considerations. Assuming that the viscosity plays a role only at very small distances z, only a logarithmic velocity distribution is possible [338]:

$$v_x(z) \approx v_\tau \left(\frac{1}{\kappa} \ln \frac{\rho z v_\tau}{\eta} + \text{const.} \right). \tag{22.15}$$

The *von Kármán's constant* κ in the *law of the wall* (22.15), as determined experimentally, is 0.41, whereas the additive constant in Eq. (22.15) is about 5.0 (for a smooth wall). The so-called *shear velocity* v_τ is connected to the friction force τ per unit area by the relation

$$\tau_{\text{fric}} = \rho v_\tau^2. \tag{22.16}$$

Eliminating v_τ from Eqs. (22.15) and (22.16) it can be seen that the friction force $\tau_{\text{fric}}(x)$ decreases slowly with the distance x. The thickness δ of the turbulent boundary layer can be estimated by observing that $\mathrm{d}\delta/\mathrm{d}x \sim v_\tau/V$, and is found to increase linearly with x:

$$\delta_{\text{turb}} \sim v_\tau x/V.$$

Flow in a pipe

The pressure loss due to the friction along a given length L of a pipe can be described by the dimensionless *friction factor*

$$C = \frac{2R\Delta p/L}{(1/2)\rho V^2}, \tag{22.17}$$

where R is the radius of the pipe and V is the average velocity of flow. Since the thickness of a boundary layer is continuously increasing downstream, at a finite

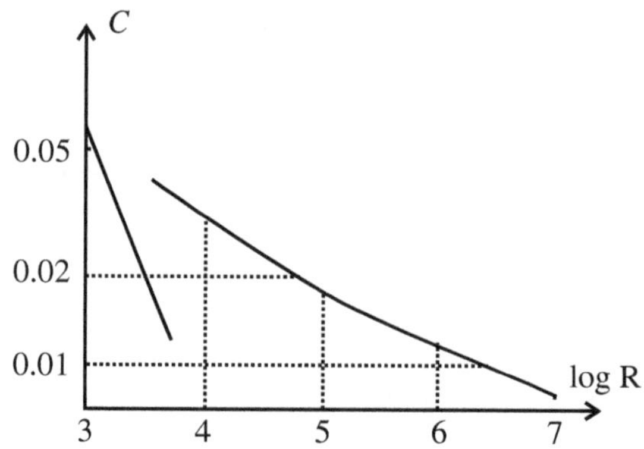

Figure 22.6 Friction factor C as a function of the Reynolds number in a circular pipe.

distance from the point of entry of the fluid the whole cross-section of the pipe will be filled by a laminar or turbulent boundary layer.

If the flux is laminar, Poiseuille's equation (22.8) implies that the factor C is inversely proportional to the Reynold's number:

$$C = 64/\mathrm{R}. \tag{22.18}$$

In a turbulent flow, the law of the wall (22.15) can be used to derive an implicit relation known as *Colebrook's equation* [60]:

$$\frac{1}{\sqrt{C}} \approx 0.88 \ln(\mathrm{R}\, C) - 0.85. \tag{22.19}$$

Figure 22.6 shows the relations between the friction factor and the Reynolds number expressed by Eqs. (22.18) and (22.19). Note the abrupt variation of C at the transition from laminar to turbulent flow, which is usually observed when $\mathrm{R} \sim 2\text{–}3 \times 10^3$.

In the case of Taylor–Couette flow (Section 22.2) it can be proven that, for large Reynolds numbers, the flow is always unstable if the two cylinders rotate in the same direction. If the cylinders rotate in opposite directions the flow is unstable only if $\Omega_2 R_2 < \Omega_1 R_1$ [284]. The first instability leads to the appearance of axisymmetric toroidal *Taylor's vortices* between the cylinders (Fig. 22.7). If the separation h between the two cylinders is very small and the outer cylinder is fixed, the vortices appear when [330]

$$\mathrm{R} \equiv \frac{\rho \Omega_1 R_1 h}{\eta} > 41.2 \sqrt{\frac{R_1}{h}}.$$

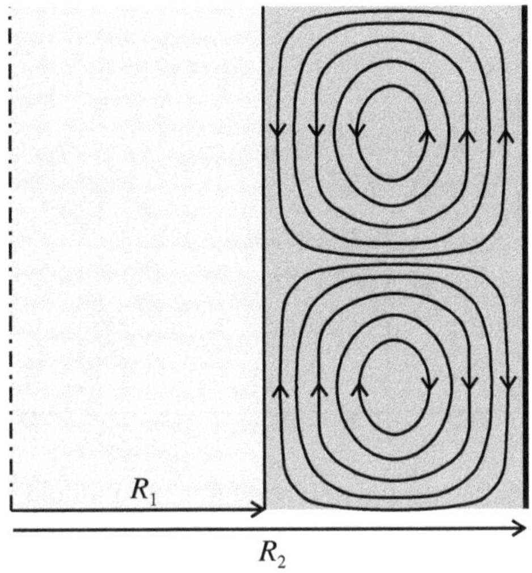

Figure 22.7 Taylor's vortices.

22.5 Drag crisis

Similarly to the definition (22.17) for a flow in a pipe, the motion of a body in a fluid can be characterized by the *drag coefficient*

$$C_{\rm d} = \frac{F_{\rm fric}/A}{(1/2)\rho V^2},$$

where V is the velocity of the body, $F_{\rm fric}$ is the drag force and A is the area of a cross-section of the body transverse to the direction of flow. The coefficient $C_{\rm d}$ is a function of the Reynolds number R and the typical dependence $C_{\rm d}(R)$ observed for a sphere is shown in Fig. 22.8. At low values of R the drag coefficient decreases according to Stokes' law, $C_{\rm d} = 24/{\rm R}$. The decrease becomes slower, and $C_{\rm d}$ almost levels off, when R $\sim 10^3$–10^5. When R ~ 2–3×10^5, the drag coefficient suddenly drops by a factor 4 or 5. This phenomenon is known as *drag crisis* and corresponds to the onset of a turbulent boundary layer. At this point a turbulent wake beyond the body, already appearing for lower values of R, suddenly becomes more narrow [176, par. 45], which reduces the value of the total drag force. If R is very high, the compressibility, parametrized by the *Mach number* $\rm M = V/c$, where c is the velocity of sound in the fluid, becomes important. As a general result R is found to increase when M increases.

22.6 Streamlined bodies

A *streamlined body* (e.g. the wing of an airplane) is an elongated object with a rounded leading edge and a sharp trailing edge (Fig. 22.9). If a streamlined body

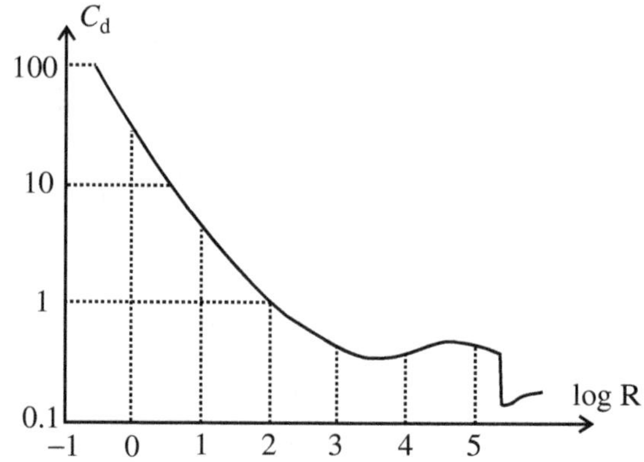

Figure 22.8 Drag coefficient for a sphere moving in a fluid as a function of the Reynolds number.

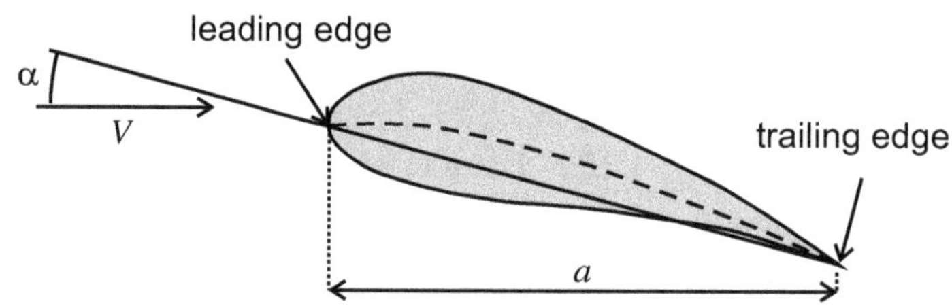

Figure 22.9 A cross-section of a thin wing.

moves in a viscous fluid with a velocity V, the skin friction is an important source of drag (which is not the case for a non-streamlined body such as a sphere).

According to a theorem first derived by Zhukovsky, the lift force on the body is given by

$$F_{\text{lift}} = \rho V \int \Gamma \, dy, \tag{22.20}$$

where ρ is the fluid density and Γ is the velocity circulation, i.e. the line integral

$$\Gamma = \oint \mathbf{v} \cdot d\mathbf{l}$$

taken along a closed contour surrounding the wing profile. If the angle of attack α in Fig. 22.9 is $\lesssim 10°$, the ratio between the lift force F_{lift} and the drag force F_{fric} can be as high as 10–100, but it decreases rapidly if α increases and the object ceases to be streamlined (Fig. 22.10).

We can also introduce the *lift coefficient*

$$C_z = \frac{F_{\text{lift}}/aL}{(1/2)\rho v^2},$$

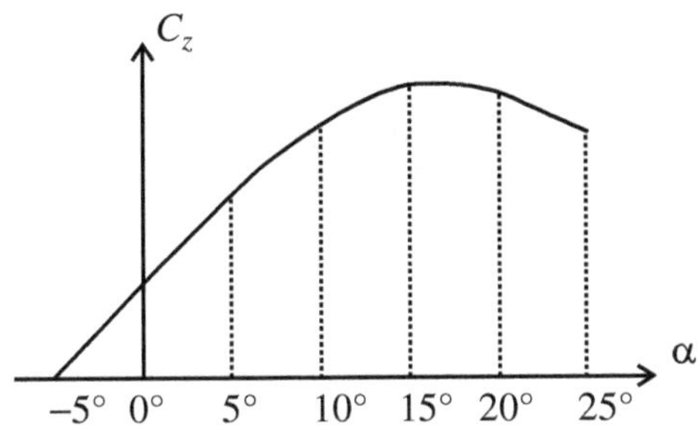

Figure 22.10 Lift coefficient versus angle of attack for a cambered airfoil.

where a and L are the width and the span of the wing, respectively. If the wing is very long, the lift coefficient is proportional to the angle of attack α, which is supposed to be small. In this case the proportionality constant depends only on the shape of the wing. For instance, if the wing is a thin plate [176, section 48] $C_z \approx 2\pi\alpha$.

An important part of the drag acting on a wing is due to the energy dissipation in the thin turbulent wake formed behind the wing (Section 22.5). This *induced drag* can be estimated using a formula derived by Prandtl [276]:

$$F_{\text{fric}} = -\frac{\rho}{4\pi}\iint \frac{\mathrm{d}\Gamma(y)}{\mathrm{d}y}\frac{\mathrm{d}\Gamma(y')}{\mathrm{d}y'} \ln\left|y - y'\right| \,\mathrm{d}y\,\mathrm{d}y'. \tag{22.21}$$

In Eq. (22.21) the integrals are taken between $y = 0$ and $y = L$, and the origin of the y axis is at one end of the wing. Note that, if the span L increases, the induced drag does not change in magnitude, whereas the lift force (22.20) increases almost linearly.

23

Lubrication

Friction between two solid surfaces can be reduced by means of a lubricant film. In hydrodynamic lubrication the films are so thick that the surfaces are prevented from coming into contact. In this case the lubrication is governed by the viscosity of the film and the frictional resistance to motion arises from the shearing of the fluid layers. The principles of fluid mechanics outlined in Chapter 22 completely determine the performance of the system. If the elastic deformation of the lubricated surfaces is significant, we speak of elastohydrodynamic lubrication. If the fluid is pumped into the gap between the two surfaces, the lubrication is hydrostatic. Solid flakes can be also intercalated to lower the coefficient of friction. In the case of boundary lubrication the surfaces touch each other on a considerable contact area. This subject is much less established and will be discussed separately in the next chapter.

23.1 Hydrodynamic lubrication

Consider a lubricant film between two solid surfaces moving with a relative velocity V as in Fig. 23.1. If the surfaces were parallel, no pressure field could be established to support a given load. This is not the case if the top surface is inclined by a certain angle. Suppose first that the lower surface is flat and the upper one has an arbitrary profile $h(x, y)$. In this case it can be proven that the pressure gradient is related to the function $h(x, y)$ by the so-called *Reynolds' equation* [286]:

$$\frac{\partial}{\partial x}\left(h^3\frac{\partial p}{\partial x}\right) + \frac{\partial}{\partial y}\left(h^3\frac{\partial p}{\partial y}\right) = 6V\eta\frac{\partial h}{\partial x}. \tag{23.1}$$

Equation (23.1) is derived from the the Navier–Stokes equation (22.1) assuming that the gap between the two surfaces is very small, the flow is laminar and the lubricant is Newtonian. Furthermore, the fluid velocity and the volumetric flow rate (per unit width) in the x direction are, respectively,

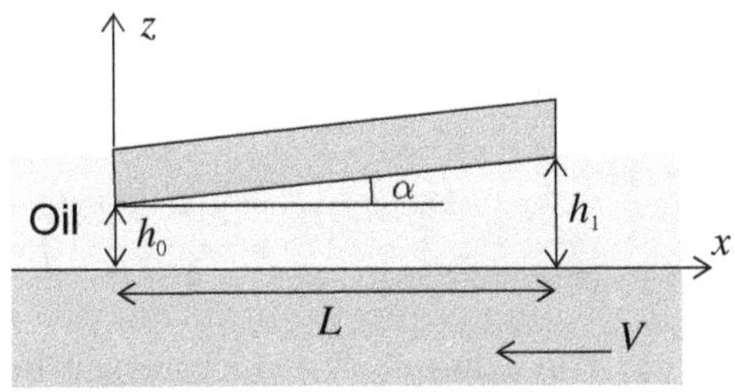

Figure 23.1 Hydrodynamic pressure generation between non-parallel surfaces.

$$v_x = -\frac{z(h-z)}{2\eta}\frac{\partial p}{\partial x} - \left(1 - \frac{z}{h}\right)V, \tag{23.2}$$

$$q_x = -\frac{h^3}{12\eta}\frac{\partial p}{\partial x} - V\frac{h}{2}. \tag{23.3}$$

The previous formulas form the basis for determining the friction force and the lubricant leakage out of the ends of the gap.[1] Further approximations can be made in two cases of practical importance.

- If the gap is very long in the y direction (perpendicular to the plane of Fig. 23.1), the pressure gradient $\partial p/\partial y$ is negligible, and Eq. (23.1) becomes

  $$\frac{dp}{dx} = -6\eta V \frac{h_{max} - h}{h^3}, \tag{23.4}$$

  where h_{max} is the separation corresponding to the maximum value of pressure in the fluid (*long bearing approximation*).
- In the opposite case, the pressure gradient in the x direction is negligible and the Reynolds' equation becomes

  $$p(x, y) = \frac{3V\eta}{h^3}\frac{dh}{dx}\left(y^2 - \frac{b^2}{4}\right), \tag{23.5}$$

  where b is the width of the bearing (*short bearing approximation*).

Pad bearings

A *pad bearing* has the simple profile shown in Fig. 23.1:

$$h(x) = h_0 + \tan\alpha = h_0 + x(h_1 - h_0)/L.$$

The pressure distribution is obtained by integrating Eq. (23.4) with the boundary conditions $p(0) = p(L) = p_{ext}$, where p_{ext} is the external pressure:

$$p(x) = p_{ext} + \frac{6V\eta L}{h_0(k-1)}\left(-\frac{1}{h(x)} + \frac{h_0}{h^2(x)}\frac{k}{k+1} + \frac{1}{h_0(k+1)}\right), \tag{23.6}$$

[1] For a derivation of these formulas see [325, section 5.1]

where $k = h_1/h_0$. The normal force that the bearing will support is obtained by integrating the pressure distribution (23.6) over the bearing area:

$$F_N = \frac{F_{N0}}{(k-1)^2}\left(-\ln k + \frac{2(k-1)}{k+1}\right), \tag{23.7}$$

where $F_{N0} = 6V\eta L^2 b/h_0^2$. The relation $F_N(k)$ expressed by Eq. (23.7) is plotted in Fig. 23.2. The normal force F_N has a maximum when $k \approx 2.19$.

The friction force is obtained by integrating the shear stress $\tau = \eta \mathrm{d}v_x/\mathrm{d}z$ using the relation (23.2) for the fluid velocity:

$$F_{\text{fric}} = \frac{V\eta Lb}{h_0}\left(\frac{6}{k+1} - \frac{4\ln k}{k-1}\right).$$

As a result, the friction coefficient $\mu = F_{\text{fric}}/F_N$ is found to be independent of the viscosity (Fig. 23.2):

$$\mu = \frac{(k-1)h_0}{L}\frac{3(k-1) - 2(k+1)\ln k}{6(k-1) - 3(k+1)\ln k}.$$

Finally, the lubricant flow is determined by the sliding speed and the film geometry:

$$Q_x = Vh_0 b\frac{k}{k+1}. \tag{23.8}$$

Note that the minimum value of μ is obtained when $k \approx 2.53$ and is usually quite small. As an example, if $\alpha = 1°$, the friction coefficient $\mu \approx 0.05$, meaning that hydrodynamic lubrication is a very efficient way to reduce friction. If $V = 10$ m/s, $h_0 = 100$ μm and $b = 10$ cm, the corresponding amount of lubricant per unit time to be supplied is $Q_x \sim 0.1$ l/s.

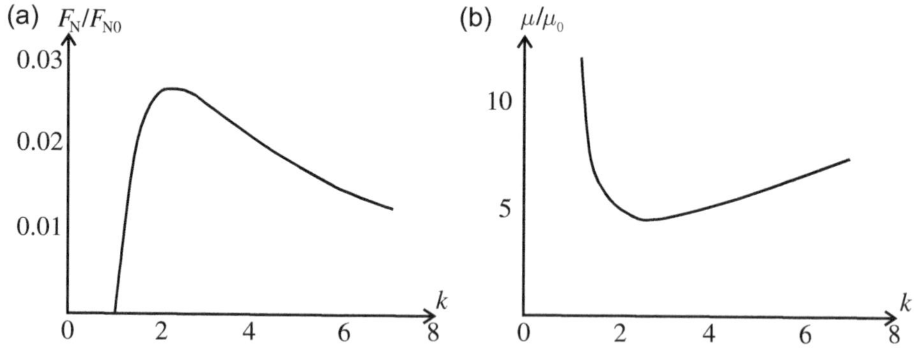

Figure 23.2 (a) Load capacity and (b) friction coefficient as a function of the ratio k between the outlet and the inlet film thickness in a pad bearing.

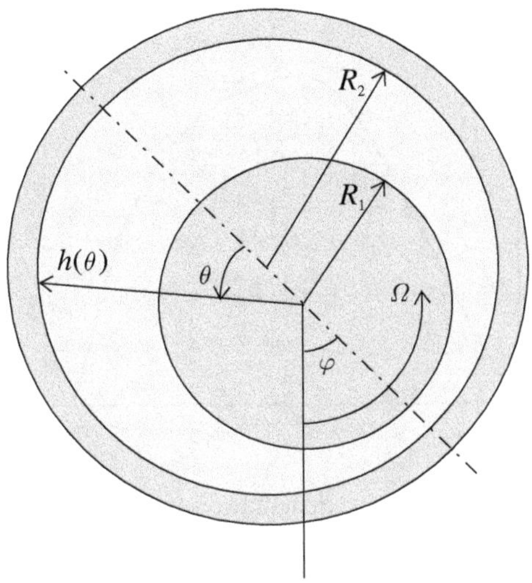

Figure 23.3 Geometry of a journal bearing.

If the angle of inclination $\alpha < 0$, the normal force (23.7) becomes negative. In this case the surfaces are sucked together by the negative pressure, and the limits of hydrodynamic lubrication (see below) are reached in a short time.

Journal bearings

A *journal bearing* is formed by a shaft of radius R_1 rotating inside a stationary bush with an angular speed Ω (Fig. 23.3). The difference between the radii R_2 and R_1 of bush and shaft is the *clearance* c. Shaft and bush are not concentric and the precise position of the shaft is determined by the load that it is carrying. Introducing the *eccentricity ratio* ε as the distance between the centers of shaft and bush divided by c, the film thickness can be written as

$$h(\theta) = c(1 + \varepsilon \cos\theta). \tag{23.9}$$

Since in most journal bearings the width b is shorter than the shaft diameter, we can use Eq. (23.5) to determine the pressure distribution. Assuming the boundary conditions (the so-called 'half Sommerfeld condition') $p(0) = p(\pi) = 0$, it can be seen that the axial variation of pressure is parabolic, and the circumferential variation is governed by a trigonometric function (Fig. 23.4):

$$p(\theta, y) = \frac{3\eta\Omega}{c^2}\left(\frac{b^2}{4} - y^2\right)\frac{\varepsilon \sin\theta}{(1 + \varepsilon\cos\theta)^3}.$$

The pressure p reaches its maximum value at the angle

$$\theta_{\max} = \arccos\frac{1 - \sqrt{1 + 24\varepsilon^2}}{4\varepsilon}.$$

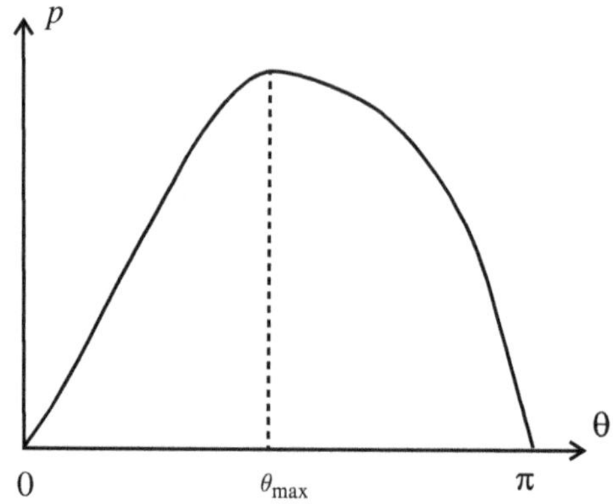

Figure 23.4 Pressure distribution in a journal bearing.

Integrating the function $p(\theta)$ around the bearing, one gets the components of the normal force parallel and perpendicular to the line of centers. This allows us to determine the load that the bearing is supporting:

$$F_N = F_{N,0} \frac{\varepsilon}{(1-\varepsilon^2)^2} \sqrt{16\varepsilon^2 + \pi^2(1-\varepsilon^2)},$$

where $F_{N,0} = \eta \Omega R b^3 / 4c^2$, and the *attitude angle* between the line of centers and the vertical:[2]

$$\varphi = \arctan \frac{\pi\sqrt{1-\varepsilon^2}}{4\varepsilon}.$$

The friction force is obtained by integrating the shear stress:[3]

$$F_{fric} = F_{fric,0} \frac{1}{\sqrt{1-\varepsilon^2}}, \tag{23.10}$$

where $F_{fric,0} = 2\pi\eta V b R_1/c$. The forces F_N and F_{fric} are plotted in Fig. 23.5 as a function of ε. Note that the load capacity F_N rises sharply with increasing ε, whereas the friction force F_{fric} is relatively insensitive to ε as long as $\varepsilon \lesssim 0.8$. In order to avoid shaft misalignments at high values and vibrations at low values, the optimum value of ε is usually chosen around 0.7.

Finally, the lubricant flow in the journal bearing is easily estimated from Eqs. (23.3) and (23.9):

$$Q_\theta(\theta) = \frac{\Omega b R_1}{2} c(1 + \varepsilon \cos\theta).$$

The rate at which the lubricant is lost due to side leakage is

$$Q = Q_\theta(0) - Q_\theta(\pi) = \Omega R c b \varepsilon.$$

[2] Note that φ is different from θ_{max}!

[3] Equation (23.10) with $\varepsilon = 0$ is known as *Petroff's law* [267].

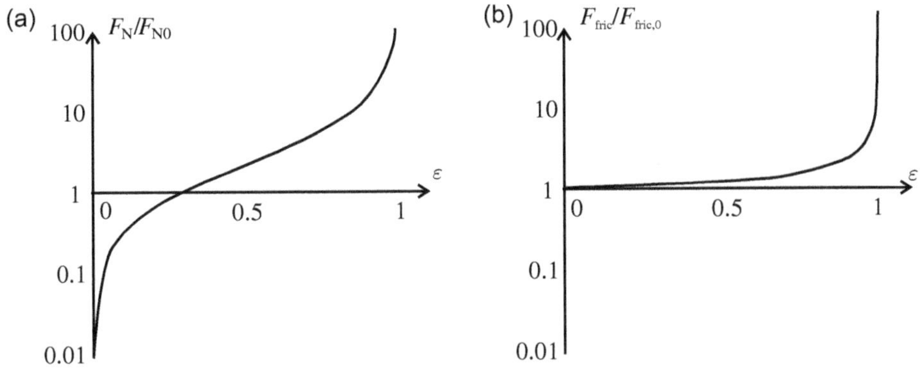

Figure 23.5 (a) Normal force and (b) friction force in a journal bearing as a function of the eccentricity ratio.

Thermal effects

The thermal energy carried away by the lubricant in unit time can be roughly estimated by multiplying the flow Q_x by $\rho C_V \Delta T$, where ρ is the fluid density, C_V is the heat capacity of the lubricant and ΔT is the temperature rise. Equating the result to the energy dissipation (per unit time) $F_{fric}V$ one gets

$$\Delta T \sim \frac{F_{fric} V}{Q_x \rho C_V}.$$

Note that values of $\Delta T \sim 100$ °C are not uncommon. For an accurate estimation, one has to solve the Reynolds equation and the heat transfer equation for the lubricant film simultaneously, which is only possible by numerical methods. The temperature dependence of the viscosity (Section 24.1) must also be taken into account.

Limits of hydrodynamic lubrication

Hydrodynamic lubrication is only effective when the sliding velocity V is relatively high. If this is not the case, Eq. (23.7) shows that the film thickness must decline to maintain the pressure field. This means that contact between surface asperities may occur, causing wear and high friction. The variation of the friction coefficient μ with the parameter $\eta V/F_N$ for a pad bearing is represented by the *Stribeck curve* (Fig. 23.6). The friction is proportional to ηV only if V is large enough. Although for perfectly smooth surfaces the separation at the lower limit of hydrodynamic lubrication can be as small as a few nanometers (see below), the theory usually breaks down in the micron range, corresponding to the roughness of most engineering surfaces.

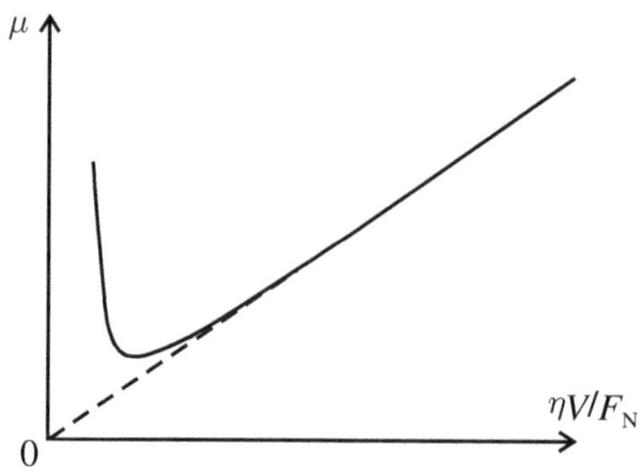

Figure 23.6 Relation between the friction coefficient and the velocity for a solid body sliding on a lubricated surface.

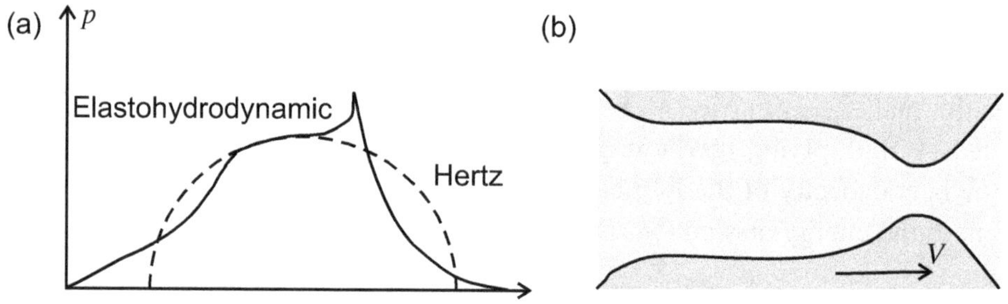

Figure 23.7 (a) Pressure distribution and (b) deformation in the elastohydrodynamic contact formed by a cylinder sliding on a flat surface.

23.2 Elastohydrodynamic lubrication

A liquid lubricant separating two surfaces forming a Hertzian contact undergoes very high pressure, and its viscosity can consequently increase by several orders of magnitude (see Section 24.1). In order to calculate the pressure distribution and the elastic deformation in the gap one has to solve the Navier–Stokes equation combined with the Barus law (introduced in the next chapter) and the Hertzian theory. This is usually done numerically in an iterative way. The result is shown in Fig. 23.7 for a cylinder sliding on a flat surface [131]. Note that the pressure peaks and quickly decreases before the outlet of the gap, where a constriction appears. This effect is even more pronounced in the contact between a sphere and a flat surface, where a characteristic 'horseshoe' constriction can be observed.

23.3 Hydrostatic lubrication

In *hydrostatic lubrication* the solid surfaces are separated by a fluid forced by an external pump. This scheme is adequate when very heavy loads and very slow

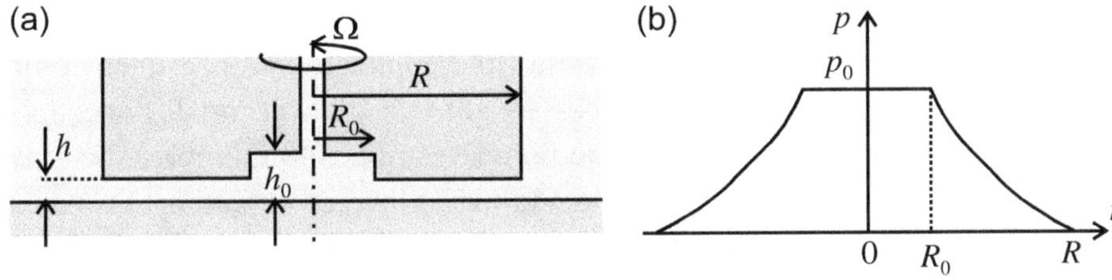

Figure 23.8 Flat circular pad bearing with central recess: (a) schematic and (b) pressure distribution.

speeds are involved (e.g. when positioning a heavy telescope). As an example, consider a circular hydrostatic pad bearing of radius R with a central recess of radius R_0, as in Fig. 23.8. The pressure distribution $p(r)$ can be easily estimated using Eq. (23.3), adapted to the radial flow, and integrating. As a result

$$p(r) = p_0 \frac{\ln(R/r)}{\ln(R/R_0)},$$

where p_0 is the recess pressure. The bearing can support a normal load

$$F_N = \frac{p_0 \pi}{2} \frac{R^2 - R_0^2}{\ln(R/R_0)},$$

independently of the fluid viscosity η. This means that any fluid circulating in the device can in principle be used as a lubricant. Assuming that the recess depth h_0 is much larger than the bearing film thickness h, the frictional torque is approximately given by

$$M_{\text{fric}} = \frac{\pi \eta \Omega}{2h}(R^4 - R_0^4),$$

where Ω is the angular velocity of the bearing. If Ω is low, the power loss $M_{\text{fric}}\Omega$ due to viscous dissipation is usually negligible with respect to the power loss $p_0 Q$ associated with pumping (Q is the volumetric fluid flow).

23.4 Solid lubrication

Consider a layered solid material. If the layers are able to slide over one another at relatively low shear stresses, the solid becomes self lubricating. This is indeed the case for graphite, MoS_2 and talc, but not for mica, where strong adhesive forces prevent smooth sliding between the sheets. These forces are due to strong electrostatic attraction between K^+ and O^{2-} ions. In talc, which is chemically and crystallographic similar to mica, only much weaker vdW forces act between the lamellae, and the friction is much lower. Note that the coefficients of static and kinetic friction in inorganic layered lubricants such as graphite or MoS_2 are quite similar: $\mu_k \approx \mu_s$. In this way, lamellar solids find application as additives in lubrication oils. When the contacting surfaces are set into motion, small flakes of the additive orient themselves parallel to the fluid streamlines to minimize the energy dissipation. In order to be used as a good additive, a layered solid must also adhere strongly to the sliding surfaces. This is the case when graphite or MoS_2, but not talc, are applied to steel [66]. Soft plastic metals such as Ag, Au, In and Pb can also reduce friction when applied as thin films.

24

Viscous phenomena in confined or spreading liquids

In this chapter we will first introduce the Eyring model to interpret the phenomenological dependence of viscosity on temperature and pressure observed in most liquids. This model can also explain important phenomena such as the similar values of static and kinetic friction in boundary lubrication with fatty acids or the shear thinning of polymers. After that, we will address specific phenomena observed when a liquid is confined in gaps with thickness comparable to the size of the liquid molecules or forms ultrathin layers on a solid surface. These phenomena include capillary condensation, flow between rough surfaces, squeezing processes and spreading of liquid droplets.

24.1 The Eyring model

The viscosity η of a liquid medium tends to decrease as the temperature T increases or the pressure p decreases. The temperature dependence of η is usually well described by the phenomenological equation

$$\eta = \eta_0 e^{\varepsilon / k_B T}, \tag{24.1}$$

where the activation energy ε depends on the molecular size. As an example, for linear hydrocarbons (alkanes) ε is of the order of 0.1 eV and increases with the number N of carbon atoms up to $N \sim 30$. Beyond this length the molecule moves in a segmental way and η remains constant. The pressure dependence of η is approximately described by the *Barus law* [13]:

$$\eta = \eta_0 e^{p/p_0}, \tag{24.2}$$

where $p_0 \sim 0.1$ GPa. The exponential dependence $\eta(p)$ implies that a contact pressure of 1 GPa, which is not uncommon in ball bearings and gears, may increase the viscosity of a mineral oil by a factor of 10^6.

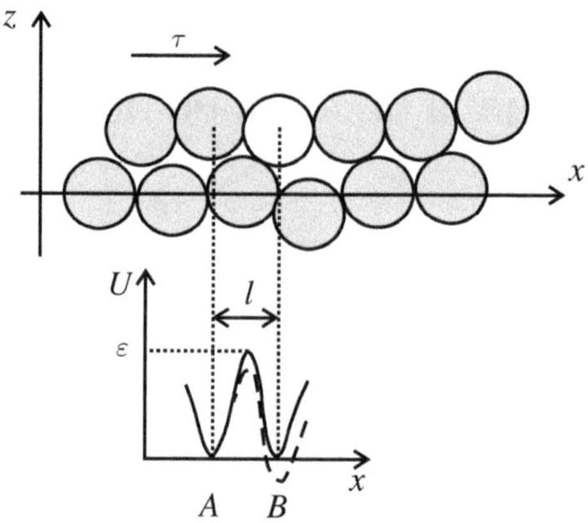

Figure 24.1 The Eyring model described in the text. Note the lowering of the energy barrier $A \to B$ caused by the shear stress (dashed curve vs. continuous curve).

Equations (24.1) and (24.2) can be understood by using the model introduced by Eyring [104]. In this model the motion of a fluid molecule is seen as a stress-aided thermally activated process. The motion occurs in the direction of the applied shear, as shown in Fig. 24.1. In order to jump from the position A to the position B the molecule has to overcome an energy barrier ε. If the molecule has a cross-section δA and moves an average distance l, the energy barrier is reduced by the amount $l\tau\delta A$ when a shear stress τ is applied. Using a result from the theory of sliding of adsorbate particles, expressed by Eq. (15.25), the viscosity of the fluid can be written as

$$\eta = \text{const.} \times \frac{e^{(\varepsilon + p\,\delta V)/k_B T}}{e^{l\tau\delta A/k_B T} - e^{-l\tau\delta A/k_B T}}, \tag{24.3}$$

where δV is the volume of the gap to be opened in order for the molecule to jump (slightly smaller than the molecular volume) and the prefactor is independent of τ. If the shear stress is low enough Eq. (24.3) reduces to

$$\eta \propto \frac{k_B T}{2l\delta A} e^{(\varepsilon + p\delta V)/k_B T},$$

and Eqs. (24.1) and (24.2) are recovered, with the parameter $p_0 = k_B T/\delta V$.

Application to boundary lubrication

Commercial oils contain additives aimed at improving the lubrication properties. The additives are usually fatty acids consisting of long hydrocarbon chains with a polar head capable of binding to metal oxides (Fig. 24.2). The adhesion is so strong

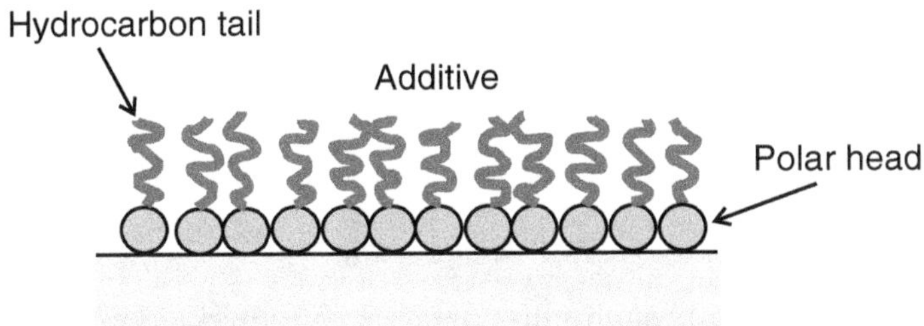

Figure 24.2 Binding of a fatty acid to a metal oxide surface.

that the adsorbate cannot fluidize and a dense film is formed with the inert hydrocarbon tails pointing away from the surface. If two flat surfaces coated by these additives are pressed together and slowly moved past each other, the sliding will take place between the ends of the hydrocarbon tails and the viscosity of the lubricant will play almost no role. The tails can in principle interdiffuse, which would lead to a slow increase of the static friction with the time of stationary contact (see Section 14.2). However, the ageing process can be inhibited if the contact pressure is large enough to significantly increase the energy barriers for interdiffusion. If this is the case the coefficients of static and kinetic friction will approximately have the same values: $\mu_{\mathrm{k}} \approx \mu_{\mathrm{s}}$.

Shear thinning

If polymers are used as additives in an otherwise Newtonian lubricant, so-called *shear thinning* may appear. This phenomenon consists of a viscosity drop caused by the fact that the long molecules of the polymer tend to align at high speed. Shear thinning is exploited in wall paints as well as in human joints, and it is usually reversible. The onset of shear thinning can also be explained by the Eyring model. Equation (24.3) implies that the viscosity starts to decrease with increasing shear stress τ when $l\tau\delta A/k_{\mathrm{B}}T \sim 1$. The corresponding critical value of the shear rate $\dot{\gamma} = \tau/\eta \sim \rho k_{\mathrm{B}}T/m\eta$, where $m \sim \rho l\delta A$ is the mass of the molecule (or of a chain segment) and ρ is the fluid density. For a 10 nm film of n-hexadecane this value is reached at a sliding velocity of the order of 100 m/s. However, the critical velocity becomes much lower if the film is squeezed.

24.2 Capillary bridges

Capillary bridges have enormous importance in granular materials, insect adhesion, head/disk systems and MEMS, where they can cause device failure. Capillary bridges can result from organic contamination at ambient conditions, from water condensation in a humid environment or from thin lubrication films. Thanks to capillary adhesion it is possible to build sand castles from humid sand but not from dry sand.

Kelvin equation

If two rough surfaces are brough into contact in a humid environment, the water will condense in the form of capillary bridges. The radius of curvature r_K of a bridge depends on the relative humidity according to the *Kelvin equation*:

$$r_K = \frac{\gamma v_m}{k_B T \ln(p/p_{sat})}. \tag{24.4}$$

In Eq. (24.4) γ and v_m are the surface tension and the molecular volume of the liquid, and p/p_{sat} is the relative vapor pressure. In the case of water $\gamma = 73$ mN/m and $v_m = 0.03$ nm^3 so that, for humidity values below 90%, r_K takes values in the nanometer range. The Kelvin equation is derived by observing that the chemical potentials of vapor and liquid are, respectively, $\mu(p) = k_B T \ln p$ and $\mu_l = \mu(p) + \Delta p \cdot v_m$ and, according to the Laplace formula, the pressure difference between vapor and liquid is $\Delta p = -\gamma/r_K$.

Thermal effects

The formation of capillary bridges in a humid environment is a thermally activated process. This means that the time required to form a bridge between two hydrophilic surfaces at nanometer distance from each other is

$$\tau = \tau_0 \exp\left(\frac{\Delta\varepsilon}{k_B T}\right),$$

where $\Delta\varepsilon$ is the energy cost of condensing the corresponding liquid volume and τ_0 is a characteristic time. Correspondingly, the adhesion force between the two surfaces is expected to increase logarithmically with the contact time t:

$$F_{adh} \propto \ln(t/\tau_0). \tag{24.5}$$

It can be proven that the prefactor in Eq. (24.5) depends on the relative humidity as $1/\ln(p_{sat}/p)$ [245, section 7.5]. For this reason, the stability of a sand castle will increase if the wet sand is kept under pressure for a while.

Equation (24.5) also implies that, in a humid environment, the friction force F_{fric} will decrease logarithmically with the sliding velocity. This logarithmic dependence was measured by Bocquet *et al.* on a paper–paper interface and on small particles jammed in a rotating drum, whereas no influence of the humidity on friction and adhesion was found on hydrophobic materials such as Teflon [26].

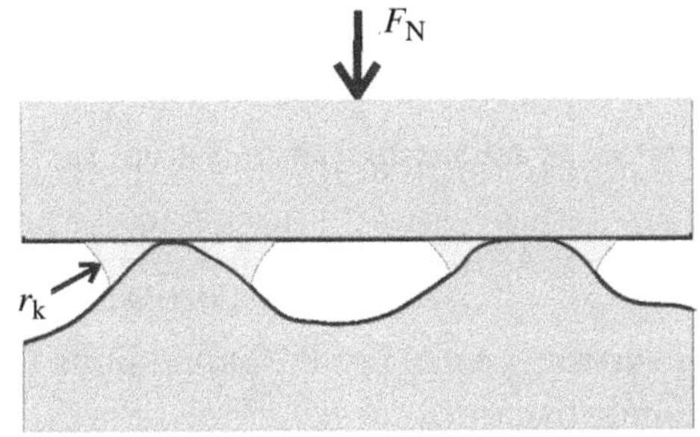

Figure 24.3 Capillary bridges formed around contacting asperities.

A logarithmic decrease of friction with sliding velocity was also reported by Riedo *et al.* on hydrophilic films using AFM [290].

Note that in AFM capillary forces are very important and usually detrimental. Since these forces are typically of the order of some tens of nanonewtons, they prevent high resolution in ambient conditions. Capillary forces can be reduced if the probing tip is covered by amphiphilic molecules. They can even be removed completely if tip and sample are immersed in a liquid (Section 18.1). On the other hand, the water meniscus formed between a tip and a hydrophilic surface can be exploited to deliver small 'ink' molecules in the so-called *dip-pen nanolithography* technique.

Capillary adhesion between rough surfaces

If a flat elastic block is pressed against a rigid rough substrate in a humid environment, the thickness of the liquid film formed at the interface (Fig. 24.3) is $d = r_K(\cos\theta_1 + \cos\theta_2)$, where θ_1 and θ_2 are the contact angles between the liquid and the solid surfaces. For water at room temperature, this thickness is in the nanometer range. If the water wets the surfaces completely, the liquid thickness is $2r_K$.

In the framework of the Persson theory the liquid covers the projected area $\Delta A = A(\zeta_K) - A(\zeta_1)$, where ζ_K is defined as the magnification at which the incipient separation $\delta_{inc} = 2r_K$ and $\zeta_1 = k_1/2\pi a$ is the highest possible magnification (Section 8.4). According to the Laplace formula the (negative) pressure in the liquid bridge is $p_{attr} = \gamma/r_K$, so that the attractive force between the two solid surfaces is:

$$F_{attr} = \frac{\gamma}{r_K}\left[A(\zeta_K) - A(\zeta_1)\right].$$

If the gap between the two surfaces is completely filled by water, the attractive pressure $p_{attr} = \gamma/r_K$ is balanced by the repulsive contact pressure $p(\delta)$ determined by Eq. (8.23). As a result, the average equilibrium separation is

$$\delta = \delta_0 \ln \frac{\beta E^* r_K}{\gamma},$$

and the corresponding work of adhesion (per unit area) is

$$W_{\mathrm{adh}} = \int_{\delta}^{\infty} \left(p_{\mathrm{attr}} - p(\delta')\right) \mathrm{d}\delta' = 2\gamma \left(1 - \frac{\delta_0}{2r_K} \ln \frac{e\beta E^* r_K}{\gamma}\right). \tag{24.6}$$

For a self-affine fractal surface we can use the expressions for δ_0 and β introduced in Section 8.4 and conclude that

$$W_{\mathrm{adh}} \approx 2\gamma \left(1 - \frac{h_{\mathrm{rms}}}{r_K} \ln \frac{k_0 h_{\mathrm{rms}} E^* r_K}{\gamma}\right), \tag{24.7}$$

in relatively good agreement with experimental results on polysilicon surfaces [67].

Note that for hard contacting materials, capillary adhesion appears only if r_K is much larger than the surface roughness. For very soft materials, capillary attraction can pull the surfaces into contact and, in this case, Eq. (24.7) is not valid any more.

24.3 Fluid flow between rough surfaces

Consider a fluid flowing between a rigid flat surface and an elastic rough solid. Suppose that the upper solid moves with a velocity $\mathbf{V}$ parallel to the lower surface. The separation between the two surfaces is $h(\mathbf{r}, t)$, where $\mathbf{r} \equiv (x, y)$, and the linear size of the nominal contact region is L. If the slope $\nabla h \ll 1$ and $h/L \ll 1$ the fluid velocity $\mathbf{v}$ varies slowly with the coordinates x and y. Assuming a slow time dependence, the Navier–Stokes equation (22.1) simplifies to

$$\eta \frac{\partial^2 \mathbf{v}}{\partial z^2} = \nabla p. \tag{24.8}$$

The fluid flow vector $\mathbf{j}$ can be calculated by solving Eq. (24.8) and integrating over the gap between the two surfaces. As a result:

$$\mathbf{j} = -\frac{h^3(\mathbf{r})}{12\eta} \nabla p + \frac{1}{2} h(\mathbf{r}) \mathbf{V}. \tag{24.9}$$

Using a perturbation treatment, the average value of $\mathbf{j}$ can be related to the roughness and the average separation for both isotropic and anisotropic surfaces [253].

Suppose now that the surfaces are in contact. In particular, we will consider the fluid flow through a rubber seal, squeezed by a pressure p_0, as shown in Fig. 24.4. The fluid is released into a region at athmospheric pressure and undergoes a pressure drop Δp. The nominal contact is assumed to occur on a rectangular area. As discussed in Section 8.3, at increasing magnification ζ a critical value ζ_c will be reached, where the non-contact regions percolate. The surface separation at this

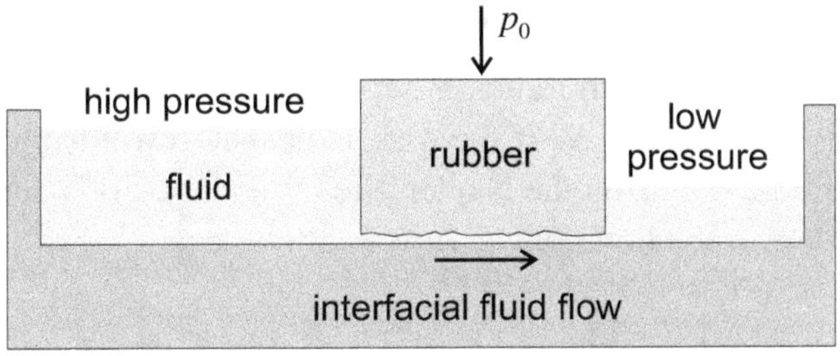

Figure 24.4 Schematic of a rubber seal.

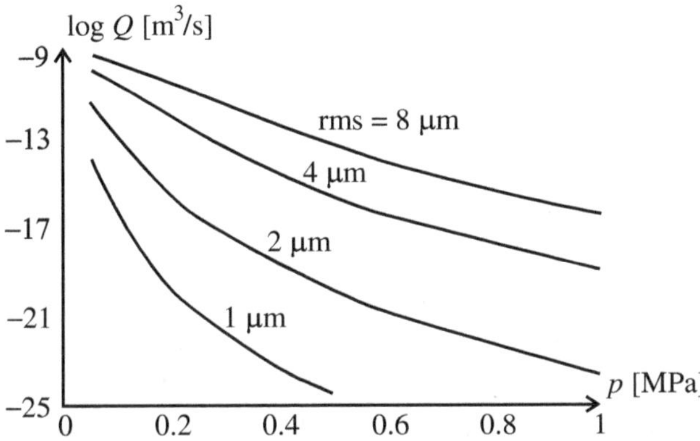

Figure 24.5 Volumetric flow of water leaking through a seal as a function of the applied pressure p_0. The simulated rubber surface has a Hurst coefficient $H = 0.8$ and rms roughness of 1, 2, 4 or 8 μm. The pressure drop across the seal is $\Delta p = 0.01$ MPa. Reproduced from [262] with permission from IOP Publishing.

point is $h_c \sim \delta_{inc}(\zeta_c)$, where δ_{inc} is defined in Section 8.4. The critical percolation channel has width and length λ_c. Assuming that all leakage occurs through it, the volume flow per unit time is given by Poiseuille's law (22.9):

$$Q \approx \frac{h_c^3}{12\eta} N \Delta p, \tag{24.10}$$

where N is the ratio between the width and length of the apparent contact area. If the magnification increases, more percolation channels with more narrow constrictions will appear but their contribution to the leak-rate will be negligible. Numeric calculations (Fig. 24.5) show that Q decreases extremely fast when p_0 increases, and increases when the surface roughness increases [262]. A good agreement with these results has been found in the case of water flowing through a silicon rubber ring pressed against sand paper [190].

24.4 Squeezed films

Consider now a rigid disk with radius R separated from a rigid flat substrate by a fluid film of thickness $h_0 \ll R$. If the disk is pushed towards the substrate by a force F_N, as in Fig. 24.6, using the Navier–Stokes equation it is not difficult to see that the thickness h of the liquid layer varies with time as

$$\frac{dh}{dt} = -\frac{2h^3 F_N}{3\pi \eta R^4} \tag{24.11}$$

or, after integrating,

$$\frac{1}{h^2(t)} - \frac{1}{h_0^2} = \frac{4F_N}{3\pi \eta R^4}. \tag{24.12}$$

The pressure distribution in the gap is

$$p(r) = p_{ext} - \frac{3\eta \dot{h}}{h^3}(R^2 - r^2), \tag{24.13}$$

where p_{ext} is the external pressure. For a steel block on a steel table lubricated by a 1 μm thick film of castor oil, this means that direct contact would be reached within few minutes. However, Eq. (24.12) breaks down when h becomes comparable to the linear size of the liquid molecules. As a result, a fluid slab of a few nanometers in thickness will remain trapped between the solid surfaces. Note that, if the surfaces are separated rapidly (so that $\dot{h} > 0$), the pressure in the gap is lower than p_{ext} and the formation of bubbles (*cavitation*) is possible.

When h reaches a value of a few nm, the liquid molecules tend to form layers parallel to the two surfaces, and the viscosity $\eta(h)$ increases rapidly. This is shown by the 'quantized' variations of many physical properties such as the thickness of the liquid slab and the yield strength k [101], the periodicity of which is close to the diameter of the molecules. Note that a finite value of k indicates that the layers must be ordered, as incommensurate structures experience a negligible pinning potential (Section 16.3).

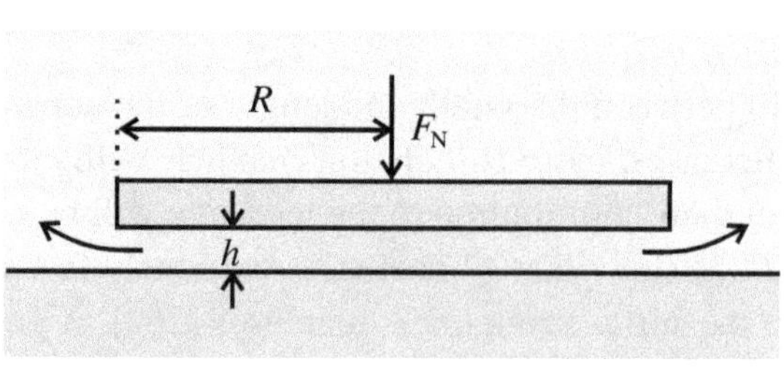

Figure 24.6 A rigid solid disk pushed against a rigid flat substrate in a liquid.

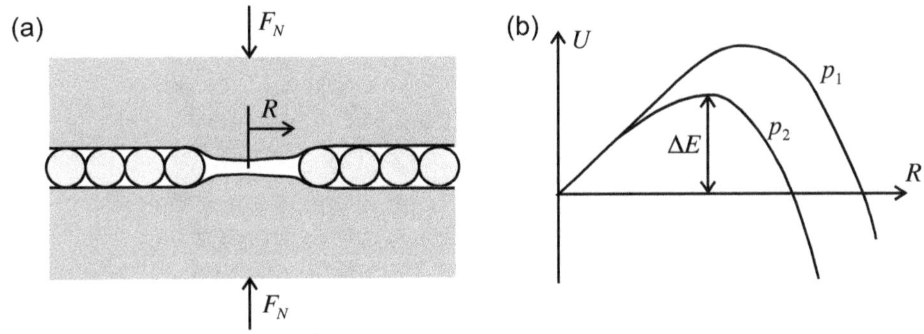

Figure 24.7 (a) The squeeze-out of the last lubricant monolayer from the contact area between two elastic solids starts with the formation of a small circular 'hole'. (b) Free energy as a function of the hole radius for two values of external pressure (with $p_2 > p_1$).

We will now focus on the squeezing of the last few layers in more detail. The squeezing process is assisted by thermal fluctuations. In particular, a fluctuation can give rise to a 'hole' in a layer. If the radius R of such hole exceeds a critical value R_c, the pressure difference Δp between the inner and the outer boundary lines will drive the layer out of the contact region, as shown in Fig. 24.7(a) for the $n = 1 \to 0$ transition. The critical radius can be estimated by starting from the expression for the adiabatic work required to form the hole [257]:

$$U(R) = 2\pi R\gamma + \pi R^2 S - \alpha R^3. \tag{24.14}$$

In Eq. (24.14) γ is the 'line tension', S is the two-dimensional spreading pressure further discussed in Section 24.5 and the last term is the relaxation energy of the walls, which are supposed to be elastic. The coefficient $\alpha \sim p_{\rm ext}^2/E$, where E is the elastic modulus of the solid materials. The dependence $U(R)$ is plotted in Fig. 24.7(b). As $p_{\rm ext}$ increases, the energy barrier $\Delta\varepsilon$ preventing the layer from squeezing out decreases. When $\Delta\varepsilon \sim 1$ eV, the layer is expelled. As a result, the critical radius R_c turns out to be of the order of 1 nm for hydrocarbons on metals or on metal oxides. As detailed in [257], the corresponding critical pressure is

$$p_c \approx 0.60\sqrt{\frac{(2\pi\gamma + 2\pi R_c S)E}{3(1-\nu^2)R_c^2}}.$$

Once a hole has nucleated, it spreads quickly through the contact area. The time evolution of the squeezed-out area A can be estimated from the Navier–Stokes and continuity equations in 2D [257]:

$$\frac{dA}{dt}\ln\left(\frac{A}{A_0}\right) = -\frac{4\pi\Delta p}{\rho\eta}, \tag{24.15}$$

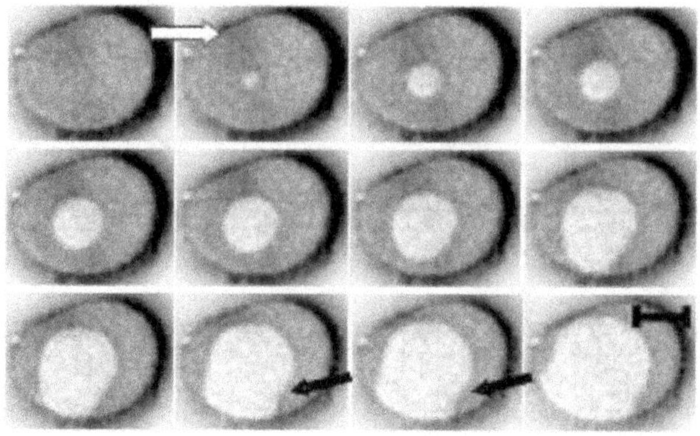

Figure 24.8 Squeeze-out of one monolayer of a silicone liquid (OMCTS) between two mica surfaces. The sequence of images was acquired at time intervals of 100 ms. Reproduced from [16] with permission from IOP Publishing.

where $A_0 = \pi R^2$ is the total contact area and ρ is the density of the liquid. Assuming that $\Delta p = p_0 =$ const., Eq. (24.15) can be easily integrated. As a result, the time required to completely squeeze out the layer is

$$t_c = \frac{\rho \eta A_0}{4\pi p_0}.$$

A better agreement with the experiments is obtained by assuming a Hertzian distribution for the contact pressure [352].

The squeeze-out of liquid layers has been demonstrated in a series of investigations by Mugele and coworkers based on the surface force apparatus [16]. An example, illustrating an $n = 3 \rightarrow 2$ transition, is shown in Fig. 24.8. Note that the circular symmetry gets broken in the late stages of the transition.

Rough surfaces

Suppose now that a cylindrical rubber block (with radius R) is squeezed against a rough rigid substrate in a liquid. Neglecting the deformation of the block caused by the non-uniform fluid pressure, the separation $\delta(\mathbf{r}, t)$ depends on the time t according to Eq. (24.11):

$$\frac{\mathrm{d}\delta}{\mathrm{d}t} \approx -\frac{2\delta^3 p(t)}{3\eta R^2}, \tag{24.16}$$

where p is the average fluid pressure in the nominal contact region. If p_0 is the pressure applied on the top surface of the cylinder,

$$p(t) = p_0 - p_{\text{con}}(t),$$

where p_{con} is the (locally averaged) asperity contact pressure. If p_0 is so small that $\delta \gg h_{rms}$ we can use Eq. (8.23) and conclude, after some simplifications, that $\delta(t)$ approaches the equilibrium separation asymptotically as [191]

$$\delta \approx \delta_\infty + \left(1 - \frac{p_{con}(0)}{p_0}\right) u_0 \exp\left[-\left(\frac{\delta(p_0)}{h_{rms}}\right)^3 \frac{t}{\tau}\right], \tag{24.17}$$

where $\delta_\infty \equiv \delta_0 \log(\beta E^*/p_0)$ and the characteristic time

$$\tau = \frac{3\eta R^2 \delta_0}{2h_{rms}^2 p_0}.$$

Equation (24.17) breaks down if the pressure p_0 is too high. In this case one has to solve the equation [191]

$$\frac{d\delta}{dt} \approx -\frac{\delta_0^2 \delta^3}{\tau h_{rms}^4} \left(1 - \frac{p_{con}}{p_0}\right)$$

together with the relation $p(\delta)$ given by Eq. (8.26).

24.5 Spreading of liquids

Finally, let us consider a small fluid drop on a flat solid surface. If the system is in equilibrium the *contact angle* θ_{eq} between droplet and substrate is determined by the solid–liquid, solid–vapor and liquid–vapor interfacial energies γ_{sl}, γ_{sv} and γ respectively (Fig. 24.9) according to *Young's equation*[1] [349]:

$$\gamma_{sv} = \gamma_{sl} + \gamma \cos\theta_{eq}. \tag{24.18}$$

If the droplet is not in equilibrium, a flow will start and only stop when Eq. (24.18) is satisfied. In case of *complete wetting* ($\theta_{eq} = 0$) the droplet spreads till it reaches a thickness defined by the van der Waals forces. This situation is typically observed on hard solids with high surface tension. As a result, the contact angle θ

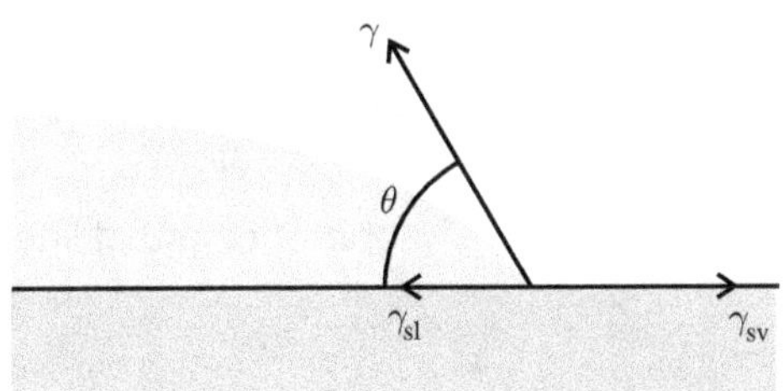

Figure 24.9 A liquid droplet in thermodynamic equilibrium with a solid and its vapor.

[1] Equation (24.18) can be proven by minimizing the free energy of the system with the condition that the volume of the fluid remains constant.

decreases with time. The speed of advance $v(t)$ is defined by the condition that the rate of decrease of the interfacial energy equals the rate of viscous dissipation [65]. When the droplet expands the variation of the interfacial energy corresponding to a section of liquid of length L is

$$\frac{\mathrm{d}U_{\mathrm{surf}}}{\mathrm{d}t} = -LSv, \tag{24.19}$$

where

$$S = \gamma_{\mathrm{sv}} - \gamma_{\mathrm{sl}} - \gamma \cos\theta \tag{24.20}$$

is the so-called *spreading pressure*. If the actual contact angle θ is small, we can assume that $S \approx S_0 + \gamma\theta^2/2$, where S_0 is the spreading pressure at $\theta = 0$. According to Eq. (22.2), the dissipation rate depends on the velocity gradient in the drop as $\eta(\mathrm{d}v/\mathrm{d}z)^2$, where η is the viscosity of the fluid. In a first approximation $\mathrm{d}v/\mathrm{d}z \approx v/h$, where $h(x) \approx \theta x$ is the droplet profile, so that

$$\frac{\mathrm{d}E_{\mathrm{diss}}}{\mathrm{d}t} \approx L\eta \int_b^R \left(\frac{v}{x\theta}\right)^2 (\theta x)\,\mathrm{d}x = L\eta v^2 \frac{\ln(R/b)}{\theta},$$

where R is the radius of the droplet and b is a cut-off distance comparable to the size of a fluid molecule. Equating the dissipation rate to the energy loss (24.19), observing that, for small values of θ, the quantity $R^3\theta$ is approximately equal to the (constant) volume V of the droplet, and neglecting S_0, one gets

$$\frac{\mathrm{d}R}{\mathrm{d}t} \ln\frac{R}{b} \approx \frac{\gamma}{\eta}\left(\frac{V}{R^3}\right)^3.$$

Finally, neglecting the slowly varying logarithmic term and integrating, we end up with the *Tanner law of wetting* [329]:

$$R(t) \approx \left(\frac{V^3\gamma}{\eta} t\right)^{1/10}. \tag{24.21}$$

The Tanner law is experimentally verified even if S_0 is not negligible. The reason is that the fluid spreads far ahead of the macroscopic contact line in a flat *precursor film* [137] of thickness between a monolayer and about 10 nm (Fig. 24.10). The 'flat

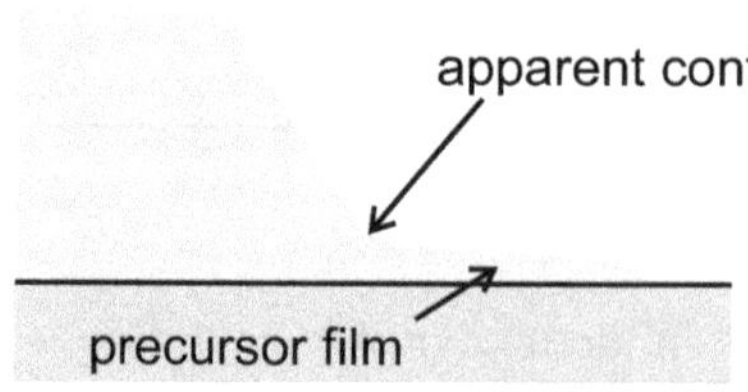

Figure 24.10 Spreading of a precursor film driven by van der Waals forces.

spreading pressure' S_0 is balanced by the energy dissipation in the precursor film, whereas the residual spreading pressure $\gamma\theta^2/2$ is balanced by the dissipation in the droplet, as described above. As shown by time-resolved ellipsometry, the radius of the precursor film advances as $\sqrt{t}$ [137]. This behavior can be reproduced using a model which does not differ substantially from that discussed in Section 24.4 [245, section 8.6.3].

Appendix A
Friction force microscopy

Similarly to standard AFM (Fig. 17.1), friction force microscopy (FFM) is based on the relative motion of a sharp tip on a solid surface. This motion is realized by a scanner formed by piezoelectric elements, which moves the surface perpendicularly to the tip with a certain time periodicity. The scanner can be also extended or retracted in order to vary the normal force F_N between tip and surface. This force is responsible for the deflection of the microcantilever supporting the tip. If F_N increases while scanning due to local variations of the surface height, the scanner is retracted by a feedback loop. If F_N decreases, the surface is brought closer to the tip by extending the scanner. In such a way, the surface topography can be reconstructed line by line from the vertical deformation of the scanner. An accurate control of the imaging process is made possible by a light beam, which is reflected from the rear of the cantilever into a photodetector. When the bending of the cantilever changes, the light spot on the detector moves up or down. This causes a variation of the photocurrent corresponding to the value of F_N to be controlled.

The scan motion is also accompanied by friction. A tangential force F with the opposite direction to the scan velocity v hinders the sliding motion. The force F causes the torsion of the cantilever, and can be recorded simultaneously with the topography if the photodetector can measure not only the normal deflection but also the torsion of the lever while scanning. In practice this is made possible by a four-quadrants photodetector, which converts the photocurrent corresponding to the lateral force into a voltage V_L. Note that the friction also causes lateral bending of the cantilever, but this effect is modest if the thickness of the lever is much less than its width.

The first atomic friction measurements by Mate *et al.* [206] were actually based on the deflection of a tungsten wire. The optical beam deflection scheme was introduced independently by two groups in 1990 [219, 205]. Other methods to measure the forces between tip and surface are capacitance detection [232], dual fiber interferometry [211], and piezoelectric levers [186]. In the first method capacitance variations of two plates close to the cantilever are measured while scanning. The second technique uses two optical fibers to detect the cantilever deflection along two orthogonal directions angled at 45° with respect to the surface normal. In the third case the normal and lateral forces are detected with two Wheatstone bridges at the base of the cantilever. The two forces are proportional to respectively the sum and the difference of the bridge signals.

Force calibration is relatively simple for rectangular cantilevers. Due to possible discrepancy with the geometric values provided by manufacturers, one should use optical or electron microscopes to determine the width, thickness and length of cantilever, w, d and L, the tip height h and the position of the tip with respect to the cantilever. The thickness of the cantilever can also be estimated from the resonance frequency f_0 using the relation

$$d = \frac{2\sqrt{12}\pi}{1.875^2}\sqrt{\frac{\rho}{E}} f_0 L^2. \tag{A.1}$$

In Eq. (A.1) ρ is the density of the cantilever and E is its Young's modulus. The normal stiffness k_{N} and the lateral stiffness k_{L} of the lever are given by

$$k_{\mathrm{N}} = \frac{Ewt^3}{4L^3}, \qquad k_{\mathrm{L}} = \frac{Gwt^3}{3h^2L},$$

where G is the shear modulus of the material. Typical values of k_{N} and k_{L} for commercial probes are few N/m and few hundreds of N/m respectively.

The next step is measuring the sensitivity S_z of the photodetector in nm/V. For beam-deflection FFMs S_z can be determined by force vs. distance curves measured on hard surfaces (e.g. Al_2O_3), where elastic deformations are negligible and the vertical elongation of the scanner equals the deflection of the cantilever. The typical response of the photodetector when the tip is brought into contact and then retracted is sketched in Fig. A.1. In the beginning no signal variations are observed until the tip jumps into contact at the distance $z = z_1$ from the surface. Further extension or retraction of the scanner results in a linear response until the tip jumps again out of contact at a distance $z_2 > z_1$. The slope of the linear part of the curve gives the required value of S_z. Assuming that the light beam is precisely positioned above the tip, the normal and lateral forces can be finally estimated as

$$F_{\mathrm{N}} = k_{\mathrm{N}} S_z V_{\mathrm{N}}, \qquad F_{\mathrm{L}} = \frac{3}{2} c_{\mathrm{L}} \frac{h}{L} S_z V_{\mathrm{L}},$$

where V_{N} is the voltage corresponding to the normal deflection of the lever.

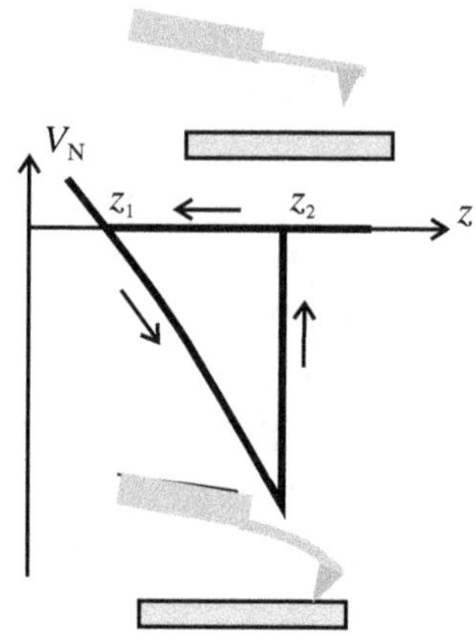

Figure A.1 A typical force–distance curve on a hard surface.

The normal stiffness k_N can also be calibrated with alternative methods. Cleveland *et al.* attached tungsten spheres to the tip, which changes the value of f_0 as

$$f_0 = \frac{1}{2\pi}\sqrt{\frac{k_N}{M + m_{eff}}}.$$

In the previous equation M is the total mass of the spheres and m_{eff} is the effective mass of the cantilever [59]. The value of k_N can be extrapolated from the frequency shifts Δf corresponding to different masses. Alternatively Hutter *et al.* observed

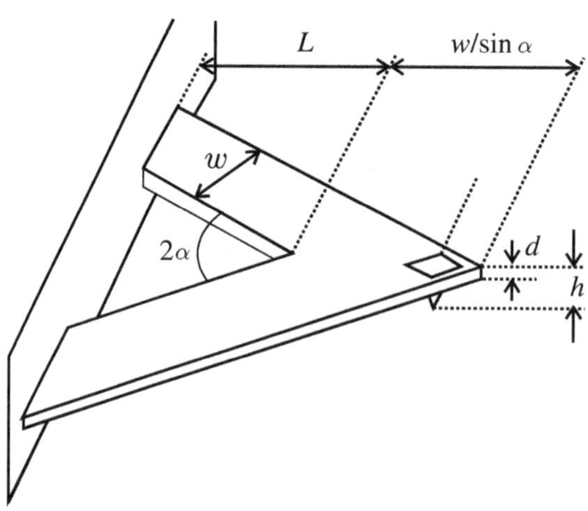

Figure A.2 Geometry of a V-shaped cantilever.

that k_N is related to the area P of the power spectrum of the thermal fluctuations of the free cantilever [143]. The correct relation is $k_N = 4k_B T/3P$, where k_B is Boltzmann's constant and T is the absolute temperature [41]. Extensions of this method have been proposed by Sader *et al.* for both normal and lateral force calibration [299, 121].

Calculations for cantilevers with different shapes require finite element analysis, although approximate formulas have been derived. In the case of V-shaped cantilevers, Neumeister *et al.* [233] have obtained an analytical approximation for the lateral stiffness:

$$k_L = \frac{Ed^3}{3(1+\nu)h^2}\left(\frac{1}{\tan\alpha}\ln\frac{w}{l\sin\alpha} + \frac{l\cos\alpha}{w} - \frac{3\sin 2\alpha}{8}\right)^{-1},$$

where the geometrical quantities L, w, α, d and h are defined in Fig. A.2. The expression for k_N is more complex and can be found in [233]. Note that V-shaped cantilevers are more sensitive to lateral forces, but also more prone to in-plane bending.

Appendix B
Viscosity of gases

The viscosity of a gas is determined by molecular diffusion. Accurate estimations can be made using the methods of kinetic theory. In this theory the fraction of molecules with momentum $\mathbf{p} = m\mathbf{v}$ close to a given position $\mathbf{r}$ at a given time t defines the *distribution function* $f(\mathbf{p}, \mathbf{r}, t)$ (m is the mass of a molecule and $\mathbf{v}$ its velocity). According to *Boltzmann's equation* this function varies with the time t as

$$\frac{\partial f}{\partial t} + \mathbf{v} \cdot \nabla f = \left.\frac{\partial f}{\partial t}\right|_{\text{coll}}, \tag{B.1}$$

where the quantity on the right hand side corresponds to the so-called *collision integral*. If collisions occur only between pairs of molecules:

$$\left.\frac{\partial f}{\partial t}\right|_{\text{coll}} = \int v_{\text{rel}}(f_1' f_2' - f_1 f_2)\, \mathrm{d}\sigma\, \mathrm{d}^3 p_2,$$

where v_{rel} is the relative velocity of two generic molecules labeled as 1 and 2, and σ is the collisional cross-section[1] and the primes indicate the values acquired in the collision.

Another key concept in kinetic theory is the *mean free path* l, which is defined as the average distance covered by a molecule between two consecutive collisions. The order of magnitude of l depends on σ and on the number n of molecules per unit volume as

$$l \sim \frac{1}{n\sigma}.$$

For air at room temperature $l \approx 70$ nm.

Considering the viscosity η of a gas as a 'diffusivity of momentum', this quantity has an order of magnitude

$$\eta \sim mnlv_T,$$

[1] If a molecular beam hits a scattering center with the same velocity, the *cross section* σ is defined as the ratio between the number of molecules scattered in unit time and the flux density of the incident molecules.

where $v_T \sim \sqrt{k_B \mathrm{T}}$ is the average velocity of a molecule at the temperature T. Since σ also depends on the temperature, a general relation for η as a function of T is not possible. However, if T is not too low, the molecules can be considered as hard elastic particles and σ varies only slightly with the velocity. In this case, $\eta \propto \sqrt{T}$, where the coefficient of proportionality is of the order of $\sqrt{mk_B}/\sigma$.

The viscosity of a *dilute gas* can be estimated using the *theory of Chapman–Enskog* [51, 86]. In this theory η is expressed as a series of polynomials. When inserted in Eq. (B.1), the series turns out to be rapidly converging. For a gas formed by rigid spheres with diameter d, the cross-section is $\sigma = \pi d^4$ and

$$\eta \approx 0.18 \frac{\sqrt{mk_B T}}{d^2}.$$

It is interesting to observe that an exact calculation of η is possible if the interaction law between the gas particles has the form $U(r) = \alpha/r^4$. The solution to this formal problem was obtained by Maxwell [208]:

$$\eta = 0.81 k_B T \sqrt{\frac{m}{\alpha}}.$$

A completely similar result holds for the thermal conductivity κ, which relates the heat flow $\mathbf{q}$ to the temperature gradient ∇T in a fluid:

$$\mathbf{q} = -\kappa \nabla T.$$

Considering the quantity κ as a 'diffusivity of energy', it is not difficult to see that its order of magnitude is $\kappa \sim cnlv_T$, where c is the molecular heat capacity.

Appendix C
Slip conditions

Consider a fluid confined by two parallel planes in relative motion as in Fig. 22.1 and suppose that the fluid sticks to the upper plane but slips on the lower one. In this case, it can be proven that the shear stress is given by Eq. (22.5), with the right hand side divided by $(1 + l/h)$ [245, section 8.6]. The *slip length* l can be seen as the fictitious distance below the surface where the no-slip boundary condition would be satisfied. Similarly, the flow rate in a cylindrical pipe is given by Poiseuille's law (22.8) with the right hand side multiplied by the factor $(1 + 4l/h)$. From dimensional considerations it is not difficult to prove that, under ideal conditions,

$$l = \frac{\eta}{\gamma m n_a},$$

where η is the viscosity, m is the mass of a liquid molecule, n_a is the adsorbate concentration and γ is the damping coefficient in Eq. (15.26). In the case of water flowing on a flat metal surface, the slip length would be thus of the order of 10 μm.

However, the slip length is influenced by many factors, e.g. surface roughness, dissolved gases, wetting, electrical properties and viscous heating. In the case of a wavy surface with amplitude h_0 and periodicity $\lambda = 2\pi/k$, Einzel *et al.* [84] derived an expression for the effective slip length l_{eff} from the Navier–Stokes equation. Introducing a critical slip length

$$l_c = \frac{\lambda^3}{8\pi^3 h_0^2},$$

they concluded that $l_{eff} \approx l_c$ if $l \gg l_c$, and $l_{eff} \approx -\lambda^2/(4\pi^2 l_c)$ if $l \ll l_c$. Thus, negative slip lengths are also possible.

References

[1] Abdurixit, A., Baratoff, A. and Meyer, E. 2000. Molecular dynamics simulations of dynamic force microscopy: applications to the Si(111)7×7 surface. *Appl. Surf. Sci.*, **157**, 355.

[2] Airy, G. B. 1862. On the strains in the interior of beams. *Phil. Trans. Roy. Soc. London*, **153**, 49.

[3] Albrecht, T. R., Grütter, P., Horne, D. and Rugar, D. 1991. Frequency modulation detection using high-Q cantilevers for enhanced force microscope sensitivity. *J. Appl. Phys.*, **69**, 668.

[4] Alireza Ghasemi, S., Goedecker, S., Baratoff, A., Lenosky, T., Meyer, E. and Hug, H. J. 2008. Ubiquitous mechanisms of energy dissipation in noncontact atomic force microscopy. *Phys. Rev. Lett.*, **100**, 236106.

[5] Amontons, G. 1699. On the resistance caused in machines (in French). *Mem. R. Acad., Paris*, 257.

[6] Archard, J. F. 1957. Elastic deformation and the laws of friction. *Proc. Roy. Soc. A*, **243**, 190.

[7] Archard, J. F. 1959. The temperature of rubbing surfaces. *Wear*, **2**, 438.

[8] Aruliah, D. A., Müser, M. and Schwarz, U. D. 2005. Calculations of the threshold force and threshold power to move adsorbed nanoparticles. *Phys. Rev. B*, **71**, 085406.

[9] Ashegi, M., Leung, Y. K., Wong, S. S. and Goodson, K. E. 1997. Phonon-boundary scattering in thin silicon layers. *Appl. Phys. Lett.*, **71**, 1798.

[10] Barabasi, A. L. and Stanley, H. E. 1995. *Fractal Concepts in Surface Growth*. Cambridge University Press.

[11] Barel, I., Urbakh, M., Jansen, L. and Schirmeisen, A. 2010. Multibond dynamics of nanoscale friction: the role of temperature. *Phys. Rev. Lett.*, **104**, 066104.

[12] Bartels, L., Meyer, G. and Rieder, K. H. 1997. Basic steps of lateral manipulation of single atoms and diatomic clusters with a scanning tunneling microscope tip. *Phys. Rev. Lett.*, **79**, 697.

[13] Barus, C. 1893. Isotherms, isopiestics and isometrics relative to viscosity. *Am. J. Sci.*, **45**, 87.

[14] Baumberger, T., Heslot, F. and Perrin, B. 1994. Crossover from creep to inertial motion in friction dynamics. *Nature*, **367**, 544.

[15] Bauschinger, J. 1881. On the the variation of yield strength and modulus of elasticity of different metals (in German). *Zivilingenieur*, **299**, 289.

[16] Becker, T. and Mugele, F. 2003. Collapse of molecularly thin lubricant layers between elastic substrates. *J. Phys.: Condens. Matter*, **15**, S321.

[17] Belak, J. and Stowers, I. F. 1990. A molecular dynamics model of the orthogonal cutting process. *Proc. Am. Soc. Precision Eng.*, 76.

[18] Bennewitz, R., Gyalog, T., Guggisberg, M., Bammerlin, M., Meyer, E. and Güntherodt, H. J. 1999. Atomic stick-slip processes on Cu(111). *Phys. Rev. B*, **60**, R11301.

[19] Bentall, R. H. and Johnson, K. R. 1967. Slip in the rolling contact of two dissimilar elastic rollers. *Int. J. Mech. Sci.*, **9**, 389.

[20] Benz, M., Rosenberg, K. J., Kramer, E. J. and Israelachvili, J. N. 2006. The deformation and adhesion of randomly rough and patterned surfaces. *J. Phys. Chem. B*, **110**, 11884.

[21] Berkovich, E. S. 1950. Three-faceted diamond pyramid for studying mechanical microhardness by indentation (in Russian). *Zavodskaya Laboratoria*, **13**, 345.

[22] Berry, M. V. and Lewis, Z. V. 1980. On the Weierstrass–Mandelbrot fractal function. *Proc. R. Soc. A*, **370**, 459.

[23] Binnig, G., Rohrer, H., Gerber, C. and Weibel, E. 1982. Surface studies by scanning tunneling microscopy. *Phys. Rev. Lett.*, **49**, 57.

[24] Binnig, G., Quate, C. F. and Gerber, C. 1986. Atomic force microscope. *Phys. Rev. Lett.*, **56**, 930.

[25] Blasius, H. 1908. Boundary layers in fluids with low friction (in German). *Z. Math. Phys.*, **56**, 1.

[26] Bocquet, L., Charlaix, E., Ciliberto, S. and Crassous, J. 1998. Moisture-induced ageing in granular media and the kinetics of capillary condensation. *Nature*, **396**, 735.

[27] Bohlein, T., Mikhael, J. and Bechinger, C. 2012. Observation of kinks and antikinks in colloidal monolayers driven across ordered surfaces. *Nature Materials*, **11**, 126.

[28] Bouchbinder, E., Goldman, T. and Fineberg, J. 2014. The dynamics of rapid fracture: instabilities, nonlinearities and length scales. *Rep. Prog. Phys.*, **77**, 046501.

[29] Bouhacina, T., Aimé, J. P., Gauthier, S., Michel, D. and Heroguez, V. 1997. Tribological behaviour of a polymer grafted in silanized silica probed with a nanotip. *Phys. Rev. B*, **56**, 7694.

[30] Boussinesq, J. V. 1885. *Applications of potentials to the study of equilibrium and motion of elastic solids (in French)*. Gauthier-Villars.

[31] Bowden, F. P. 1953. Friction on snow and ice. *Proc. Roy. Soc. London A*, **217**, 462.

[32] Bowden, F. P. and Tabor, D. 1986. *Friction and Lubrication*. Clarendon Press.

[33] Brace, W. F. and Byerlee, J. D. 1966. Stick slip as a mechanism for earthquakes. *Science*, **153**, 990.

[34] Bradley, R. S. 1932. The cohesive force between solid surfaces and the surface energy of solids. *Phil. Mag.*, **13**, 853.

[35] Braun, O. M. and Kivshar, Y. S. 2004. *The Frenkel–Kontorova Model: Concepts, Methods, and Applications*. Springer.

[36] Braun, O. M. and Naumovets, A. G. 2006. Nanotribology: microscopic mechanisms of friction. *Surf. Sci. Rep.*, **60**, 79.

[37] Bufler, H. 1959. On the theory of rolling friction (in German). *Ing. Arch.*, **27**, 137.

[38] Buldum, A., Ciraci, S., and Batra, I. P. 1998. Contact, nanoindentation, and sliding friction. *Phys. Rev. B*, **57**, 2468.

[39] Burridge, R. and Knopoff, L. 1967. Model and theoretical seismicity. *Bull. Seismol. Sec. Am.*, **57**, 341.

[40] Bush, A. W., Gibson, R. D., and Thomas, T. R. 1975. The elastic contact of a rough surface. *Wear*, **35**, 87.

[41] Butt, H. J. and Jaschke, M. 1995. Calculation of thermal noise in atomic-force microscopy. *Nanotechnology*, **6**, 1.

[42] Cannara, R. J., Brukman, M. J., Cimatu, K., Sumant, A., Baldelli, S., and Carpick, R. 2007. Nanoscale friction varied by isotopic shifting of surface vibrational frequencies. *Science*, **318**, 780.
[43] Car, R. and Parrinello, M. 1985. Unified approach for molecular dynamics and density-functional theory. *Phys. Rev. Lett.*, **55**, 2471.
[44] Carlson, J. M. and Batista, A. A. 1996. Constitutive relation for the friction between lubricated surfaces. *Phys. Rev. E*, **53**, 4153.
[45] Carlson, J. M. and Langer, J. S. 1989. Properties of earthquakes generated by fault dynamics. *Phys. Rev. Lett.*, **62**, 2632.
[46] Carpick, R. W., Agrait, N., Ogletree, D. F., and Salmeron, M. 1996. Measurement of interfacial shear (friction) with an ultrahigh vacuum atomic force microscope. *J. Vac. Sci. Technol. B*, **14**, 1289.
[47] Carpick, R. W., Ogletree, D. F. and M. Salmeron. 1999. A general equation for fitting contact area and friction vs load measurements. *J. Coll. Int. Sci.*, **211**, 395.
[48] Carter, F. W. 1926. On the action of a locomotive driving wheel. *Proc. Roy. Soc. A*, **112**, 151.
[49] Cattaneo, C. 1938. On the contact of two elastic bodies: local distribution of stresses (in Italian). *Rend. R. Acc. Naz. dei Lincei*, **27**, 342, 434, 474.
[50] Cerruti, V. 1882. *Memoirs of Mathematical Physics (in Italian)*. Acc. Lincei.
[51] Chapman, S. 1916. On the law of distribution of molecular velocities, and on the theory of viscosity, in a non-uniform single monatomic gas. *Phil. Trans. Roy. Soc. A*, **216**, 279.
[52] Chen, J., Ratera, I., Park, J. Y. and Salmeron, M. 2006. Velocity dependence of friction and hydrogen bonding effects. *Phys. Rev. Lett.*, **96**, 236102.
[53] Chen, L. B., Zukoski, C. F., Ackerson, B. J. *et al.* 1992. Structural changes and orientational order in a sheared colloidal suspension. *Phys. Rev. Lett.*, **69**, 688.
[54] Childs, T. H. C. 1973. The persistence of asperities in indentation experiments. *Wear*, **25**, 3.
[55] Choi, J. S., Kim, J. S., Byun, I. S. *et al.* 2011. Friction anisotropy-driven domain imaging on exfoliated monolayer graphene. *Science*, **333**, 607.
[56] Chumak, A. A., Milonni, P. W. and Berman, G. P. 2004. Effects of electrostatic fields and Casimir force on cantilever vibrations. *Phys. Rev. B*, **70**, 085407.
[57] Ciavarella, M. and Decuzzi, P. 2001. The state of stress induced by the plane frictionless cylindrical contact. II. The general case (elastic dissimilarity). *Int. J. Sol. Struc.*, **38**, 4525.
[58] Cleveland, J. P., Anczykowski, B. and Elings, V. B. 1991. Energy dissipation in tapping-mode atomic force microscopy. *Appl. Phys. Lett.*, **59**, 2171.
[59] Cleveland, J. P., Manne, S., Bocek, D. and Hansma, P. K. 1993. A nondestructive method for determining the spring constant of cantilevers for scanning force microscopy. *Rev. Sci. Instrum.*, **64**, 403.
[60] Colebrook, C. F. 1939. Turbulent flow in pipes, with particular reference to the transition regime between smooth and rough pipe laws. *J. Inst. Civ. Eng. London*, **11**, 133.
[61] Cumings, J. and Zettl, A. 2000. Low-friction nanoscale linear bearing realized from multiwall carbon nanotubes. *Science*, **289**, 602.
[62] Das, B. M. 2006. *Principles of Geotechnical Engineering*. Thompson.
[63] Daw, M. S. and Baskes, M. I. 1984. Embedded-atom method: derivation and application to impurities, surfaces, and other defects in metals. *Phys. Rev. B*, **29**, 6443.
[64] Dayo, A., Alnasrallah, W. and Krim, J. 1998. Superconductivity-dependent sliding friction. *Phys. Rev. Lett.*, **80**, 1690.

[65] de Gennes, P. G. 1985. Wetting: statics and dynamics. *Rev. Mod. Phys.*, **57**, 827.

[66] Deacon, R. F. and Goodman, J. F. 1958. Lubrication by lamellar solids. *Proc. Roy. Soc. London A*, **243**, 464.

[67] Del Rio, F. W., Dunn, M. L., Phinney, L. M., Bourdon, C. J. and de Boer, M. P. 2007. Rough surface adhesion in the presence of capillary condensation. *Appl. Phys. Lett.*, **90**, 163104.

[68] Denk, W. and Pohl, D. 1991. Local electrical dissipation imaged by scanning force microscopy. *Appl. Phys. Lett.*, **59**, 2171.

[69] Deresiewicz, H. 1954. Contact of elastic spheres under an oscillating torsional couple. *Trans. ASME, J. App. Mech.*, **21**, 52.

[70] Deresiewicz, H. 1968. A note on Hertz impact. *Acta Mech.*, **6**, 110.

[71] Derjaguin, B. V., Muller, V. M. and Toporov, Y. P. 1975. Effect of contact deformations on the adhesion of particles. *J. Colloid Interface Sci.*, **53**, 314.

[72] Dienwiebel, M., Verhoeven, G. S., Pradeep, N., Frenken, J. W. M., Heimberg, J. A. and Zandbergen, H. W. 2004. Superlubricity of graphite. *Phys. Rev. Lett.*, **92**, 126101.

[73] Dieterich, J. H. 1979. Modelling of rock friction: 1. Experimental results and constitutive equations. *J. Geophys. Res.*, **84**, 2161.

[74] Dieterich, J. H. and Kilgore, B. D. 1994. Direct observation of frictional contacts: new insights for state-dependent properties. *Pure Appl. Geoph.*, **143**, 283.

[75] Dietzel, D., Feldmann, M., Fuchs, H., Schwarz, U. D. and Schirmeisen, A. 2009. Transition from static friction to kinetic friction of metallic nanoparticles. *App. Phys. Lett.*, **95**, 053104.

[76] Dong, Y., Li, Q. and Martini, A. 2013. Molecular dynamics simulation of atomic friction: a review and guide. *J. Vac. Sci. Technol. A*, **31**, 030801.

[77] Doolittle, A. K. 1951. Studies in newtonian flow II, The dependence of the viscosity of liquids on free-space. *J. Appl. Phys.*, **22**, 1471.

[78] Dorofeyef, I., Fuchs, H., Wenning, G. and Gotsmann, B. 1999. Brownian motion of microscopic solids under the action of fluctuating electromagnetic fields. *Phys. Rev. Lett.*, **83**, 2404.

[79] Drucker, D. C. and Prager, W. 1952. Soil mechanics and plasticity analysis of limit design. *Q. Appl. Math.*, **10**, 157.

[80] Dürig, U. 1999. Conservative and dissipative interactions in dynamic force microscopy. *Surf. Interface Anal.*, **27**, 467.

[81] Dürig, U. 2000. Extracting interaction forces and complementary observables in dynamic probe microscopy. *Appl. Phys. Lett.*, **76**, 1203.

[82] Eason, G. and Shield, R. T. 1960. The plastic indentation of a semi-infinite solid by a perfectly rough punch. *ZAMP*, **11**, 33.

[83] Einstein, A. 1906. A new determination of molecular dimensions (in German). *Annalen der Phys.*, **19**, 289.

[84] Einzel, D., Panzer, P. and Liu, M. 1990. Boundary condition for fluid flow: curved or rough surfaces. *Phys. Rev. Lett.*, **64**, 2269.

[85] Elkaakour, Z., Aime, J. P., Bouhacina, T., Odin, C. and Masuda, T. 1994. Bundle formation of polymers with an atomic force microscope in contact mode: a friction versus peeling process. *Phys. Rev. Lett.*, **73**, 3231.

[86] Enskog, D. 1917. The kinetic theory of phenomena in fairly rare gases (in German). PhD Thesis, University of Uppsala.

[87] Erlandsson, R., Olsson, L. and Martensson, P. 1996. Inequivalent atoms and imaging mechanisms in ac-mode atomic-force microscopy of Si(111)7×7. *Phys. Rev. B*, **54**, R8309.

[88] Fajardo, O. Y., Gnecco, E. and Mazo, J. J. 2014. Out-of-plane and in-plane actuation effects on atomic-scale friction. *Phys. Rev. B*, **89**, 075423.

[89] Falvo, M. R., Taylor, R. M., Helser, A. *et al.* 1999. Nanometre-scale rolling and sliding of carbon nanotubes. *Nature*, **397**, 236.

[90] Filippov, A., Dienwiebel, M., Frenken, J. W. M., Klafter, J. and Urbakh, M. 2008. Torque and twist against superlubricity. *Phys. Rev. Lett.*, **100**, 046102.

[91] Filippov, A. E., Popov, V. L. and Urbakh, M. 2011. Mechanism of wear and ripple formation induced by the mechanical action of an atomic force microscope tip. *Phys. Rev. Lett.*, **106**, 025502.

[92] Filleter, T. and Bennewitz, R. 2010. Structural and frictional properties of graphene lms on SiC(0001) studied by atomic force microscopy. *Phys. Rev. B*, **81**, 155412.

[93] Flamant, A. 1892. On the pressure distribution in a rectangular solid transversely loaded (in French). *Comptes Rendus Acad. Sci. Paris*, **114**, 1465.

[94] Frank, F. C. and Read, W. T. Jr. 1950. Multiplication processes for slow moving dislocations. *Phys. Rev.*, **79**, 722.

[95] Frenkel, Y. I. and T. Kontorova. 1938. On the theory of plastic deformation and doubling (in Russian). *Zh. Eksp. Teor. Fiz.*, **8**, 1340.

[96] Freund, L. B. 1998. *Dynamic Fracture Mechanics*. Cambridge University Press.

[97] Fusco, C. and Fasolino, A. 2005. Velocity dependence of atomic-scale friction: a comparative study of the one- and two-dimensional Tomlinson model. *Phys. Rev. B*, **71**, 045413.

[98] Gao, J. P., Luedtke, W. D., Gourdon, D., Ruths, M., Israelachvili, J. N. and Landman, U. 2004. Frictional forces and Amontons law: from the molecular to the macroscopic scale. *J. Phys. Chem. B*, **108**, 3410.

[99] Gauthier, M. and Tsukada, M. 1999. Theory of noncontact dissipation force microscopy. *Phys. Rev. B.*, **60**, 11716.

[100] Gauthier, M. and Tsukada, M. 2000. Damping mechanism in dynamic force microscopy. *Phys. Rev. Lett.*, **85**, 5348.

[101] Gee, M. L., McGuiggan, P. M. and Israelachvili, J. N. 1990. Liquid to solidlike transitions of molecularly thin films under shear. *J. Chem. Phys.*, **93**, 1895.

[102] Giessibl, F. 1997. Forces and frequency shifts in atomic-resolution dynamic-force microscopy. *Phys. Rev. B.*, **56**, 16010.

[103] Gladwell, G. M. L. 1976. On some unbounded contact problems in plane elasticity theory. *Trans. ASME, Series E, J. App. Mech.*, **43**, 263.

[104] Glasston, S., Laidler, K. J. and Eyring, H. 1941. *The Theory of Rate Processes*. McGraw Hill.

[105] Gnecco, E. 2010a. A collisional model for AFM manipulation of rigid nanoparticles. *Beilstein J. Nanotechnol.*, **1**, 158.

[106] Gnecco, E. 2010b. Quasi-isotropy of static friction on hexagonal surface lattices. *Europhys. Lett.*, **91**, 66008.

[107] Gnecco, E., Bennewitz, R., Gyalog, T. *et al.* 2000. Velocity dependence of atomic friction. *Phys. Rev. Lett.*, **84**, 1172.

[108] Gnecco, E., Bennewitz, R. and Meyer, E. 2002. Abrasive wear on the atomic scale. *Phys. Rev. Lett.*, **88**, 215501.

[109] Gnecco, E., Socoliuc, A., Maier, S. *et al.* 2009a. Dynamic superlubricity on insulating and conductive surfaces in ultra-high vacuum and ambient environment. *Nanotechnology*, **20**, 025501.

[110] Gnecco, E., Riedo, E., King, W. P., Marder, S. R. and Szoszkiewicz, R. 2009b. Linear ripples and traveling circular ripples produced on polymers by thermal AFM probes. *Phys. Rev. B*, **79**, 235421.

[111] Gnecco, E., Rao, A., Mougin, K., Chandrasekar, G. and Meyer, E. 2010. Controlled manipulation of rigid nanorods by atomic force microscopy. *Nanotechnology*, **21**, 215702.

[112] Gnecco, E., Roth, R. and Baratoff, A. 2012. Analytical expressions for the kinetic friction in the Prandtl–Tomlinson model. *Phys. Rev. B*, **86**, 035443.

[113] Goldman, T., Livne, A. and Fineberg, J. 2010. Acquisition of inertia by a moving crack. *Phys. Rev. Lett.*, **104**, 114301.

[114] Goldstein, H., Poole, C. P. Jr. and Safko, J. L. 2001. *Mechanics*. Third edn. Pearson International Editions.

[115] Goodman, L. E. 1962. Contact stress analysis of normally loaded rough spheres. *Trans. ASME, Series E, J. App. Mech.*, **29**, 515.

[116] Goodman, L. E. and Keer, L. M. 1965. The contact stress problem for an elastic sphere indenting an elastic cavity. *Int. J. Sol. Struc.*, **1**, 407.

[117] Goryl, M., Budzioch, J., Krok, F. *et al.* 2012. Probing atomic-scale friction on reconstructed surfaces of single-crystal semiconductors. *Phys. Rev. B*, **85**, 085308.

[118] Gotsmann, B. and Dürig, U. 2004. Thermally activated nanowear modes of a polymer surface induced by a heated tip. *Langmuir*, **20**, 1495.

[119] Gotsmann, B. and Fuchs, H. 2001. Dynamic force spectroscopy of conservative and dissipative forces in Al-Au(111) tip–sample system. *Phys. Rev. Lett.*, **86**, 2597.

[120] Gotsmann, B., Seidel, B., Anczykowski, B. and Fuchs, H. 1999. Conservative and dissipative tip–sample interaction forces probed with dynamic AFM. *Phys. Rev. B*, **60**, 11051.

[121] Green, C. P., Lioe, H., Cleveland, J. P., Proksch, R., Mulvaney, P. and Sader, J. E. 2004. Normal and torsional spring constants of atomic force microscope cantilevers. *Rev. Sci. Instr.*, **75**, 1988.

[122] Greenwood, J. A. and Williamson, J. B. 1966. Contact of nominally flat surfaces. *Proc. Roy. Soc. A*, **295**, 300.

[123] Griffith, A. A. 1921. The phenomena of rupture and flow in solids. *Phil. Trans. Roy. Soc. A*, **221**, 163.

[124] Guggisberg, M., Bammerlin, M., Loppacher, C. *et al.* 2000. Separation of interactions by noncontact force microscopy. *Phys. Rev. B*, **61**, 11151.

[125] Gürgöze, M. 2005. On the representation of a cantilevered beam carrying a tip mass by an equivalent springmass system. *J. Sound Vib.*, **202**, 538.

[126] Gutenberg, B. and Richter, C. F. 1956. Magnitude and energy of earthquakes. *Ann. Geofis.*, **9**, 1.

[127] Gysin, U., Rast, S., Ruff, P., Lee, D. W., Vettiger, P. and Gerber, C. 2004. Temperature dependence of the force sensitivity of silicon cantilevers. *Phys. Rev. B*, **69**, 045403.

[128] Hadamard, J. 1911. Permanent slow motion of a liquid viscous sphere in a viscous liquid (in French). *Comptes Rendus*, **152**, 1735.

[129] Haddow, J. B. 1967. On a plane strain wedge indentation paradox. *Int. J. Mech. Sci.*, **9**, 159.

[130] Hamilton, G. M. 1983. Explicit equations for the stresses beneath a sliding spherical contact. *Proc. Inst. Mech. Eng. C*, **197**, 53.

[131] Hamrock, B. J. and Dowson, D. 1981. *Ball Bearing Lubrication, the Elastohydrodynamics of Elliptical Contacts*. John Willey & Sons.

[132] Hardy, C., Baronet, C. M. and Tordion, G. V. 1971. Elastoplastic indentation of a half-space by a rigid sphere. *J. Num. Meth. Eng.*, **3**, 451.

[133] Haucke, H., Liu, X., Vignola, J. F., Houston, B. H., Marcus, M. H. and Baldwin, J. W. 2005. Effects of annealing and temperature on acoustic dissipation in a micromechanical silicon oscillator. *Appl. Phys. Lett.*, **86**, 181903.

[134] Heathcote, H. L. 1921. The ball bearing: in the making, under test and on service. *Proc. Inst. Automobile Engineers*, **15**, 569.
[135] Hencky, H. 1923. On some statically determinate cases of equilibrium in plastic bodies (in German). *Z. angew. Math. Mech.*, **3**, 241.
[136] Hertz, H. 1882. On the contact of elastic solids (in German). *Z. reine angew. Math. Mech.*, **92**, 156.
[137] Heslot, F., Cazabat, A. M. and Levinson, P. 1989. Dynamics of wetting of tiny drops: ellipsometric study of the late stages of spreading. *Phys. Rev. Lett.*, **62**, 1286.
[138] Hill, R. 1950. *Mathematical Theory of Plasticity*. Oxford University Press.
[139] Hill, R., Lee, E. H. and Tupper, S. J. 1947. Theory of wedge-indentation of ductile metals. *Proc. R. Soc. A*, **188**, 273.
[140] Hirano, M., Shinjo, K., Kaneko, R. and Murata, Y. 1991. Anisotropy of frictional forces in muscovite mica. *Phys. Rev. Lett.*, **67**, 2642.
[141] Hirano, M., Shinjo, K., Kaneko, R. and Murata, Y. 1997. Observation of superlubricity by scanning tunneling microscopy. *Phys. Rev. Lett.*, **78**, 1448.
[142] Howald, L., Lüthi, R., Meyer, E. and Güntherodt, H. J. 1995. Atomic-force microscopy on the Si(111)7 $\times$ 7 surface. *Phys. Rev. B*, **51**, 5484.
[143] Hutter, J. L. and Bechhoefer, J. 1993. Calibration of atomic- force microscope tips. *Rev. Sci. Instrum.*, **64**, 1868.
[144] Hyun, S., Pei, L., Molinari, J.-F. and Robbins, M. O. 2004. Finite-element analysis of contact between elastic self-affine surfaces. *Phys. Rev. E*, **70**, 026117.
[145] Inglis, C. E. 1913. Stresses in a plate due to the presence of cracks and sharp corners. *Proc. Inst. Naval. Arch.*, **55**, 163.
[146] Irwin, G. 1957. Analysis of stresses and strains near the end of a crack traversing a plate. *J. App. Mech.*, **24**, 361.
[147] Ishikawa, M., Okita, S., Minami, N. and Miura, K. 2000. Load dependence of lateral force and energy dissipation at NaF(001) surface. *Surf. Sci.*, **445**, 488.
[148] Israelachvili, J. N. 1991. *Intermolecular and Surface Forces*. Academic Press.
[149] Israelachvili, J. N. 1992. Adhesion forces between surfaces in liquids and condensable vapours. *Surf. Sci. Rep.*, **14**, 109.
[150] Jacobs, T. D. B. and Carpick, R. 2013. Nanoscale wear as a stress-assisted chemical reaction. *Nature Mat.*, **8**, 108.
[151] Jansen, L., Hölscher, H., Fuchs, H. and Schirmeisen, A. 2010. Temperature dependence of atomic-scale stick-slip friction. *Phys. Rev. Lett.*, **104**, 256101.
[152] Jarvis, S. P., Uchihashi, T., Ishida, T., Tokumoto, H. and Nakayama, Y. 2000. Temperature dependence of atomic-scale stick-slip friction. *J. Phys. Chem.*, **104**, 6091.
[153] Johnson, K. L. 1955. Surface interaction between elastically loaded bodies under tangential forces. *Proc. Roy. Soc. A*, **230**, 521.
[154] Johnson, K. L. 1961. Energy dissipation at spherical surfaces in contact transmitting oscillating forces. *J. Mech. Eng. Sci.*, **3**, 362.
[155] Johnson, K. L. 1968. Deformation of a plastic wedge by a rigid flat die under the action of a tangential force. *J. Mech. Phys. Sol.*, **16**, 395.
[156] Johnson, K. L. 1987. *Contact Mechanics*. Cambridge University Press.
[157] Johnson, K. L., Kendall, K. and Roberts, A. D. 1971. Surface energy and the contact of elastic solids. *Proc. Roy. Soc. A*, **324**, 301.
[158] Johnson, K. L., Greenwood, J. A. and Higginson, J. G. 1985. The contact of elastic regular wavy surfaces. *Int. J. Mech. Sci.*, **27**, 383.
[159] Johnson, W., Mahtab, F. U. and Haddow, J. B. 1964. Indentation of a semi-infinite block by a wedge of comparable hardness. *Int. J. Mech. Sci.*, **6**, 329.

[160] Jost, H. P. 1966. *Lubrication (Tribology) Education and Research. A Report on the Present Position and Industry's Needs*. Dept. of Education and Science, Her Majesty's Stationery Office.

[161] Kalker, J. J. 1967a. On the rolling contact of two elastic bodies in the presence of dry friction. PhD Thesis, Technical University Delft.

[162] Kalker, J. J. 1967b. A strip theory for rolling with slip and spin. *Proc. Kon. Ned. Akad. van Wetenschappen.*, **B70**, 10.

[163] Kato, K. 1915. Mathematical investigation on the mechanical problems of transmission line (in Japanese). *J. Jap. Soc. Mech. Eng.*, **19**, 41.

[164] Kawai, S., Canova, F. F., Glatzel, T., Foster, A. S. and Meyer, E. 2011. Atomic-scale dissipation processes in dynamic force spectroscopy. *Phys. Rev. B*, **84**, 115415.

[165] Kawai, S., Koch, M., Gnecco, E. *et al.* 2014. Quantifying the atomic-level mechanics of single long physisorbed molecular chains. *PNAS*, **111**, 3968.

[166] Kisiel, M., Gnecco, E., Gysin, U., Marot, L., Rast, S. and Meyer, E. 2011. Suppression of electronic friction on Nb films in the superconducting state. *Nat. Mat.*, **10**, 119.

[167] Koch, S., Stradi, D., Gnecco, E. *et al.* 2013. Elastic response of graphene nanodomes. *ACS Nano*, **7**, 2927.

[168] Kopta, S. and Salmeron, M. 2000. The atomic scale origin of wear on mica and its contribution to friction. *J. Chem. Phys.*, **113**, 8249.

[169] Kramers, H. A. 1940. Brownian motion in a field of force and the diffusion model of chemical reactions. *Physica*, **7**, 284.

[170] Krim, J. and Widom, A. 1988. Damping of a crystal oscillator by an adsorbed monolayer and its relation to interfacial viscosity. *Phys. Rev. B*, **38**, 12184.

[171] Krylov, S. Y., Jinesh, K. B., Valk, H., Dienwiebel, M. and Frenken, J. W. M. 2005. Thermally induced suppression of friction at the atomic scale. *Phys. Rev. E*, **71**, 65101.

[172] Lakes, R. S. 1987. Foam structures with a negative Poisson's ratio. *Science*, **235**, 1038.

[173] Lamb, H. 1911. On the uniform motion of a sphere through a viscous fluid. *Phil. Mag.*, **21**, 112.

[174] Lamb, H. 1932. *Hydrodynamics*. Cambridge University Press.

[175] Landau, L. D. and Lifshitz, E. M. 1976. *Mechanics*. Third edn. Butterworth-Heinemann.

[176] Landau, L. D. and Lifshitz, E. M. 1987. *Fluid Mechanics*. Second edn. Butterworth-Heinemann.

[177] Landau, L. D., Pitaevskii, L. P., Lifshitz, E. M. and Kosevich, M. 1986. *Theory of Elasticity*. Third edn. Butterworth-Heinemann.

[178] Landman, U., Luedtke, W. D. and Ringer, E. M. 1992. Atomistic mechanisms of adhesive contact formation and interfacial processes. *Wear*, **153**, 3.

[179] Langer, M., Kisiel, M., Pawlak, R. *et al.* 2014. Giant frictional dissipation peaks and charge-density wave slips at the $NbSe_2$ surface. *Nature Mat.*, **13**, 173.

[180] Langewisch, G., Falter, J., Fuchs, H. and Schirmeisen, A. 2013. Giant frictional dissipation peaks and charge-density wave slips at the $NbSe_2$ surface. *Phys. Rev. Lett.*, **110**, 036101.

[181] Lantz, M. A., OShea, S. J., Welland, M. E. and Johnson, K. L. 1997. Atomic-force-microscope study of contact area and friction on $NbSe_2$. *Phys. Rev. B*, **55**, 10776.

[182] Lantz, M. A., Wiesmann, D. and Gotsmann, B. 2009. Dynamic superlubricity and the elimination of wear on the nanoscale. *Nature Nanotech.*, **4**, 586.

[183] Lee, D. W., Kang, J. H., Gysin, U. *et al.* 2005. Fabrication and evaluation of single-crystal silicon cantilevers with ultra-low spring constants. *J. Micromech. Microeng.*, **15**, 2179.

[184] Leung, O. M. and Goh, M. C., 1992. Orientational ordering of polymers by atomic force microscope tip–surface interaction. *Science*, **255**, 64.

[185] Lifshitz, R. and Roukes, M. 2000. Thermoelastic damping in micro- and nanomechanical systems. *Phys. Rev. B*, **61**, 5600.

[186] Linnemann, R., Gotszalk, T., Rangelow, I. W., Dumania, P. and Oesterschulze, E. 1996. Atomic force microscopy and lateral force microscopy using piezoresistive cantilevers. *J. Vac. Sci. Technol. B*, **14**, 856.

[187] Livshits, A. I. and Shluger, A. L. 1997. Self-lubrication in scanning-force-microscope image formation on ionic surfaces,. *Phys. Rev. B*, **56**, 12482.

[188] Lockett, F. 1963. Indentation of a rigid/plastic material by a conical indenter. *J. Mech. Phys. Sol.*, **11**, 345.

[189] Loppacher, C., Bammerlin, M., Guggisberg, M. *et al.* 1999. Phase variation experiments in non-contact dynamic force microscopy using phase locked loop techniques. *Appl. Surf. Sci.*, **140**, 287.

[190] Lorenz, B. and Persson, B. N. J. 2009. Leak rate of seals: comparison of theory with experiment. *Europhys. Lett.*, **86**, 44006.

[191] Lorenz, B. and Persson, B. N. J. 2010. Time-dependent fluid squeeze-out between solids with rough surfaces. *Eur. Phys. J. E*, **32**, 281.

[192] Lorenz, B., Persson, B. N. J., Dieluweit, S. and Tada, T. 2011. Rubber friction: comparison of theory with experiment. *Eur. Phys. J. E*, **34**, 129.

[193] Love, A. E. H. 1888. On the small free vibrations and deformations of elastic shells. *Phil. Trans. Roy. Soc.*, **17**, 491.

[194] Love, A. E. H. 1929. Stress produced in a semi-infinite solid by pressure on part of the boundary. *Phil. Trans. Roy. Soc. A*, **228**, 377.

[195] Love, A. E. H. 1939. Boussinesq's problem for a rigid cone. *Quart. J. Math. (Oxford series)*, **10**, 161.

[196] Luan, B. and Robbins, M. O. 2006. The breakdown of continuum models for mechanical contacts. *Nature*, **435**, 929.

[197] Lubkin, J. L. 1951. Torsion of elastic spheres in contact. *Trans. ASME, Series E, J. App. Mech.*, **18**, 183.

[198] Luebbe, J., Troeger, L., Torbruegge, S. *et al.* 2010. Achieving high effective Q-factors in ultra-high vacuum dynamic force microscopy. *Meas. Sci. Technol.*, **21**, 125501.

[199] Lüthi, R., Meyer, E., Haefke, H., Howald, L., Gutmannsbauer, W. and Güntherodt, H. J. 1994. Sled-type motion on the nanometer scale: determination of dissipation and cohesive energies of C_{60}. *Science*, **266**, 1979.

[200] Lüthi, R., Meyer, E., Bammerlin, M. *et al.* 1997. Ultrahigh vacuum atomic force microscopy: true atomic resolution. *Surf. Rev. Lett.*, **4**, 1025.

[201] Maier, S., Sang, Y., Filleter, T. *et al.* 2005. Fluctuations and jump dynamics in atomic friction experiments. *Phys. Rev. B*, **72**, 245418.

[202] Maier, S., Gnecco, E., Baratoff, A. and Meyer, E. 2008. Atomic-scale friction modulated by a buried interface: combined atomic and friction force microscopy experiments. *Phys. Rev. B*, **78**, 045432.

[203] Mandel, J. 1967. Rolling resistance of an undeformable cylinder on a perfectly plastic solid. *Friction and Wear* (in French). GAMI, Paris.

[204] Marder, M. 1991. New dynamical equation for crack. *Phys. Rev. Lett.*, **66**, 2484.

[205] Marti, O., Colchero, J. and Mlynek, J. 1990. Combined scanning force and friction microscopy of mica. *Nanotechnology*, **1**, 141.

[206] Mate, C. M., McClelland, G. M., Erlandsson, R. and Chiang, S. 1987. Atomic-scale friction of a tungsten tip on a graphite surface. *Phys. Rev. Lett.*, **59**, 1942.

[207] Maugis, D. 1992. Adhesion of spheres – the JKR-DMT transition using a Dugdale model. *J. Colloid Interface Sci.*, **150**, 243.

[208] Maxwell, J. C. 1866. On the viscosity or internal friction of air and other gases. *Phil. Trans. Roy. Soc. Lond.*, **156**, 249.

[209] Maxwell, J. C. 1868. On the dynamical theory of gases. *Phil. Mag.*, **35**, 129.

[210] May, W. D., Morris, E. L. and Atack, D. 1959. Rolling friction of a hard cylinder over a viscoelastic material. *J. App. Phys.*, **30**, 1713.

[211] McClelland, G. M. and Glosli, J. N. 1992. Friction at the atomic scale. *NATO ASI Series E*, eds. L. Singer and H. M. Pollock, vol. 220. Kluwer, Dordrecht.

[212] McGuiggan, P. M. and Israelachvili, J. N. 1990. Adhesion and short-range force between surfaces. Part II: Effects of surface lattice mismatch. *J. Mater. Res.*, **5**, 2232.

[213] Mecke, K. R. and Dietrich, S. 1999. Effective Hamiltonian for liquid–vapor interfaces. *Phys. Rev. E*, **59**, 6766.

[214] Medyanik, S. N., Liu, W. K., Sung, I. H. and Carpick, R. W. 2006. Predictions and observations of transitions in atomic-scale stick-slip friction. *Phys. Rev. Lett.*, **97**, 136106.

[215] Melan, E. 1938. On the plasticity of the spatial continuum (in German). *Ing. Arch.*, **9**, 116.

[216] Merkle, A. P. and Marks, L. D. 2007. Friction in full view. *Appl. Phys. Lett.*, **90**, 064101.

[217] Meyer, E., Hug, H. and Bennewitz, R. 2004. *Scanning Probe Microscopy: The Lab on the Tip*. Springer.

[218] Meyer, E., Lüthi, R., Howald, L., Bammerlin, M., Guggisberg, M. and Güntherodt, H. J. 1996. Site-specific friction force spectroscopy. *J. Vac. Sci. Technol. B*, **14**, 1285.

[219] Meyer, G. and Amer, N. 1990. Simultaneous measurement of lateral and normal forces with an optical-beam-deflection atomic force microscope. *Appl. Phys. Lett.*, **57**, 2089.

[220] Mindlin, R. D. and Deresiewicz, H. 1953. Elastic spheres in contact under varying oblique forces. *Trans. ASME E, J. App. Mech.*, **20**, 327.

[221] Mindlin, R. D., Mason, W. P., Osmer, J. F. and Deresiewicz, H. 1952. Effects of an oscillating tangential force on the contact surfaces of elastic spheres. *Proc. 1st US Nat. Congress of App. Mech.*, E. Sternberg (ed.).

[222] Minor, A. M., Morris, J. W. and Stach, E. A. 2001. Quantitative in-situ nanoindentation in an electron microscope. *Appl. Phys. Lett.*, **79**, 1625.

[223] Moriwaki, T. and Okuda, K. 1989. Machinability of copper in ultra-precision micro diamond cutting. *Ann. CIRP*, **38**, 115.

[224] Mougin, K., Rao, A., Cuberes, M. T. *et al.* 2008. Manipulation of gold nanoparticles: influence of surface chemistry, temperature, and environment (vacuum versus ambient atmosphere). *Langmuir*, **24**, 1577.

[225] Murakami, Y. (ed.). 1987. *Stress Intensity Factor Handbook*. Pergamon Press.

[226] Müser, M. H. and Robbins, M. O. 2000. Conditions for static friction between flat crystalline surfaces. *Phys. Rev. B*, **61**, 2335.

[227] Müser, M. H. 2004. Structural lubricity: role of dimension and symmetry. *Europhys. Lett.*, **66**, 97.

[228] Müser, M. H., Wenning, L. and Robbins, M. O. 2001. Simple microscopic theory of Amontons's laws for static friction. *Phys. Rev. Lett.*, **86**, 1295.

[229] Napolitano, S., DAcunto, M., Baschieri, P., Gnecco, E. and Pingue, P. 2012. Ordered rippling of polymer surfaces by nanolithography: influence of scan pattern and boundary effects. *Nanotechnology*, **23**, 475301.

[230] Navier, C. L. M. H. 1822. Memoir on the laws of fluid motion (in French). *Mem. Acad. Sci. Inst. France*, **6**, 380.

[231] Nayak, P. R. 1971. Random process model of rough surfaces. *ASME J. Lubr. Technol.*, **93**, 398.

[232] Neubauer, G., Cohen, S. R., McClelland, G. M., Horn, D. E. and Mate, C. M. 1990. Force microscopy with a bidirectional capacitance sensor. *Rev. Sci. Instrum.*, **61**, 2296.

[233] Neumeister, J. M. and Ducker, W. A. 1994. Lateral, normal, and longitudinal spring constants of atomic-force microscopy cantilevers. *Rev. Sci. Instrum.*, **65**, 2527.

[234] Nieminen, J. A. 1992. Mechanism of lubrication by a thin solid film on a metal-surface. *Modelling Simul. Mater. Sci. Eng.*, **1**, 83.

[235] Nita, P., Casado, S., Dietzel, D., Schirmeisen, A. and Gnecco, E. 2013. Spinning and translational motion of Sb nanoislands manipulated on MoS_2. *Nanotechnology*, **24**, 325302.

[236] Nita, P., Pimentel, C., Luo, F. *et al.* 2014. Molecular resolution friction microscopy of Cu phthalocyanine thin films on dolomite (104) in water. *Nanoscale*, **6**, 8334.

[237] Oliver, W. C. and Pharr, G. M. 1992. An improved technique for determining hardness and elastic modulus using load and displacement sensing indentation experiments. *J. Mater. Res.*, **7**, 1564.

[238] Oseen, C. W. 1910. On the Stokes' formula and a related problem in hydrodynamics (in German). *Ark. Mat. Astron. Fys.*, **6**, 1.

[239] Park, J. Y., Ogletree, D. F., Thiel, P. A. and Salmeron, M. 2006. Electronic control of friction in silicon pn junctions. *Science*, **313**, 186.

[240] Peressadko, A. G., Hosoda, N. and Persson, B. N. J. 2005. Influence of surface roughness on adhesion between elastic bodies. *Phys. Rev. Lett.*, **95**, 124301.

[241] Persson, A. 1964. On the stress distribution of cylindrical elastic bodies in contact. Doctoral dissertation, Chalmers Tekniska Hogskola, Göteborg.

[242] Persson, B. N. J. 1991. Surface resistivity and vibrational damping in adsorbed layers. *Phys. Rev. B*, **44**, 3277.

[243] Persson, B. N. J. 1993. Theory of friction and boundary lubrication. *Phys. Rev. B*, **48**, 18140.

[244] Persson, B. N. J. 1996. On the origin of the stick-slip motion of lubricated surfaces. *Chem. Phys. Lett.*, **254**, 114.

[245] Persson, B. N. J. 2000. *Sliding Friction: Physical Principles and Applications*. Second edn. Springer.

[246] Persson, B. N. J. 2001a. Theory of rubber friction and contact mechanics. *J. Chem. Phys.*, **155**, 3840.

[247] Persson, B. N. J. 2001b. Theory of time-dependent plastic deformation in disordered solids. *Phys. Rev. B*, **61**, 5949.

[248] Persson, B. N. J. 2002. Adhesion between an elastic body and a randomly rough hard surface. *Eur. Phys. J. E.*, **8**, 385.

[249] Persson, B. N. J. 2003. On the mechanism of adhesion in biological systems. *J. Chem. Phys.*, **118**, 7614.

[250] Persson, B. N. J. 2006a. Contact mechanics for randomly rough surfaces. *Surf. Sci. Rep.*, **61**, 201.

[251] Persson, B. N. J. 2006b. Rubber friction: role of the flash temperature. *J. Phys.: Condens. Matter*, **18**, 7789.

[252] Persson, B. N. J. 2007. Relation between interfacial separation and load: a general theory of contact mechanics. *Phys. Rev. Lett.*, **99**, 125502.

[253] Persson, B. N. J. 2010. Fluid dynamics at the interface between contacting elastic solids with randomly rough surfaces. *J. Phys.: Condens. Matter*, **22**, 265004.

[254] Persson, B. N. J. 2012. Contact mechanics for layered materials with randomly rough surfaces. *J. Phys.: Condens. Matter*, **24**, 095008.

[255] Persson, B. N. J. and Brener, E. A. 2005. Crack propagation in viscoelastic solids. *Phys. Rev. E*, **71**, 036123.

[256] Persson, B. N. J. and Rydberg, R. 1985. Brownian motion and vibrational phase relaxation at surfaces: CO on Ni(111). *Phys. Rev. B*, **32**, 3586.

[257] Persson, B. N. J. and Tosatti, E. 1994. Layering transition in con ned molecular thin lms: nucleation, growth and analogy with spreading of wetting drop. *Phys. Rev. B*, **50**, 5590.

[258] Persson, B. N. J. and Tosatti, E. 1999. Theory of friction: elastic coherence length and earthquake dynamics. *Solid State Commun.*, **109**, 739.

[259] Persson, B. N. J. and Tosatti, E. 2000. Qualitative theory of rubber friction and wear. *J. Chem. Phys.*, **112**, 2021.

[260] Persson, B. N. J. and Tosatti, E. 2001. The effect of surface roughness on the adhesion of elastic solids. *J. Phys. Chem.*, **115**, 5597.

[261] Persson, B. N. J. and Volokitin, A. I. 1995. Electronic friction of physisorbed molecules. *J. Chem. Phys.*, **103**, 8659.

[262] Persson, B. N. J. and Yang, C. 2008. Theory of the leak-rate of seals. *J. Phys.: Condens. Matter*, **20**, 315011.

[263] Persson, B. N. J., Albohr, O., Creton, C. and Peveri, V. 2004. Contact area between a viscoelastic solid and a hard, randomly rough, substrate. *J. Chem. Phys.*, **120**, 8779.

[264] Persson, B. N. J., Albohr, O., Heinrich, G. and Ueba, H. 2005a. Crack propagation in rubber-like materials. *J. Phys.: Condens. Matter*, **17**, R1071.

[265] Persson, B. N. J., Albohr, O., Tartaglino, U., Volokitin, A. I. and Tosatti, E. 2005b. On the nature of surface roughness with application to contact mechanics, sealing, rubber friction and adhesion. *J. Phys.: Condens. Matter*, **17**, R1.

[266] Persson, B. N. J., Lorenz, B. and Volokitin, A. I. 2010. Heat transfer between elastic solids with randomly rough surfaces. *Eur. Phys. J. E*, **31**, 3.

[267] Petroff, N. P. 1883. Friction in machines and the effect of lubricant (in Russian). *Inzh. Zh. St. Petersburg*, **1**, 71.

[268] Peyrard, M. and Aubry, S. 1983. Critical behaviour at the transition by breaking of analyticity in the discrete Frenkel–Kontorova model. *J. Phys. C*, **16**, 1593.

[269] Pfeiffer, O., Loppacher, C., Wattinger, C. *et al.* 2000. Using higher flexural modes in non-contact force microscopy. *Appl. Surf. Sci.*, **157**, 337.

[270] Pietrement, O. and Troyon, M. 2001. Study of the interfacial shear strength pressure dependence by modulated lateral force microscopy. *Langmuir*, **17**, 6540.

[271] Pimentel, C., Pina, C. M. and Gnecco, E. 2013. Epitaxial growth of calcite crystals on dolomite and kutnahorite (104) surfaces. *Cryst. Growth Des.*, **13**, 2557.

[272] Pina, C. M., Miranda, R. and Gnecco, E. 2012. Anisotropic surface coupling while sliding on dolomite and calcite crystals. *Phys. Rev. B*, **85**, 073402.

[273] Poiseuille, J. L. M. 1840. Experimental research on the motion of liquids in tubes of very small diameter (in French) *Computes Reduces*, **11**, 961.

[274] Polaczyk, C., Schneider, T., Schöfer, J. and Santner, E. 1998. Microtribological behavior of Au(001) studied by AFM/FFM. *Surf. Sci.*, **402**, 454.

[275] Prandtl, L. 1903. On the torsion of prismatic rods (in German). *Phys. Zeitschr.*, **4**, 758.

[276] Prandtl, L. 1918. Theory of lifting surfaces (in German). *Nachr. Ges. Wiss. Göttingen. Math.-Phys. Kl.*, 151.

[277] Prandtl, L. 1928. A conceptual model for the kinetic theory of solids (in German). *Z. angew. Math. Mech.*, **8**, 85.

[278] Qi, Y., Cheng, Y. T., Cagin, T. and Goddard III, W. A. 2002. Friction anisotropy at Ni(100)/(100) interfaces: molecular dynamics studies. *Phys. Rev. B*, **66**, 085420.

[279] Rabe, U., Janser, K. and Arnold, W. 1996. Vibrations of free and surface-coupled atomic force microscope cantilevers: theory and experiment. *Rev. Sci. Instrum.*, **67**, 3281.

[280] Radok, J. R. M. 1957. Viscoelastic stress analysis. *Q. App. Math.*, **15**, 198.

[281] Rao, A., Gnecco, E., Marchetto, D. *et al.* 2009a. The analytical relations between particles and probe trajectories in atomic force microscope nanomanipulation. *Nanotechnology*, **20**, 115706.

[282] Rao, A., Wille, M. L., Gnecco, E., Mougin, K. and Meyer, E. 2009b. Trajectory fluctuations accompanying the manipulation of spherical nanoparticles. *Phys. Rev. B*, **80**, 193405.

[283] Rast, S., Gysin, U., Ruff, P., Gerber, C., Meyer, E. and Lee, D. W. 2006. Force microscopy experiments with ultrasensitive cantilevers. *Nanotechnology*, **17**, 189.

[284] Rayleigh, L. 1917. On the dynamics of revolving fluids. *Proc. Roy. Soc. London A*, **93**, 148.

[285] Reynolds, O. 1883. An experimental investigation of the circumstances which determine whether the motion of water shall be direct or sinuous, and of the law of resistance in parallel channels. *Phil. Trans. Roy. Soc. London*, **174**, 935.

[286] Reynolds, O. 1886. On the theory of lubrication and its application to Mr Beauchamp Tower's experiments including an experimental determination of the viscosity of olive oil. *Phil. Trans. Roy. Soc. London*, **177**, 157.

[287] Ribeiro, R., Shan, Z., Minor, A. M. and Liang, H. 2007. In situ observation of nano-abrasive wear. *Wear*, **263**, 1556.

[288] Rice, J. R. 1993. Spatio-temporal complexity of slip on a fault. *J. Geophys. Res.*, **98**, 9885.

[289] Richmond, O., Morrison, H. L. and Devenpeck, M. L. 1974. Sphere indentation with application to the Brinell hardness test. *Int. J. Mech. Sci.*, **16**, 75.

[290] Riedo, E., Levy, F. and Brune, H. 2002. Kinetics of capillary condensation in nanoscopic sliding friction. *Phys. Rev. Lett.*, **88**, 185505.

[291] Riedo, E., Gnecco, E., Bennewitz, R., Meyer, E. and Brune, H. 2003. Interaction potential and hopping dynamics governing sliding friction. *Phys. Rev. Lett.*, **91**, 084502.

[292] Risken, H. 1989. *The Fokker–Planck Equation*, 2nd edn. Springer.

[293] Ritter, C., Heyde, M., Schwarz, U. D. and Rademann, K. 2002. Controlled translational manipulation of small latex spheres by dynamic force microscopy. *Langmuir*, **18**, 7798.

[294] Roth, R., Fajardo, O. Y., Mazo, J. J., Meyer, E. and Gnecco, E. 2014. Lateral vibration effects in atomic-scale friction. *Appl. Phys. Lett.*, **104**, 083103.

[295] Rubio, G., Agrait, N. and Vieira, S. 1996. Atomic-sized metallic contacts: mechanical properties and electronic transport. *Phys. Rev.Lett.*, **76**, 2302.

[296] Rugar, D., Budakian, R., Mamin, H. J. and Chui, B. W. 2004. Single spin detection by magnetic resonance force microscopy. *Nature*, **43**, 329.

[297] Ruina, A. I. 1983. Slip instability and state variable friction laws. *J. Geophys. Res.*, **88**, 10359.

[298] Rybczynski, W. 1911. On the progressive motion of a liquid sphere in a viscous medium (in German). *Bull. Acad. Sci. Cracovie A*, **40**.

[299] Sader, J. E., Chon, J. W. M. and Mulvaney, P. 1999. Calibration of rectangular atomic force microscope cantilevers. *Rev. Sci. Instr.*, **70**, 3967.

[300] Saitoh, K., Hayashi, K., Shibayama, Y. and Shirahama, K. 2010. Gigantic maximum of nanoscale noncontact friction. *Phys. Rev. Lett.*, **105**, 236103.

[301] Samuels, L. E. and Mulhearn, T. O. 1956. The deformed zone associated with indentation hardness impressions. *J. Mech. Phys. Sol.*, **5**, 125.

[302] Sang Y., Dubé, M. and Grant, M. 2001. Thermal effects in atomic friction. *Phys. Rev. Lett.*, **87**, 174301.

[303] Sasaki, N. and Tsukada, M. 2001. Theoretical evaluation of the frequency shift and dissipated power in noncontact atomic force microscopy. *Appl. Phys. A*, **72**, S39.

[304] Schmidt, R. H., Haugstad, G. and L. Gladfelter W. 1999. Correlation of nanowear patterns to viscoelastic response in a thin polystyrene melt. *Langmuir*, **15**, 317.

[305] Schwarz, U. D., Zwörner, O., Köster, P. and Wiesendanger, R. 1997. Quantitative analysis of the frictional properties of solid materials at low loads. *Phys. Rev. B*, **56**, 6987.

[306] Sharvin, Y. V. 1965. A possible method for studying Fermi surfaces. *Sov. Phys. JETP*, **21**, 655.

[307] Sheehan, P. E. and Lieber, C. M. 1996. Nanotribology and nanofabrication of MoO_3 structures by atomic force microscopy. *Science*, **272**, 1158.

[308] Sih, J. C. 1973. *Handbook of Stress-Intensity Factors*. Inst. of Fracture and Solid Mech., Lehigh University.

[309] Sneddon, I. N. 1946. Boussinesq's problem for a flat-ended cylinder. *Proc. Cambridge Phil. Soc.*, **42**, 29.

[310] Sneddon, I. N. 1965. The relation between load and penetration in the axisymmetric Boussinesq problem for a punch of arbitrary profile. *Int. J. Eng. Sci.*, **3**, 47.

[311] Socoliuc, A., Gnecco, E., Bennewitz, R. and Meyer, E. 2003. Ripple formation induced in localized abrasion. *Phys. Rev. B*, **68**, 115416.

[312] Socoliuc, A., Bennewitz, R., Gnecco, E. and Meyer, E. 2004. Transition from stick-slip to continuous sliding in atomic friction: entering a new regime of ultralow friction. *Phys. Rev. Lett.*, **92**, 134301.

[313] Socoliuc, A., Gnecco, E., Maier, S., Pfeiffer, O., Baratoff, A. and Meyer, E. 2006. Atomic-scale control of friction by actuation of nanometer-sized contacts. *Science*, **313**, 207.

[314] Sørensen, M. R., Jacobsen, K. W. and Stoltze, P. 1996. Simulations of atomic-scale sliding friction. *Phys. Rev. B*, **53**, 2101.

[315] Spence, D. A. 1973. An eigenvalue problem for elastic contact with finite friction. *Proc. Cambridge Phil. Soc.*, **73**, 249.

[316] Steele, W. A. 1973. The physical interaction of gases with crystalline solids. *Surf. Sci.*, **36**, 317.

[317] Steiner, P., Roth, R., Gnecco, E., Glatzel, T., Baratoff, A. and Meyer, E. 2009. Modulation of contact resonance frequency accompanying atomic-scale stick-slip in friction force microscopy. *Nanotechnology*, **20**, 495701.

[318] Steiner, P., Roth, R., Gnecco, E., Baratoff, A. and Meyer, E. 2010. Angular dependence of static and kinetic friction on alkali halide surfaces. *Phys. Rev. B*, **82**, 205417.

[319] Steiner, P., Gnecco, E., Krok, F., *et al.* 2011. Atomic-scale friction on stepped surfaces of ionic crystals. *Phys. Rev. Lett.*, **106**, 186104.

[320] Stillinger, F. H. and Weber, T. A. 1985. Computer simulation of local order in condensed phases of silicon. *Phys. Rev. B*, **31**, 5262.

[321] Stilweel, N. A. and Tabor, D. 1961. Elastic recovery of conical indentations. *Proc. Phys. Soc*, **78**, 169.

[322] Stipe, B. C., Mamin, H. J., Stowe, T. D., Kenny, T. W. and Rugar, D. 2001. Magnetic dissipation and fluctuations in individual nanomagnets measured by ultrasensitive cantilever magnetometry. *Phys. Rev. Lett.*, **86**, 2874.

[323] Stokes, G. G. 1845. On the theories of internal friction of fluids in motion. *Trans. Cambridge Phil. Soc.*, **8**, 287.

[324] Stokes, G. G. 1851. On the effect of the internal friction of fluids on the motion of pendulums. *Trans. Cambridge Phil. Soc.*, **9**, 8.

[325] Stolarski, T. A. 1990. *Tribology in Machine Design*. Butterworth-Heinemann Ltd.

[326] Such, B., Krok, F. and Szymonski, M. 2008. AFM tip-induced tipple pattern on AIII-BV semiconductor surfaces. *Appl. Surf. Sci.*, **254**, 5431.

[327] Tabor, D. 1948. A simple theory of static and dynamic hardness. *Proc. Roy. Soc. A*, **192**, 247.

[328] Tabor, D. 1977. Surface forces and surface interactions. *J. Colloid Interface Sci.*, **58**, 2.

[329] Tanner, L. 1979. The spreading of silicon oil drops on horizontal surfaces. *J. Phys. D*, **12**, 1473.

[330] Taylor, G. I. 1923. Stability of a viscous liquid contained between two rotating cylinders. *Phil. Trans. Roy. Soc. London A*, **233**, 289.

[331] Ternes, M., Lutz, C. P., Hirjibehedin, C. F., Giessibl, F. J. and Heinrich, A. J. 2008. The force needed to move an atom on a surface. *Science*, **319**, 1066.

[332] Tomlinson, G. A. 1929. A molecular theory of friction. *Phil. Mag.*, **7**, 905.

[333] Tripathi, M., Paolicelli, G., D'Addato, S. and Valeri, S. 2012. Controlled AFM detachments and movement of nanoparticles: gold clusters on HOPG at different temperatures. *Nanotechnology*, **23**, 245706.

[334] Vanossi, A., Manini, N., Urbakh, M., Zapperi, S. and Tosatti, E. 2013. Modeling friction: from nanoscale to mesoscale. *Rev. Mod. Phys.*, **85**, 529.

[335] Verhoeven, G. S., Dienwiebel, M. and Frenken, J. W. M. 2004. Model calculations of superlubricity of graphite. *Phys. Rev. B*, **70**, 165418.

[336] Volokitin, A. I., Persson, B. N. J. and Ueba, H. 2006. Enhancement of noncontact friction between closely spaced bodies by two-dimensional systems. *Phys. Rev. B*, **73**, 165423.

[337] von Kármán, T. 1921. On the laminar and turbulent friction (in German). *ZAMM*, **1**, 233.

[338] von Kármán, T. 1930. Mechanical similitude and turbulence (in German). *Nach. Ges. Wiss. Göttingen, Math.- Phys. Kl.*, **5**, 58.

[339] von Mises, R. 1913. Mechanics of solid bodies in plastic deformable state (in German). *Nach. Ges. Wiss. Göttingen, Math.- Phys. Kl.*, **1**, 582.

[340] von Schlippe, B. and Dietrich, R. 1941. Shimmying of a pneumatic wheel (in German). *Ber. Lilienthal Ges.*, **140**, 35.

[341] Westergaard, H. M. 1939. Bearing pressures and cracks. *J. App. Mech.*, **6**, 49.

[342] Williams, M. L., Landel, R. F. and Ferry, J. D. 1955. The temperature dependence of relaxation mechanisms in amorphous polymers and other glass-forming liquids. *J. Amer. Chem. Soc.*, **77**, 3701.

[343] Wyder, U., Baratoff, A., Meyer, E. *et al.* 2007. Interpretation of atomic friction experiments based on atomistic simulations. *J. Vac. Sci. Technol. B*, **25**, 1547.

[344] Yang, C. and Persson, B. N. J. 2008. Contact mechanics: contact area and interfacial separation from small contact to full contact. *J. Phys.: Condens. Matter*, **20**, 215214.

[345] Yang, J. 2000. Surface effects and high quality factors in ultra-thin single-crystal silicon cantilevers. *Appl. Phys. Lett.*, **77**, 3860.

[346] Yang, W. H. 1966. The contact problem for viscoelastic bodies. *Trans. ASME, Series E, J. App. Mech.*, **33**, 395.

[347] Yanson, A. I., Rubio Bollinger, G., Van den Brom, H. E., Agrait, N. and van Ruitenbeek, J. M. 1998. Formation and manipulation of a metallic wire of single gold atoms. *Nature*, **395**, 783.

[348] Yoshizawa, H., Chen, Y. L. and Israelachvili, J. 1993. Recent advances in molecular level understanding of adhesion, friction and lubrication. *Wear*, **168**, 161.

[349] Young, T. 1805. An essay on the cohesion of fluids. *Phil. Trans. Roy. Soc. London*, **95**, 65.

[350] Zener, C. 1938. Interfacial friction in solids III. Experimental demonstration of thermoelastic internal friction. *Phys. Rev.*, **53**, 100.

[351] Zhao, X., Hamilton, M., Sawyer, W. G. and Perry, S. S. 2007. Thermally activated friction. *Tribol. Lett.*, **27**, 113.

[352] Zilberman, S., Becker, T., Mugele, F., Persson, B. N. J. and Nitzan, A. 2003. Dynamics of squeeze-out: theory and experiments. *J. Chem. Phys.*, **118**, 11160.

[353] Zwörner, O., Hölscher, H., Schwarz, U. D. and Wiesendanger, R. 1998. The velocity dependence of friction forces in point-contact friction. *Appl. Phys. A*, **66**, 263.

Index

For EU product safety concerns, contact us at Calle de José Abascal, 56–1°,
28003 Madrid, Spain or eugpsr@cambridge.org.

www.ingramcontent.com/pod-product-compliance
Lightning Source LLC
LaVergne TN
LVHW062117200926
848484LV00007B/1696

* 9 7 8 1 1 0 7 0 0 6 2 3 2 *